Physical Geography: The Canadian Context

Physical Geography:
The Canadian Context

Allan Falconer, Ph.D.

Barry D. Fahey, Ph.D.

Russell D. Thompson, Ph.D.

Department of Geography
University of Guelph
Guelph, Ontario

McGraw-Hill Ryerson Limited
Toronto   Montreal   New York   London   Sydney
Johannesburg   Mexico   Panama   Düsseldorf   Singapore
São Paulo   Kuala Lumpur   New Delhi

PHYSICAL GEOGRAPHY: THE CANADIAN CONTEXT

ISBN 0-07-077556-7

1 2 3 4 5 6 7 8 9 10 D 3 2 1 0 9 8 7 6 5 4
Printed and bound in Canada

CONTENTS

SECTION ONE   INTRODUCTION   *1*

Chapter 1 The Nature of Physical Geography and Its Significance for Canada   *3*
Chapter 2 Practical Knowledge of Physical Geography in Southern Ontario During the Nineteenth Century   *10*
Chapter 3 Physical Geography in the Early Twentieth Century   *19*
Chapter 4 The Component Parts of Contemporary Physical Geography   *30*

SECTION TWO   GEOMORPHOLOGY   *37*

Chapter 5 Geomorphology in Canada   *39*
Chapter 6 The Changing Interpretation of Landforms   *58*
including the following reprints:
On the Ridges, Elevated Beaches, Inland Cliffs and Boulder Formations of the Canadian Lakes and Valley of St. Lawrence   *58*
Late-Pleistocene Lakes in the Ontario and the Erie Basins   *62*
Examination of the Carbonate Content of Drift in the Area of Foxe Basin, N. W. T.   *71*
Models of Hillslope Development Under Mass Failure   *78*
Landform Material Science—Rock Control in Geomorphology   *94*
Analysis of Scallop Patterns by Simulation Under Controlled Conditions   *98*

SECTION THREE   WEATHER AND CLIMATE   *113*

Chapter 7 The Status of Canadian Climatology   *115*
Chapter 8 The Climate of the Canadian Arctic   *119*
including the following reprints:
The Arctic   *120*
Influence of Vegetation on Permafrost   *142*
Chapter 9 The Climatology of Snow and Ice   *152*
including the reprint:
Micrometeorological Observations of the Snow and Ice Section, Division of Building Research, National Research Council   *153*
Chapter 10 Agroclimatology   *168*
Chapter 11 Urban Climatology   *175*
including the reprint:
Comparison of Urban/Rural Counter and Net Radiation at Night   *180*

SECTION FOUR   **SOILS AND VEGETATION**   *205*

Chapter 12   Soils in the Canadian Landscape   *207*
             including the following reprints:
                    The Canadian Soils System   *211*
                    The Soils of the Rankin Inlet Area, Keewatin, N. W. T.,
                    Canada   *219*
                    A Comparison of Luvisolic Soils from Three Regions in
                    Canada   *231*
                    Properties and Geomorphic Relationships of Soils Developed
                    on a Quartzite Ridge near Kingston, Ontario   *248*
Chapter 13   Studies in the Vegetation of Canada   *260*
             including the following reprints:
                    Forest Regions of Canada   *263*
                    Bogland Ecosystems: Some Biogeographical Units   *272*

SECTION FIVE   **OVERVIEW**   *285*

Chapter 14   Physical Geography: A Canadian Context   *287*

             Appendix   *303*
             Author Index   *308*
             Subject Index   *311*
             Index of Contributing Authors   *313*
             Index of Reprinted Papers   *313*

ACKNOWLEDGEMENTS

The preparation of a volume of this type requires a great deal of co-operation and support from many people. We gratefully acknowledge the owners of copyright of the papers reproduced herein and thank them for their support of this venture. Detailed credits of author, source, and publisher are included as footnotes in the text where appropriate and here we wish to acknowledge the reprint permissions received from the International Association for Great Lakes Research, Information Canada, Ohio State University Press, University of Chicago Press, Royal Meteorological Society, National Academy of Sciences and National Research Council, Atmospheric Environment Service, D. Reidel Publishing Company, Institute of Arctic and Alpine Research, University of Colorado, Elsevier Publishing Company, Agricultural Institute of Canada, *The Canadian Geographer,* Pergamon of Canada Limited, and University of Wisconsin.

We appreciate the assistance and support from the University of Guelph and from our colleagues in the Department of Geography, particularly Fredric A. Dahms, Chairman, and Eiju Yatsu. Additional comments and advice from D. M. Brown and D. W. Hoffman of the Department of Land Resource Science were valuable and we thank them for devoting their time to reading various drafts of the manuscript. The efforts of Michal J. Bardecki in compiling the bibliographies and Fred Adams in redrawing some diagrams have been especially helpful. Typing of the manuscript by Mrs. L. Cawthra and Miss B. Nixon was done with a great deal of patience and forebearance and we are particularly grateful for their assistance with this difficult work.

A. F.

B. D. F.

R. D. T.

PREFACE

This book is not intended to be a completely integrated treatise. Rather, it represents a volume of material which is not available in physical geography texts at the introductory level. The material we consider to be lacking falls into two categories: contextual and developmental. The contextual category is not usually included as a component of physical geography courses. However, such material shows what physical geography is, how it was pertinent to the growth of the country, and why it is once again becoming important in planning the use of our land.

In the first section of the book, we present a generalized picture of the context of physical geography which, we hope, will provide light reading for students in the early portion of a physical geography course. The material is not a definitive statement of context nor is it a documented history of Canada. The paper by K. Kelly entitled *Practical Knowledge of Physical Geography in Southern Ontario During the Nineteenth Century* was specially commissioned to provide an academic base for this section of the book. The area of study he develops is fascinating and such an approach should provide an important ingredient in studies of Canadian history. Physical geography as an influence on settlement is inextricably linked with knowledge of physical geography in all its aspects. Such a perception of physical geography is equally pertinent to present day urban and regional planning. The environmental impact of city growth is becoming more obvious but little is being done to alleviate this problem and integrate cities with the natural environment.

The developmental category contains studies which go beyond (in some cases far beyond) the limits of introductory level courses. Developmental material has two purposes: to illustrate the evolution of physical geography and to make our readers realize that knowledge of physical geography is not complete.

Sections 2 to 4 of this volume contain primarily developmental material and they are divided into the major segments of physical geography for ease of reference. Each section surveys Canadian research and progress in the appropriate subject areas. No segment is regarded as complete or comprehensive but each strives to illustrate the type of work undertaken in Canada, and the reprinted papers have been chosen for this reason. The chapter by J. T. Parry is ideally suited to this purpose and the geomorphology section benefits from his overview of Geomorphology in Canada which is a special revision of a paper initially prepared for the centennial edition of *The Canadian Geographer* (1967). Such overviews are not available for climate or vegetation and soils and so the style and content of these sections differ from that adopted in the geomorphology section.

In order to fully understand many of the reprinted papers, readers will require some understanding of physical geography. Thus sections 2 to 4 are quite different in character from section 1 and they are intended to be more advanced than that section.

We hope that these sections will motivate students to delve more deeply into the material on which their introductory texts are based. Perhaps this will provide a challenge and stimulate further study of the physical geography complex. We hope also that students will have a more complete understanding of Canadian studies in physical geography and thus become more aware of the Canadian environment.

There is no reason to be satisfied with the content of introductory courses in physical geography; the competent, intelligent student will look beyond the required material, asking "so what?" and "where do we go from here?" Consequently when sections 2 to 4 fail to perplex the reader, the volume will have served its purpose.

Section 5 is a necessary and important component of the book because it provides an overview of the subject area and complements the material in sections 2, 3 and 4. We are honoured that J. Brian Bird agreed to write the concluding chapter and we thank him for providing a comprehensive overview of a subject in which he has been active for many years.

We hope that this volume provides a useful compendium of comment, examples and challenge for students of physical geography. It is not a book with a single purpose but it is intended to be more than a convenient reprint service and we hope it will foster some curiosity about Canadian physical geography. There are many aspects of Canada's natural environment which are yet to be explained. We anticipate that future research and explanation in these topics will expand the source material available for subsequent editions of this book.

A. F.
B. D. F.
R. D. T.

Guelph 1973

Section 1
Introduction

Chapter 1

The Nature of Physical Geography And Its Significance for Canada

The nature of physical geography may be deduced from the content of books with titles indicating that they introduce the subject, from the activities of physical geographers, and from the university and college courses which teach "physical geography." One definition of physical geography is "the physical basis of geography" (Strahler 1960). Further expanding this definition, Strahler claims that physical geography consists of a body of basic principles of earth science which have been selected to include primarily environmental influences which vary from place to place over the earth's surface (Strahler 1960, p. 1). This view of environmental influences is implicit in Hare's introduction to Bird's recently published book about the natural landscapes of Canada (Bird 1972). Hare comments on the present-day interest in ecology and wider relationships between plant and animal communities and the rest of nature. He goes on to indicate that the rest of nature includes landforms, and these "together with wildlife, forests, and the prairies are part of the natural environment of Canadian society." Hare also notes the significance of landforms in the history of Canada and indicates that greater knowledge of landforms "is one of the keys to a richer and more appealing future" (Bird 1972, p. 5).

A centennial publication prepared under the auspices of the Canadian Association of Geographers (Warkentin 1968) entitled "Canada, A Geographical Interpretation" contains a very similar message; indeed, in that volume the influence of the natural environment on the whole growth of the nation is a major theme. While these comments can be substantiated from several other such works, it still remains true that no volume adequately integrates the growth of our knowledge in physical geography with the development of the country. We acknowledge the role played by the natural environment in the development of the nation. Our knowledge of the natural environment is applied to increase our understanding of human geography but to this point no one has attempted to apply knowledge of the cultural and economic growth of the nation to increase our understanding of physical geography.

In the limited space available here, the cultural and economic influences visible in the development of physical geography cannot be fully explored. There are several apparent links between the development of knowledge of physical geography and major historical events. For ease of presentation we can use the history of the emergence of Canada as an analogue to introductory physical geography courses.

If we examine the descriptions of introductory physical geography courses and the text books used for them, we find that the initial portions of the course include consideration of the world as a globe; the concepts of latitude and longitude; such matters as time zones, great circles, early navigation practices; and the orbit of the earth about the sun. All these items have a significance for Canada and for all other places on the face of the earth. It is appropriate that these items are in the introductory portions of such a course because they serve as an introduction to Canada. We should take note that the voyages of Columbus, Cabot, etc., were an expression of the very items contained in the *early* portion of a present-day physical geography course. World knowledge had expanded to a point where men were able to test the hypothesis that the world was not flat. Subsequent advances in navigation and the ability of early mariners to plot their position on a graticule of some form enabled voyages to be recorded and repeated.

All these ventures depended upon some understanding of the earth's orbit around the sun and they implanted for many centuries the use of local solar time. Only in the late nineteenth century did Canada adopt a system of standard time zones to resolve the chaos resulting from the rapid transfer of passengers by rail. The earliest railroad timetables gave time of departure from the station of origin and used the erroneous premise that a twenty-four-hour journey westwards or eastwards resulted in a time of arrival *precisely* one day later at destination. Let us take as an example a train leaving Calgary at noon and travelling east with a speed of 30-40 m.p.h. Twenty-four hours later the time would again be noon in Calgary and the train would be steaming eastwards into Winnipeg. The time in Winnipeg would be 1:28 p.m. (solar time) and thus by the timetable the noon-hour train would be eighty-eight minutes late! When travel was at a more leisurely pace, details of this type were of little consequence. If physical geography had been taught even as recently as one century ago here in Canada, it would not have included information about standard time zones because they did not exist.

The early explorers were untroubled by standard time zones but they would be considerably troubled by the duration of their voyages. If we can visualize these ancient mariners moving erratically across the ocean, we can see that they would be anxious for sight of land. Land would give temporary stability, fresh supplies of water and perhaps meat, fruit and vegetables, and above all it would be the turning point in the voyage. After visiting these new shores and claiming them for God, King and Country, the explorers had achieved much of their purpose. It only remained for them to return as heroes to their native country or their sponsoring monarchs and relate the details of the distant lands.

In many ways the introduction to North America was based on the very ingredients of introductory physical geography, the simplest example being the concept of the spheroidal world and a knowledge of its relationship to the sun. Trans-Atlantic travel in those times also led to an increase in man's perception of oceanic and atmospheric circulation. The generalized understanding of wind circulation of the world which developed from these early impressions finally led to the tricellular model of global circulation of the atmosphere still displayed in many introductory texts. The importance of this model through three centuries when all trans-ocean communication was by sailing vessel should not be under-rated. In a general way these items were of importance to the pioneers of the sixteenth and seventeenth centuries. World knowledge at that time advanced through a growth in our understanding of latitude and longitude, navigational techniques, and atmospheric circulation.

Knowledge in the sixteenth and seventeenth centuries was frequently expressed in terms of religious dogma and superstition. That knowledge appears now to have been inadequate; indeed, by our twentieth century standards the advances and discoveries of that era were things which we now tend to regard as "obvious." It is instructive to consider that to-day our knowledge remains incomplete and we could, with good reason, assert that it is frequently expressed in terms of quantitative dogma and the much vaunted scientific method. Rather than looking patronizingly back to the advances of earlier centuries we should instead consider that there has been change in technology, in the fashion of reporting investigations, and in overall levels of educational attainment.

The Emergence of Canada

We can consider the impact of sixteenth and seventeenth century knowledge on North America. It was early in the sixteenth century that Canada was "discovered." Discovery is a totally inappropriate term because the indigenous peoples had already "discovered" the North American continent centuries before this. Canada did not exist as a dominion until 1867 and certainly did not have its present form until Newfoundland joined Confederation in 1949. We must therefore extend the use of "Canada" into the pre-Confederation era to designate the present territory even before the political birth of the nation.

Carefully avoiding the disputed claims to the "discovery" of the New World and directing our attention to what is now Canada, it is interesting to consider the needs of the explorers and the colonists arriving here from Western Europe. It matters little which visitors first capture our attention. Perhaps we should consider the Viking settlements; or the indigenous peoples; or the national and political fortunes of the French, English, and Portuguese who visited the eastern shores of Canada and attempted to settle there. Whichever historical niche we choose, there can be no doubt that these visitors would be enormously relieved to see land.

This itself establishes the basic structure of physical geography. Once we acknowledge that the earth has a surface, we thereby permit it to have a surface form or morphology. It is quite certain that the morphology of the land was not the initial concern of the explorers who would be more intent on replenishing their supplies of fresh water and food. However, land of any form was the first quest of the explorers since supplies of fresh water in those times would be an important item for a ship's crew and the supplies of fruits and vegetables an item of similar significance. The meat to be obtained from the indigenous animals would be a considerable bonus. Any supplies obtained from indigenous peoples could be an unexpected addition.

These items were an expression of the explorers' perception of physical geography. Streams would provide fresh water, berries and fruits would be more readily obtained in heathland, and forest animals would not usually be found on barren uplands. Native settlements would be sited with maximum protection from the climatic hazards and would therefore be found in sheltered areas probably in regions of well-drained soils. Thus there would be a vital significance to the total complex of landform, vegetation, soils, and climate which would provide support for the life system represented by the explorers' vessel. The "natural environment" or the "physical environment" was, at that time, understood as an ecosystem in a very basic way and evaluated as a life support system by the early explorers who were concerned that it should support *their* lives.

Political reality also loomed large as the successful explorers returned to their sponsors, patrons, benefactors or monarchs to report on the findings of their journeys. We will here accept that any explorer succeeded if he returned alive, since a definition of "success" detracts from our main purpose. The successful explorer described in detail the nature of the land visited, its peoples, wildlife, vegetation, and potential wealth. The dominant aim of the sixteenth century explorers was to find a sea route to China and to establish trade in silks, spices, and the supposed extreme riches of the Orient. Such was his perception of the Orient that Cabot believed he had attained his goal when he landed in Newfoundland (or Cape Breton or Labrador) in the summer of 1497. Similarly Cartier, on discovering the St. Lawrence estuary, felt he would have access to the Orient by sailing westwards through it.

During the sixteenth century, evaluation of the resources of a continent was conditioned by the requirements of the sponsor. Champlain's early travels and the details of the attempts to establish colonies indicate clearly the impact of the physical environment, but his sponsors were interested in furs and territorial expansion rather than an evaluation of the quality of the new land. They required that the land could support the garrison forces to maintain their trading position and a few settlers so that the country could properly be regarded as a colony. Similar evaluation of the environment is to be found in the early records of the Hudson's Bay Company and the chronicles of the early British settlements.

The Nature of the New World

The explorers returning to their port of origin had to satisfy many inquiries about the new lands which had been claimed in the name of the monarch. It was out of this need to convey information that we have our early maps of the New World, which, in fact, were frequently dedicated to nobility or royalty. Taken in sequence these maps show the development of cartographic techniques, the improved perception of map projections and their properties, and the progressively deeper penetration of explorers into the interior of the continent. Through a period of three centuries from 1600 to 1900, these maps also show the increased perception of the nature of the land and its resources.

Perception of the New World as a resource area is perhaps the earliest manifestation of European interest in North America. The fishermen had discovered the Grand Banks quite early in the sixteenth century. British explorers in 1527 attempted to find the elusive north-west passage to the Orient, and their expedition landed briefly in Newfoundland where they recorded observation of fourteen fishing vessels of Portuguese and French origin.

The detailed record of the visitors to these shores is more properly a part of historical and cultural geography. We can here note the settlements which developed along the Nova Scotia shore where fishermen established homes during their summer season's fishing. Many of these fishermen bought furs from the local Indians for knives, ribbon, or cooking pots. Furs became valued possessions in Western Europe and eventually were decreed to be fashionable. The demand for furs became considerable, and much of the competition between the French and English in Canada had its roots in the desire to control the lucrative fur trade. This, however, was an animal resource and as such is not usually considered in the content of physical geography at the introductory level.

Merchants in control of the fishing, but themselves comfortably established in Britain, attempted to prohibit settlement in Newfoundland. It is clear that they were not completely successful in this venture; they did, however, inhibit settlement there until the late eighteenth century. A recently published study (Head 1972) indicates that other factors were of importance in delaying the establishment of permanent settlements in Newfoundland, notably no satisfactory source of carbohydrates and an unattractive physical environment. When permanent settlements were established, Head notes that there was a rapid deforestation of the coastal lowlands. This resulted from a demand for timber for construction and firewood and the inherent difficulties in reestablishing seedlings in areas exposed by fire and the cutting of timber. Head concludes that the delay in the colonization of Newfoundland was because the conditions necessary for successful year-round habitation were lacking.

The late fifteenth and early sixteenth centuries were a period of frequent visits by explorers; for example, in 1497 John Cabot landed in Newfoundland believing that he had landed in China. The years 1501 and 1502 are notable for the visits of Gasper and Miguel Cortereal, two Portuguese mariners who also visited the coast of Labrador and ventured as far south as the Bay of Fundy. The date Sebastian Cabot (son of John) sailed into Hudson Bay is considered to be 1509. In 1524, the French explorer Cartier formed the opinion that there was a substantial gulf to the west of the island. In the context of the current world knowledge, such a channel clearly led to the east and the Orient. Cartier returned in 1534 to sail along the southern coast of Labrador and the following year (1535) Cartier again explored the Gulf of St. Lawrence with a three-ship expedition. Penetrating further westward he was saddened to find the two shores converging—he had discovered merely a river! He reached the site of present-day Quebec (it was then the Huron-Iroquois village of Stadacona) and leaving two ships and the bulk of his party there, Cartier continued in his smallest vessel and reached the site on which Montreal is now built. He returned to Stadacona where his party had built a fort.

The winter at this site (Quebec City) was far more severe than the explorers anticipated. The latitude was equivalent to Lyons (France) and some three degrees further south than Paris. Four feet of snow and the freeze-over of a salt-water estuary accompanied by a shortage of supplies forced the explorers to trade with the Indians who, for a large number of hatchets and knives, supplied a few smoked eels. Twenty-five men from the group of seventy-four died of scurvy and in May, when the river finally cleared of ice, survivors set sail for France.

On his return to France, Cartier provided a glowing account of the new lands, and even though he was reluctant to return, he eventually did so and wintered once more on the St. Lawrence. He was followed by Roberval who had been appointed Viceroy of New France and charged to establish this colony in the New World. Roberval returned to France without establishing a colony as great as the New Spain to the south of the Great Lakes.

To this point physical geography had played an important role in the development of Canada. The conditions off the east coast provided a rich fishing ground which attracted the vessels of many western European nations. In the summer months the adjacent land areas appeared to offer a favourable environment for settlement. These areas also were rich in furs. The climate which was responsible for the excellence of the furs was not fully understood by potential settlers. This is reflected in the severe difficulties encountered by Cartier's party when it attempted to winter at Quebec City. Likewise, the failure of settlement in Newfoundland in this era can be attributed to an adverse natural environment (Head 1971, 1972).

During the half-century following Cartier's return to New France, the demand for fur as a fashion fabric in Europe rendered the fur trade a most lucrative undertaking. The very conditions of severe climate that made the furs of North America so desirable were the very conditions which inhibited settlement.

In 1603 a further expedition arrived in Canada and the observer and cartographer of this expedition was Samuel de Champlain. Champlain, a dedicated settler, made an alliance with the Algonquin Indians who occupied lands along the St. Lawrence River which was then the major communications channel. Champlain explored this river in a manner which was to determine the pattern of Canadian development.

On reaching the Quebec narrows, he commented on the beauty of the country. He noted that the soil would, if tilled, be as good as the soil of France and he described the vegetation association of oaks, cypress, birch, firs, aspens, fruit trees, and vines. As Champlain progressed along the river, he evaluated the geomorphology and noted that the site of Three Rivers provided a good site for a permanent fort. He explored the Richelieu River but only as far as the rapids of St. Ours and then returned to the St. Lawrence. Geomorphic control of his journeys of exploration was clearly evident here. A similar problem was posed by the Lachine Rapids. He obtained further information by questioning the local tribes and gathered a verbal "picture" of Lakes Ontario, Erie, and Huron. He formed the impression that these lakes were the great western sea beyond which lay the Orient.

Champlain's return engendered more enthusiasm for Canada. The King (Henry) agreed with Champlain's thesis that a monopoly of the fur trade was a necessary thing which would be bestowed by governmental gift so that the recipient could be pressured into establishing colonies in Canada. On his return to the New World, Champlain established a settlement on Ste. Croix, a disastrous settlement where thirty-five from a party of seventy-nine died during the first winter.

The following winter, however, saw a settlement established at Port Royal in western Nova Scotia (now the Port Royal National Historic Park). For this winter, the settlers were more adapted to their environment, and they caught rabbits and other small animals to keep their settlement well provisioned. Champlain produced very competent and detailed maps of the areas he had visited and these showed hills, rivers, and vegetation. He also cultivated a garden and thus began the process of evaluating land for crop production. Bishop (1972) in his study of this era sums up the relevance of Champlain's activity when he notes that "Champlain elected to stay in America in order to join the exploring party. This was his duty as a geographer."

There was no doubt about the role of a geographer in the seventeenth century. He was an explorer and recorder of unknown lands and his *duty* was to record the coasts and rivers, hills, valleys, mountains, and forest for his patrons and for future settlers.

It is not at this point appropriate to discuss in detail the remainder of Champlain's exploits. He was a remarkable man and a devoted colonist who did much to promote viable settlements in eastern Canada. The influence of politics in this era of discovery can, however, be paralleled by more recent developments. In the above discussion it is apparent that knowledge of physical geography expanded as a consequence of the French desire to exploit the fish and fur resources of Canada. We shall see below how the desire to exploit the agricultural resources of Ontario led to a better understanding of soils and vegetation in that area. Similar events can be traced through the era of forest exploitation and the settlement of the Prairies, and it is interesting to note that at the present time developments in the Arctic are following a similar pattern. In all cases there is an initial period of exploitation which seldom is based on a study of the natural environment. In the past this has approached crisis point before a knowledge of physical geography has been developed and applied, e.g., the deforestation of the Newfoundland coastal lowlands.

An additional example would be the complete destruction of eastern Canada's hardwood and mixed forests and subsequent soil erosion. The continuous cropping of soils until they become exhausted has been a feature of exploitive agriculture across the country, and the development of badlands in Alberta or incipient dust-bowl conditions elsewhere in the Prairies are well-known examples. These events reflect a poor understanding of the processes at work in the natural environment. The present-day advances into the Arctic appear to be a little more soundly based in that some outline information about the physical geography of the Arctic is available. However, detailed knowledge of all processes acting in the Arctic and northern areas does not exist. It seems that physical geography will expand as Canada's north is exploited but, once again, much of the knowledge will be gained as a consequence of economic exploitation.

To provide a clearer insight into this situation the following paper has been written. It shows how settlement in Ontario during the nineteenth century was undertaken by people who did not necessarily understand the nature of the soil or the climatic requirements of the crops.

Chapter 2

Practical Knowledge of Physical Geography in Southern Ontario During The Nineteenth Century*

Kenneth Kelly

This paper outlines the growth of practical knowledge of physical geography in southern Ontario during the nineteenth century and indicates some of the ways in which physical conditions affected the pattern and character of agriculture. During the nineteenth century there were romantically motivated, aesthetically based discussions of the physical geography of the province and "scientific enquiries" concerned largely with the origin of phenomena. The paper ignores these except where they contributed to, or became part of, practical knowledge. It is concerned solely with the development of a practical common lore built from the observations of many settlers and widely employed in the development of agriculture. This lore, to which the paper now turns, was the result of a trial and error learning process designed to lead to farmer awareness of, and adaptation to, physical conditions.

Early Awareness of Vegetation and Soil Patterns

During the first half of the nineteenth century the development of knowledge of vegetation patterns proceeded on two scales. Travellers sketched in the basic distribution of the three major vegetation types of southern Ontario: forest, marsh, and "plains land." At the same time the government surveyors appointed to peg out farm lots produced very detailed information on the tree cover of the land. The travel literature showed marshes, sometimes referred to as wet prairies, flanking Lake St. Clair, the Detroit River, and the lower reaches of the Thames. It described "plains land" as savanna-like dry grassland with stunted oak or pine growing singly or in groves. "Plains land" occurred south of Rice Lake and close to the Lake Ontario shore in the vicinity of Oakville. But it reached its greatest extent in Norfolk County with a large outlier just west of Brantford and smaller ones stretching into Elgin County. The published literature did not offer a clear picture of the forest which covered the rest of southern Ontario. It was more concerned with the forest as an obstacle to progress and with aesthetics and romantic imagery than with species composition. But it was precisely in the detailed description of forest vegetation that the surveyors' reports excelled. The data on the occurrence of tree species which the surveyors recorded never were mapped but were available at Toronto in the form of traverse notations.

Using these data the settler could discover the character of the vegetation on any lot which interested him. Sellers could also advertise their lands as covered by

* © Kenneth Kelly, Department of Geography, University of Guelph, Guelph, 1973. Reprinted by permission of the author.

a particular type of forest, and the classification of forest types was quite sophisticated as will be seen shortly. Thus the descriptions of vegetation in the published literature convey only a poor impression of the settler's knowledge of his surroundings. Although he had at most a crude image of patterns and species composition on a broad scale, he had a detailed picture of the forest communities in the small area in which he would live and work.

The settler had at his disposal rules of thumb learned from neighbours or read in the numerous and inexpensive settlers' guides, which allowed the use of vegetation as an indicator of the quality of soil and site. These rules of thumb had been developed largely in the northeastern United States and Quebec, where settlement had begun earlier. The ready-made system of land classification was based on the belief that soil and drainage conditions exercised a dominant (or even exclusive) influence on the pattern and character of vegetation; as will be discussed later, the settler had no concept of man as an agency of vegetational change or of plant succession. However, given a knowledge of the forest cover of any lot the settler was able immediately to apply the rules of thumb and gauge, with varying degrees of accuracy, the capability of the land for wheat cultivation[1] and the scale of investment necessary to bring it into production.[2]

The settlers' guides written during the first half of the nineteenth century clearly document the classification of soils in terms of their capability for wheat.[3] According to the rules of thumb, pure stands of pine were underlain by the very worst soils, infertile and sandy. These soils were developed on what later were recognized, as the glacial and postglacial history of the province became clearer, to be the most thoroughly drained portions of sand plains or, in places, kame moraines.[4] The savanna-like "plains land," which also was judged poor for wheat, similarly occupied sand plains. Where hardwood stands had an admixture of pine, or where beeches grew in association only with hemlock, settlers deemed the land at best mediocre. They held both pine and hemlock to be indicative of a high proportion of sand in the soil. Several of the hardwoods when growing alone or in certain combinations were diagnostic of a poor or mediocre wheat soil. Oak and chestnut together dominating a site (commonly on sand plains), or beech accompanied only by maple (on well-drained uplands, often on kame moraines where these were not occupied by pine) indicated light, sandy, and dry soils. While these lands would produce good quality crops for a few years, they could not sustain successful long-term wheat cultivation. Most agricultural writers considered the swamp lands under cedar, fir, spruce, and tamarack to be of high fertility. This judgement was essentially sound. Although they sometimes grew on low, waterlogged areas of sand plain, these trees commonly characterized what were discovered later to be muck soils, glacial spillways, and the lower lying portions of clay plains. But, for reasons to be outlined below, few settlers located on these lands. There was general agreement, however, that first class wheat land was found under a mixed hardwood cover which included maple, basswood, elm, and beech. These mixed hardwood forests occupied the loams and clay loams of the till plains and better drained sections of the clay plains.

Settlers also classified land in terms of development costs, the amount of immediately useable space these investments produced, and site characteristics affecting the time of sowing and harvesting. In large part, these matters were a function of drainage conditions. The surveyors' reports provided a base of information. They employed, but did not define, the terms marsh, swamp, wet, and springy to describe some of the lands they traversed. Settlers used the type of forest cover as an indicator of drainage conditions where these were not specified by the surveyors. According to popular lore, the sandy soils under pine suffered from a permanent water deficiency. On the somewhat better oak-chestnut and beech-maple land, plant roots had difficulty getting moisture once the humic mould had gone from the soil surface. However, for a few years these sites offered great advantages. Settlers immediately could (hand) cultivate all of the land except that occupied by the stumps of the relatively widely spaced trees; there were no wet patches on these sites. Settlers with very little capital commonly chose this land for their first farm because of the ease of bringing it into production coupled with the acceptable, but not bountiful, wheat yields over the short term. Conversely, very few settlers located on lands under cedar, tamarack, fir and spruce. In contrast to other forested sites, clearing alone was insufficient here to permit successful cultivation. These sites were swampy or very wet and required artificial drainage. To accomplish this, settlers had not only to clear but also to stump and deep plough the land and then dig drainage ditches. Because the development costs were too high, most settlers avoided these lands despite their potentially high productivity. Settlers could cultivate the mixed hardwood lands immediately after clearing, but realized a smaller useable area per acre than on the oak-chestnut or beech-maple sites. Generally speaking, the tree cover was denser on the mixed hardwood lands so that more stumps remained on the field. Furthermore, seasonally wet patches occurred which were underlain by a relatively impermeable clay subsoil. Nevertheless, because water always was available to plant roots and because they produced high wheat yields, the lands under mixed hardwoods were the preferred sites for agricultural settlement during the first half of the nineteenth century.

The settlers' evaluations of land sometimes were inaccurate, for the rules of thumb were based on generalizations on the site tolerance of trees. For example, maple and elm were regarded only as indicators of a well drained site. In fact they often also grew on till with an impermeable clay subsoil and which, therefore, was wet and cold in spring. On such land, and on the clay plains, spring cultivation was delayed until the standing water evaporated from the surface. Unless farmers invested in artificial drainage, harvests were late and yields poor. Furthermore, the assessment of land sometimes was erroneous because the popular folklore viewed wild vegetation patterns as static and, on the small scale, as the result only of soil and drainage conditions. Initially it had no place for the concepts of plant succession and of changing mosaics of vegetation following fires and the destruction of dominant trees.

The Emergence of an Awareness of Plant Succession

Although the literature, with very few exceptions, ignored the problem, individual settlers as they cleared their land rapidly developed an awareness of plant succession. Catherine Parr Traill noted with some surprise that the young of the original dominant vegetation did not immediately re-occupy a site following clearing.[5] She, like most settlers, found that formerly minor constituents of the forest flora rapidly spread into the clearings. On lands cleared of hardwoods, fireweed (also known as wild lettuce) first invaded the fields; and this was followed by choke cherry or sometimes sumach. Where pines had once stood, the raspberry was the first colonist, succeeded often by aspen poplar.[6] These and doubtless other successions occurred on abandoned clearings and threatened the farmers' fields, choking out their crops. Recolonization constituted Ontario's first and greatest weed problem. It forced farmers with little capital to resort to the biennial naked fallow which they harrowed repeatedly (or after stumping, ploughed) to control weed growth. Where settlers attempted to take a crop from each field every year, they found their plants overtopped by "underbrush" and the cost of an intensive agriculture was a great labour investment in weeding.

During the 1830s several writers had called attention to the role of fire as a factor explaining vegetation patterns. Patrick Shirreff observed that on the savanna-like "plains land" the vegetation did not necessarily reflect soil conditions. In his opinion, fires were responsible for the sparseness of the tree cover and for the areal dominance of grasses.[7] Mrs. Traill, describing the Rice Lake plains, noted that the Indians frequently had burned over the land to improve the forage for deer.[8] These comments evoked no echo in the general literature until the 1880s, by which time both government and lumbermen had become concerned about the effects of fires on the province's timber resources. They commented frequently on, and bemoaned, the replacement of good merchantable pine—after fires—by a worthless second growth of birch and poplar.[9] A general awareness of fire as an agent of vegetational change and of plant succession finally emerged strongly in published form and began to shape government policy on timber management on the Canadian Shield.

Knowledge of Soil Depletion and Erosion

The early agricultural literature warned settlers—on the strength of the experience of older lands—that unless they practised mixed farming, their soils rapidly would become exhausted. However, the great majority of settlers conceded the likelihood of rapid exhaustion only on the fragile sandy lands under oak-chestnut and beech-maple. They believed that most of Ontario's soils, especially the loams and clay loams, were much more durable, capable of producing satisfactory crops of wheat for many years without manure and with no crop rotation other than the alternation of wheat with a naked fallow. Indeed, the settlers' guides assured them that most of their soils were "strong" and that the mixed hardwood lands "would yield wheat for ever."[10]

Although they denied the likelihood of exhaustion on most soils, farmers soon became aware of soil depletion. They saw their fall wheat yields drop—on mixed hardwood land—from an initial average of thirty-five or even forty down to about twenty-five bushels per acre. They associated this decline with the dissipation of the rich organic mould inherited from most kinds of deciduous forest. By mid-century, farmers had begun to turn to manure and improved crop rotations as a means to increase yields. This course of action was influenced in part by the published reports on the benefits of crop rotation derived from advances in soil chemistry, but to a larger extent by the opening of markets for livestock and for crops other than wheat. New market opportunities made it possible for farmers to combat depletion while at the same time increasing their immediate income. However, farmers and the literature ascribed the loss of soil nutrients solely to upward removal via crops; the significance of leaching in general and specifically the process of podzolization remained unrecognized.

By mid-century farmers sought to counteract soil depletion through changes in land management, but they did not even recognize the problem of soil erosion until close to the end of the nineteenth century. During the 1880s they observed that large scale deforestation had caused a change in the run-off:percolation ratio and saw that, with greater run-off, top soil was carried away.[11] Soil erosion was most evident in the 1880s in the older cleared lands close to the shore of Lake Ontario but it also was noticeable, and a cause for concern, in the more recently opened areas. For example, farmers in St. Vincent township (Grey County) lamented the loss of valuable top soil which was caused by the clearing of the heights of land and steeper slopes.[12] Observers also noted erosion along the banks of streams after the forest had been stripped away. The farmer response to the erosion problem was a part of the general reforestation drive which gained momentum towards the end of the nineteenth century. Reforestation remains, of course, a major method of erosion control at the present day.

Climate

Settlers had no rules of thumb for classification of climate; their knowledge grew slowly through experience and observation. Many gentlemen settlers brought thermometers with them and took delight also in measuring rain and snow fall. However, the accumulation of such data made little contribution to practical common lore. Because they did not know the temperature and rainfall requirements of a crop, farmers could not judge from such statistics whether or not it would grow.

Through experience, settlers built a store of knowledge of general climatic conditions which had a direct bearing on agriculture. Compared to England, the spring sowing season in Ontario was late and short. The ground was "bound up" by frost until the middle of March and some of the fields remained too wet and heavy to cultivate until mid-April.[13] During the 1830s, before the artificial draining of wet patches of land had begun, farmers near Toronto seldom were able to sow their barley, oats, or peas (or indeed spring wheat) before the twentieth of May.[14] In the same area the hay season began on about the tenth of July and the general harvest in early August. But harvest, like spring sowing, was rushed. A "parching heat" made the straw dry and brittle, and this caused the loss of heads of grain unless farmers harvested their crops quickly. Where farmers raised spring crops, harvest times were particularly hectic. All the spring crops matured simultaneously and demanded farmers' attention at the same time. These climatic problems, together with those of seasonal waterlogging, slowed down the intensification of Ontario's agriculture and demanded the acceptance of labour-saving machinery, which in turn called for the removal of stumps, roots, and rocks from the fields.

Farmers' awareness of climatic patterns relevant to particular crops came in a direct way. They sowed crops, some perhaps speculatively and others for their own subsistence; several years later, they decided, after an examination of average yields and as a commercial market opened, to incorporate permanently into, or drop from, their agriculture a specific crop. In this manner farmers discerned critical boundary lines and transition zones and delimited crude climatic regions. They could not describe the climate of a region, but they did know that the crop would grow well in it (assuming that soil and drainage conditions were appropriate). The first climatic transition line could be sketched in by the 1850s. South of it the cultivation of fall wheat was profitable; to the north farmers had to make shift with the less remunerative spring varieties. No publication ever described the course of this transition line across southern Ontario. But an examination of the agricultural society reports makes it clear that its location was well known at the county level. Its position was a part of a local, and not provincial, folk knowledge.

The classification of regions according to the suitability of their climates for specific crops continued through but by no means was complete by the end of the nineteenth century.[15] The growing awareness of climatic transitions and regions cannot be documented convincingly at the present time because of the lack of studies. However, the climatic regions best suited for fruit production were found by the end of the 1870s. For example, farmers in Simcoe County had discovered the pockets of favourable climate for orchard fruit production around Lake Simcoe and between Georgian Bay and the Niagara escarpment near Collingwood. At about the same time farmers in parts of the Niagara peninsula began to capitalize on the favourable fruit climate there.

The last quarter of the nineteenth century also witnessed the expanding knowledge of the occurrence of climatic change and of the relationship between vegetation and the climate near the ground. The early settlers looked for—and much of the promotional settlement literature promised—great changes (always for the better) as they cleared the land. These climatic improvements never materialized. Instead, by the 1880s, farmers noticed a deterioration of the climate insofar as it affected agriculture. They related this deterioration essentially to deforestation. Following overclearance, the winds in winter swept without obstacle across the country, blowing the snow from the fields planted with fall wheat. (Commonly this snow was dumped in drifts across the highways making winter travel extremely difficult.) Deprived of its insulating snow cover, the wheat frequently was killed in late winter or early spring as alternate freezing and thawing heaved the young plants out of the ground. While in earlier decades many farmers had come to believe that soil depletion would destroy Ontario's advantage as a fall wheat producer, now they held winter kill to be the major threat. The cold winds sweeping over wide stretches of land also implied higher costs of livestock production. Not only were the barns now colder during winter than when more forest protected the land, so that animals required more feed to keep in good shape, but also the lack of shelter brought problems to hay production. Nor were fall wheat production and livestock raising the only sectors of the agricultural economy affected. The emerging orchard fruit growing industry suffered losses because of the lack of shelter in winter.

By the 1880s farmers also complained about the deterioration of the summer climate. They now experienced prolonged and often severe summer droughts where none had been noticeable when more trees clothed the land. Again, they saw a causal relationship between the change in the microclimate and deforestation. Many farmers believed that the destruction of the forest had caused a reduction in rainfall and a change in its distribution through the year. For example, residents of Markham township agreed that "the rainfall is much more variable and distributed in a manner less useful to the farmer than when larger forests existed through the land."[16] However, more perceptive observers saw increased summer drought as a function not of diminished rainfall but of a change in the run-off:percolation ratio and the fact that winds now came sweeping over the land without hinderance.[17]

In their recognition of a change in the ratio of run-off to percolation, farmers saw that forest cover functioned to "store" run off and to regulate the discharge of water. Although the concept of the hydrological cycle was not yet formulated, farmers now understood much of what happened to water between the time it fell as rain or snow and the time it reached the Great Lakes. A farmer in Bruce County expressed this comprehension well when he wrote that after deforestation there was no reduction of rainfall:[18]

> But there is quite a change in the manner of it (rainwater) getting into creeks and rivers. Formerly when this country was covered with forests, the rain very gradually soaked its way into the ground and thus slowly raised the springs, creeks, and rivers. Now, after rain, the creeks receive it much sooner, as a

great deal of water does not penetrate the soil at all, but flows over the surface and gets at once into the creeks, and raises them sometimes very suddenly. . . . And for the same reason we have more floods and also more times of low water.

This kind of awareness, coupled with the recognition of the problem of erosion, the fears of the impending collapse of water-powered industries, and the growing shortage of fuel and building timber, produced a new farmer attitude to forests.

Farmers attributed much of the deterioration of their environment to deforestation. To them and to the provincial government, then, the means to rebuild a productive landscape seemed clear. The substantial reforestation of southern Ontario was called for. An image of the new landscape which would help sustain the prosperity of rural southern Ontario was developed. The "sandy wastes" (sand plains and kame moraines), the heights of land, and the steep slopes were to be planted with trees. Sheltering rows of trees were necessary between fields, along highways, and around farm buildings. Stream banks were to be marked and protected by a regrowth of woodland. Most farmers subscribed to this ideal, and many began replanting trees. The shelter planting and the farm woodlot became significant features on the landscape. However, the majority of farmers, while grasping the nature of the environmental problem which confronted them, were unwilling to invest time or money in improvements which would bring direct economic benefit only after many years.

The Classification of Land at the End of the Century

By the end of the nineteenth century judgements of land quality resembled those current at the present day. Farmers not only classified land in terms of its capability for wheat but also for a range of other crops. For example, the recognition of wet, heavy, clay soils as good for dairy farming had come during the late 1860s and early 1870s with the spread of cheese factories. During the 1870s both soil and climatic regions suitable for fruit production were demarcated. Towards the end of the nineteenth century farmers adjusted more closely than ever before to soil quality in their land use patterns. The patterns of agricultural specialization which are presently observable had begun to take shape.

Similarly, the areas now considered marginal for physical reasons were largely recognized as such by the end of the nineteenth century. For example, the wave of enthusiasm for the agricultural colonization of the southern part of the Canadian Shield had subsided. The large proportion of its area occupied by rock outcrops and swamps, the generally thin soils, and isolation from markets in combination helped divert the tide of settlement to the western provinces. The provincial government, supported by lumber interests, inclined more and more to the view that the Shield should be developed as a sustained yield forestry territory. Furthermore, it was generally conceded that the sandy plains and kame moraines of southern Ontario also offered a poor base for a prosperous agriculture. These were destined in the popular judgement for abandonment or reforestation—despite small scale attempts at truck farming—for tobacco had not yet become a significant commercial crop. Drumlin fields either were abandoned or converted to low grade pastures. During the 1870s and 1880s swamps and marshes had briefly seemed to be the new internal frontiers of southern Ontario. They were described as potentially "the very best of farmland" which could be reclaimed at relatively low cost. In these years significant areas of swamp and marsh were reclaimed or improved, especially in the southwest of the province. However, by the end of the 1880s, partly as a function of agricultural depression, the enthusiasm for large scale drainage had waned. Swamps then were perceived as permanent gaps in the agricultural settlement pattern.

Conclusion

Throughout the nineteenth century the physical setting exercised a strong influence on the character and pattern of agriculture. Adaptation was a basic theme in the agricultural evolution of southern Ontario. However, farmers adapted to an entire complex of total environmental factors, of which physical conditions represented but one set. Farmers' activities also were shaped by factors such as the range of market demand and the distribution of agencies expressing it, transportation networks and costs, and capital availability. The physical milieu shaped early agriculture with a heavy hand. The pattern of early farming reflects that of lands judged both good for wheat and relatively easy to bring into production (although ease of access to markets also played a part). The extensive character of most early agriculture can be ascribed to the demands for labour for non-farming operations such as clearing, stumping, and the removal of stones from fields sited on certain kinds of till deposit. The biennial naked fallow clearly represented a low labour input attempt to control the recolonization of fields by forest plants.[19]

Farmers' adaptation to their physical geography became more subtle and sophisticated during the second half of the nineteenth century. During this period farmers saw more clearly and in greater detail the physical character of the province and became aware of the processes shaping it, and they also recognized their own role in modifying the land. Adaptation to these newly perceived characters and processes can be seen on all scales in the agricultural geography of southern Ontario. In improved crop rotations and the greater emphasis on manure, shelter plantings and woodlots; in the areal specialization of agriculture; in the abandonment of large areas; and in the establishment of Algonquin Park: in these, one discovers portions of the adaptive "strategy" of the latter part of the nineteenth century. However, it must be made clear that this adaptation was to economic as well as to physical conditions. Thus improved crop rotations were adopted to counter-act soil depletion, but their acceptance by farmers seeking maximum immediate profit was made possible by a broadened range of market demand. The planting of trees on the farm and the establishment of woodlots was in part a response to a new awareness of soil erosion and to the recognition that the climate near the ground and the run-off:percolation ratio had changed for the worse following deforestation. However, it also was a reflection in part of the fear that living costs would rise if coal had to replace locally produced wood as a source of heat and energy. The areal specialization of agriculture allowed farmers to capitalize on optimum conditions for the raising of certain products. But these conditions were not entirely physical. Transportation, processing, and marketing facilities also were important factors explaining the new pattern of agriculture. The establishment of Algonquin Park not only reflected the new evaluation of Shield land as marginal for agriculture, but also the desire to preserve an area of wilderness within southern Ontario and to capitalize on the growing demand for recreation.

REFERENCES

1. Wheat was the only crop raised in southern Ontario for which there was a significant commercial demand during the first half of the nineteenth century.

2. In other areas, in Britain for example, observers were beginning to classify and evaluate soils by means of the bed rock on which they were developed. This approach was not followed in Ontario in part because of poor knowledge of local geology and in part because patterns of "natural" vegetation were readily observable. Nor was the Ontario farmers' approach to the classification of land unreasonable. To many observers still, vegetation represents the most subtle guide to the sum total of environmental factors. It should be noted also that in the nineteenth century, Ontario practical folk lore was not concerned with explaining the character of soils, but only with recognizing good ones.

3. The material for this and the following paragraph is taken from Kenneth Kelly, "The Evaluation of Land for Wheat Cultivation in Early Nineteenth Century Ontario," *Ontario History*, Vol. LXII (1970), pp. 57-64.

4. The physiographic terms employed here are those used in L. J. Chapman and D. F. Putnam, *The Physiography of Southern Ontario* (Second edition, University of Toronto Press: 1966).

5. Catherine Parr Traill, *The Backwoods of Canada*, New Edition (1846), p. 87.

6. Samuel Thompson, *Reminiscences of a Canadian Pioneer*, (Toronto: 1884), p. 74.

7. Patrick Shirreff, *A Tour of North America* (1835), p. 160.

8. Catherine Parr Traill, op cit., p. 86.

9. See, for example, the *Ontario Forestry Report for 1891*, pp. 8 and 16.

10. Thomas Radcliffe (ed.), *Authentic Letters from Upper Canada* (Toronto: Macmillan, 1953), p. 90—reprint of the 1835 edition.

11. See, for example, Thomas Beall, "Personal Observations on the Effects of the Removal of our Forests," 1st Prize Essay, *Farmer's Advocate*, Vol. XXII (1887), p. 37.

12. *Ontario Forestry Report for 1886*, p. 27.

13. William Hutton, *Letters on the Prospects of Agricultural Settlers in Upper Canada* (1835), p. 107; Frederick Widder, Pamphlet on the Canada Company (1843), p. 2.

14. William Hutton, op. cit., p. 107.

15. In fact nothing approaching a full awareness of climatic patterns across the whole of southern Ontario relevant to agriculture developed until the publication of Putnam and Chapman's "The Climate of Southern Ontario" in *Scientific Agriculture* (pp. 401-446) in 1938.

16. *Ontario Forestry Report for 1886*, p. 17.

17. Thomas Beall (1887), p. 37.

18. *Ontario Forestry Report for 1884*, p. 19.

19. See Kenneth Kelly, "Wheat Farming in Simcoe County in the mid-Nineteenth Century," *The Canadian Geographer*, Vol. XV (1971), pp. 95-112.

Chapter 3

Physical Geography in The Early Twentieth Century

Kelly's illustration of the growth of knowledge about Ontario's physical geography in the nineteenth century suggests items which are of significance to twentieth century Ontario. The first Ontario Soil survey was not undertaken until 1914 and the first map was not published until 1923. A quarter century later Chapman and Putnam published their definitive study "The Physiography of Southern Ontario" which was an excellent record of the physical geography of southern Ontario. This piece of work considers geomorphology, soils, climate, vegetation, and agricultural activity in a grand overview of the complex of phenomena which constitute a landscape. Their descriptive evaluation of an extended area has now become widely used by municipalities, planning and conservation authorities, and many other levels of government. It has also been applied in evaluating the whole of southern Ontario for many purposes. The fact that such a volume has only been available for a comparatively short time is very important. Such studies are not available for most of Canada and settlement and development of the country has proceeded without the aid of such integrated studies. The result has been an over-extension of the cultivated area, parts of which are now being allowed to revert to rough pasture. Results on a personal level include the failure of pioneer farms which were subsequently abandoned and the demise of the small businesses which were established to serve such settlements.

The great areas of Prairie and the northlands had been opened to settlement by the construction of the transcontinental railway by the latter part of the nineteenth century and less than seventy years ago a select committee of the Canadian Senate was considering evidence about Canada's fertile northland. Much of this evidence provides an interesting insight into the perception of physical geography when many present-day citizens were in their early childhood. Two generations ago the evidence of Mr. Fred Durnford was recorded as follows:

EVIDENCE OF FRED G. DURNFORD, C.E., OF THE DEPARTMENT OF THE INTERIOR, DELIVERED BEFORE THE SELECT COMMITTEE FEBRUARY 12, 1907. *

Mr. Durnford explained the map last mentioned by Mr. Young, the preceding witness, and which he (Mr. Durnford) had prepared. This map was on a scale of twelve and one-half miles to the inch, an enlargement of the map made by the geographer of the department and extending from north latitude 54 up to 60, and from 93 degrees up to 120 degrees, west longitude. Each of a number of squares shown upon the map contained approximately 2,600 square miles or 1,664,000 acres.

Land which, from the information in the department and from other sources (a large number, over one hundred authorities con-

* Reprinted from *Canada's Fertile Northland* edited by Capt. E. J. Chambers (Ottawa: Government Printing Bureau, 1907), pp. 30-32, 34, 35. Reprinted by permission of Information Canada.

sulted) is suitable for cultivation, was shown on the map in red. Land about which there is very little information was shown in yellow, land which is muskeg or rocky, or generally unsuitable, in brown.

Areas which contained spots of good land per cent. All the land coloured red, including were indicated on the map in mixtures of brown and red.

Mr. Durnford explained that he had taken out some figures giving the relative proportions of these various classes of land in the different provinces.

Taking the part of the Northwest Territories shown on the map (southern Keewatin and the southern fringe of Mackenzie), the red land gives about 59,800 square miles, equal to 38,272,000 acres. The yellow or unknown land in that same territory 54,600 square miles, equal to 34,944,000 acres. The brown and water areas in the same territory 23,000 square miles, equal to 14,720,000 acres. That gives a total area for the Northwest Territories, in this portion under consideration of 137,400 square miles, or 87,936,000 acres. The proportion of the red to the whole is 43 per cent. The yellow in that portion of the Northwest Territories is 40 per cent, and the brown and water 17 per cent. The yellow and the red together gives 83 that in the neighbourhood of Fort Churchill appears to be suitable for agricultural and grazing purposes, and generally for settlement.

Witness thought a farmer might do well in the neighbourhood of Fort Churchill. Witness based this opinion on information obtained from about one hundred authorities. In the portion of the province of Saskatchewan under discussion, the red area was 31,200 square miles, equivalent to 19,968,000 acres; the yellow 52,000 square miles, equivalent to 33,280,-000 acres. The brown and water area 41,800 square miles, equivalent to 26,752,000 acres, a total for Saskatchewan of 125,000 square miles, or 80,000,000 acres. The relative proportion of the red land in Saskatchewan was smaller than in the Northwest Territories, being 25 per cent; the yellow 42 per cent. Adding these together it gave 67 per cent for yellow and red, the brown and water combined amounted to 33 per cent.

In the northern part of the province of Alberta the red gave 65,000 square miles, equal to 41,600,000 acres; the yellow 77,910 square miles, equivalent to 49,862,400 acres; the brown and water areas 12,910 square miles, equal to 8,262,400 acres; total 155,820 square miles, equivalent to 99,724,800 acres; grand total for area under consideration, 418,220 square miles, equivalent to 276,660,800 acres; and for this the proportion of red was 37 per cent, the proportion of yellow 44 per cent, and the proportion of brown and water areas 19 per cent for the whole area under consideration. The proportions in Alberta were 42 per cent red, 50 per cent yellow, making 92 per cent of the whole in yellow and red, and 8 per cent of brown and water. Hudson bay was not included.

The yellow represented land on which Mr. Durnford could find no information. It is possible that there is information, but he had not been able to lay his hands on it. The greater portion of the land coloured yellow has not been explored or surveyed, and inasmuch as a good deal of that land seemed to be in the middle of the country which is known to be fit for settlement, the witness thought it should be explored and surveyed at once.

Mr. Durnford mentioned that he had often spoken about this section of country with the son of Professor Agassiz. It is known by the name of the Ancient Lake Agassiz, and the whole of this area, as far as one can gather from geological information, is the ancient bed of Lake Agassiz. The territory lying along the Nelson and the Churchill rivers forms part of this ancient lake. The theory is that as glaciers disappeared they deposited over this surface large quantities of humus, and that the ancient lakes spread southwards, and all the land which formed the ancient bed of Lake Agassiz is fertile.

Some Ancient Authorities

The most ancient authority Mr. Durnford consulted with reference to the land along the Churchill river was David Thompson, the man who towers above all the great men of his period, the man whom we first hear of as a boy of fourteen at Churchill, and who took astronomical and meteorological observations all over this country, traversed from the south of the Indian lakes, by Jasper lake, and Columbia river to the Pacific ocean, and to whose maps the Dominion of Canada is indebted for much information, for outside of David Thompson, who surveyed as far north as the southern portion of the Indian lakes, and Peter Fiddler, who completed this work, we are altogether in the dark about the country west and northwest of Churchill. David Thompson's note on the Nelson river from its mouth is to the effect that for 137 statute miles of the river's length the banks are clay and suitable for cultivation. We have not so much information upon the country north of Churchill river, because we are dependent on Thompson and Fiddler, and Thompson's notes, which the witness had consulted, only extend as far as the southern portion of Indian lake. Peter Fiddler's notes are unavailable—as they cannot be found at the present time—but they speak of this portion as very rocky, but from what he says about the character of the soil, one is to believe that it is a Huronian formation and likely to be mineralized.

Agriculture at Churchill

One of the best authorities upon Churchill in ancient days was a man of the name of Robson, a civil engineer, who constructed Fort Churchill, who was there at various periods from 1733 onwards, and appears to have been a very careful observer as well as a good engineer. He spoke of the vegetables which he had raised there, and also of the horses which had been employed for several years, and also the cattle at the fort. He said that in spite of the cold winds on Esquimaux Point he was able to produce excellent vegetables. He dug down in the soil—it was in the month of July—and found that he had to dig down a depth of three feet six inches before he came to the frost, represented by a sheet of eight inches of ice, and he makes the note that this thin stratum of ice below does not in any way affect the vegetation. He went on to speak of the horses that were used in drawing stones and other material for the fort, and the fine butter that was made, and spoke of it generally as a good agricultural

country round about there. That was in 1733 to 1747.

Mr. Durnford thought it was 1784 when David Thompson first started his diary. It extended on to 1850. He was one of the first men to cross the Rocky mountains and the discoverer of several passes. Howe's Pass should have been named after him. He went very near the Yellow Head Pass, but did not go through, passing by what he calls the Athabaska Portage. Right across the continent from Churchill to the mouth of the Columbia river, he has left a very valuable series of meteorological observations taken every winter during the time he was with the Hudson Bay Company, for seven years, and later, from 1797 to 1814 with the Northwest Company. He observed at Split Lake; he observed at Sepiwesk Lake, also at Cumberland House, at York Factory, at the South Indian Lake, at Reed Lake, Peace river, &c., &c., and left a series of meteorological tables which are of great value. The opinion that one must gather from his writings is that the principal reason agriculture was not carried on was because the mouths to be fed did not appreciate the benefits of eating vegetables. The Indians being all meat eaters it was thought superfluous on the part of the companies to attempt to raise vegetables or grain for them.

Mr. Durnford had caused to be written on the map described the different points where barley and wheat have been grown. Barley has been grown at York Factory and at Nelson House, wheat and other cereals have been raised there and at other points along the Nelson river cereals have been raised, so in his opinion there is very little doubt that in this country, if the necessity arose, cereals of all sorts could be grown.

Forestry

Mr. Durnford proceeded to refer to what authorities he had consulted had written regarding the forestry of this northwestern district.

Mr. Ellis back in 1748 gave a drawing which is very interesting, showing the size of the timber, and the houses they constructed while wintering at the mouth of the Nelson river.

Mr. McInnes, who is connected with the Geological Survey, had made a very valuable report. He was through this country, going from the head of Lake Winnipeg, as far east as Split lake last year, and he states that just north of Burntwood river near the Hart and Nelson rivers, on the 56th parallel, he found spruce timber as large as twenty inches in diameter, growing at the present time where the fires had not swept through. All this country seems to have been devastated by fire at different periods, but the information we have going back for a century and a half, shows that large timber grew at the mouth of the Nelson river, and large timbers are still growing near Split lake on the Odie river. Spruce and poplar are found, and wherever poplar grows in the northern region you may be sure the soil is good.

The witness read a short article by the United States Consul at London, Ont., speaking about the timber areas of this northern country, and he makes the statement that from the east coast of Labrador north of the fiftieth parallel in a north-westerly direction to Alaska there is a belt of timber about 3,000 miles long, and about 500 miles broad which he terms the spruce area. In a very strong article he speaks about the impossibility of exhausting that timber. He refers to Dr. Bell's statement in which he says that the area of our northern forests is forty-four times as large as England and that one such area will supply the present population of Canada, as showing that the timber in that belt may be said to be practically inexhaustible. In the southern border the timber is large enough for lumber, and in the northern it is good enough for pulp.

Climate

Mr. Durnford pointed out that the climate varies considerably. He drew attention to the fact that the further south we go the better the stamina of the men we find there. The witness had travelled quite largely in India, and found the nearer he approached the Himalayas the finer the class of men. The men from the mountains, the Sikhs, are men of magnificent physique. You find this applies also as regards the Esquimaux, who appear to be a fine race physically, kindly in their disposition and nature, not cruel to the same extent as those of more southern latitudes, and you find the same thing down in Patagonia. Towards the limit, as you may say, at which men or cereals can be grown you find the best. That had been brought to the notice of the witness very strongly living out in India. Rice is the staple grain of that country, and grows well, yet we find in Carolina a much better quality. The nearer to the poles it is possible for plants or the human species to survive, there the best of their species are found, and so, though the northern climate is rigorous it is habitable.

Mr. Durnford quoted the experience of Mr. Hanbry, who started from Churchill and went north and along the Chesterfield Inlet up to the Arctic ocean, travelled west along the Arctic ocean and up to the Coppermine river to Great Bear lake, passing two years amongst the Eskimos in 1904 and 1905. He collected some very valuable information as regards the climate of that northern country. Its people, of course, have been used to the rigours of the climate. He says that new-born children are laid on the snow by their mothers, without their receiving injuries, and he makes a statement which would at first seem almost a fairy story did we not know that he had been living among the Esquimaux in their snow houses. He says that a temperature in that very dry climate of 23 degrees is equivalent to 60 degrees in a more humid one, and that when the temperature reached 28 above zero, they had to cut a hole in the snow house because they found it uncomfortably warm. It is a strange but very valuable statement as tending to show that though the first persons to go into our north country, for instance, natives of the old country might suffer through ignorance, those who learn how to live there would undergo no greater inconvenience than they would in a climate such as we have in Ottawa.

Settlements

Throughout that northern country the only settlements are Hudson bay settlements. There is a small settlement at Churchill and a Hudson bay post on Split lake, one at Nelson House and one at Red lake—they are scattered all over that territory. The people grow only what is necessary for their own use. The people who are dependent upon them have to be fed, and as they have meat it matters not whether they raise large quantities of vegetables or not. There are comparatively few cattle kept. At Cumberland and Norway Houses they have a few, and at one time they used to have a large number. Now that they are nearer markets, they probably purchase their supplies. They are not an agricultural people and do not care to keep cattle. Of course in that country cattle would need shelter, whether at Churchill or Norway House, in the winter season. You have to go considerably further west before you could let them remain out during the whole winter.

Means of Communication

The means of communication between Lake Winnipeg and Churchill hitherto has been by boats, as also to York Factory. The Canadian Northern railroad is now extending its line from Ekomami to the Pas, a distance of about ninety miles, and have run their survey some 75 or more miles north of that. The distance from Churchill to the Pas on their route map is 450 miles. The distance from Churchill to Liverpool is 2,926 geographical miles, as compared with 2,931 from Montreal, via Cape Race, to Liverpool, a difference of five miles in favour of Churchill. The distance from Montreal to Liverpool, by the Straits of Belle Isle, is 2,763 miles, and from New York to Liverpool, 3,079 miles. The distance from Winnipeg to Churchill is approximately 650 miles, and about the same distance between Prince Albert and Churchill.

Clearly the settlement of Canada's northern and western frontiers was of considerable concern to the leaders of the country. It is obvious from the evidence of Fred Durnford that maps were of value at that time and furthermore the maps were considered to be definitive! The "evidence" centres around the areas of different colours used on the map to denote three categories: red for land suitable for cultivation, brown for rocky or muskeg land, and yellow for areas about which little was known.

Durnford's evidence has been on record for less than seventy years. The first section consists of areal estimates of the colouring on the map (eight paragraphs) and a note about Lake Agassiz (one paragraph). A further two paragraphs on minerals show the total appreciation of geology/geomorphology (three paragraphs) as less than fifty per cent of the appreciation of the colouring on the map (eight paragraphs). Such was the comprehension of geomorphology. Similarly, attention must be drawn to the comments on climate which are best described as "qualitative" although a little quantification is found in the penultimate sentence. Agriculture is again the focus for remarks which may be considered as integrated physical geography, and some information about the soils and vegetation can be gained from the agriculture and forestry sections.

While repetition is of no intrinsic merit, it would be a sad thing to omit the early example of quantitative fluvial geomorphology contained in the fourth paragraph of Elihu Stewart's remarks (reprinted below). Indeed, the whole of Mr. Stewart's remarks constitute a fine piece of integrated geography and the reader is referred to the full text as the extracts reprinted here do not do justice to his work.

EVIDENCE OF MR. ELIHU STEWART, OF THE CITY OF OTTAWA, AT THE TIME SUPERINTENDENT OF FORESTRY FOR THE DOMINION GOVERNMENT, SINCE RETIRED FROM THE PUBLIC SERVICE TO ENGAGE IN PRIVATE BUSINESS.*

Mr. Stewart explained that his knowledge of the country beyond the Saskatchewan has principally been derived from two trips that he made, one in 1902 to the Peace river, and during the past season (1906) down the Athabaska, down the Slave river, and down the

* Reprinted from *Canada's Fertile Northland* edited by Capt. E. J. Chambers (Ottawa: Government Printing Bureau, 1907), pp. 44, 47. Reprinted by permission of Information Canada.

Mackenzie to the Delta, and thence across to the Yukon, and back by the way of Dawson. In 1902 he made the journey from Edmonton to Peace river by way of Athabaska Landing, and thence up the Athabaska river to the mouth of Little Slave lake, through the whole length of Lesser Slave lake to the end of that lake, about 75 miles in length, the distance from Athabaska Landing to this point being about 215 miles. From Edmonton he drove to Athabaska Landing, about 100 miles thence up the stream to the junction of the Little Slave river, and up that river to Lesser Slave lake and then to the end of the lake, Buffalo bay. Thence he drove across (80 miles) to the Peace river crossing.

As to the area of what is broadly considered the Great Mackenzie basin, but which really includes the basins of the Mackenzie's tributaries, including the Athabaska, the Peace, the Liard, &c., Mr. Stewart computed it as 451,000 square miles, larger by over 100,000 square miles than the basin of the St. Lawrence and all the great lakes, and nearly three times the area drained by both of its branches, and the main Saskatchewan river. The Mackenzie, from information Mr. Stewart gained there during the summer of 1906, opens the latter part of May and closes the latter part of October. That does not mean that you can take a boat and go from McMurray down and across Lake Athabaska, and also across Slave lake as early as that. There is often ice in the lakes when the rivers are open. This distance from Athabaska Landing to Fort McPherson is 854 miles. He took the Athabaska river to Lake Athabaska, hence down the Slave river, passing the junction of the Peace, down to where he had the advantage of a steamer. There he had to make a portage of 16 miles, and then get a steamer that took him 1,300 miles, or to be exact, 1,299 miles, according to Mr. Ogilvie's survey, down to Fort McPherson, at the delta of the Mackenzie.

The steamer that took Mr. Stewart from Fort Smith, down the Slave, and across the Slave lake and all the rest of the way that 1,300 miles drew five and one-half feet of water.

The average width of the Mackenzie is about a mile. Of course, there are some bars occasionally. After it receives the Peace it is an immense volume of water, but there are no falls below the Slave river; none on the Mackenzie at all, but there is very swift water. The Mackenzie river has not the drainage area of the Mississippi, but it has a greater drainage area than the St. Lawrence above Montreal. Taking the St. Lawrence down to the Gulf it is a little larger than the Mackenzie, taking all the tributaries. For nearly the whole 10,000 miles of the Mackenzie it is so rapid that it is impossible to row a boat against the current. Mr. Stewart did not think it would be six miles an hour, but he pointed out it is very hard to row even against four miles an hour and make any headway. The banks of the Mackenzie are wooded all the way.

Forestry

After returning from his first trip in 1902, Mr. Stewart wrote his annual report to the Department, and he read a few sentences therein written, as he had prepared his report when the matter was fresh in his memory.

'The principal tree between the Rocky mountains and the plains is the spruce, mostly the white spruce, and from its position near the prairie there is no doubt that it will be more sought after to meet the increasing demands from that quarter.

'The country along the upper waters north of the Saskatchewan and the Athabaska and Peace river is partly prairie and partly wood. The varieties of timber are principally aspen and balsam poplar, the former predominating, and white spruce. The poplars as we go north seem to increase in size and height, and as we approach Lesser Slave lake and between this lake and the crossing of the Peace river. Below the junction of the Smoky they grow very clean and straight trees, not over a foot or fourteen inches, but reaching a height of 17 or 18 feet, making excellent building timber, as well as fencing and fuel. In some parts there are stretches of good spruce well adapted for lumbering purposes. There has so far been but little destruction from fire in this quarter. The land is mostly level, soil excellent, and if the summer frosts do not prevent it, the country will begin soon to settle up and there will be an ample supply of timber for local uses, if not for export to the adjoining prairie regions.'

Mr. Stewart followed the reading of the preceding extract from his report with the remark:—

'I never saw as fine poplar as I saw there. A considerable number of poplars were over a foot, but a foot would be a fair average. I have seen poplar in all parts of the prairie country, but never saw any growing up as straight. The wheat from Vermilion, it is said, took the first prize at the Chicago exhibition.'

The Hon. Mr. Lougheed added: 'Yes, I saw it there myself.'

Mr. Stewart explained that spruce suitable for commercial purposes grows to the Arctic sea. He was astonished to find that the limit of tree growth extended as far north as it does.

He thought it extended probably ten degrees further north in this district than in Labrador. The different kinds of trees that we have in the Mackenzie Basin include white spruce, black spruce, the larch or tamarack, which is found as far north as the spruce, the jack pine and the balsam. Mr. Stewart did not see any balsam in the Arctic circle; aspen, white poplar, balm of Gilead and birch are all found down as far as Fort MacPherson. The natives make their canoes out of birch bark at Fort Macpherson. The size of the timber becomes less as you get towards the north. There is timber growing near the junctions of the Peace and Slave rivers, probably 14 inches in diameter. Below Fort Good Hope the timber is smaller. Some of it has been made into flooring and lumber is made from the timber there. There is a large supply of spruce suitable for pulp.

The forestry section of Mr. Stewart's evidence perhaps requires clarification since the fourth paragraph should not be erroneously construed to give the impression that the wheat at Vermillion was as straight or as substantial as the poplar trees. In order to more fully represent the status of physical geography at this time the evidence of William McInnes is also reprinted. A member of the Geological Survey, McInnes gave a more detailed and reasoned account of the regions in which he had worked. Rather than oblique reference to the theories of Professor Agassiz's son, there is a clear acceptance of glacial theory and a description of glacial depositional features. The rivers are also commented on and the soils and vegetation are related to these comments in some cases, particularly under the forestry heading.

EVIDENCE OF WILLIAM McINNES, M.A., GEOLOGIST, OF THE GEOLOGICAL SURVEY, GIVEN BEFORE THE SELECT COMMITTEE, FEBRUARY 20, 1907.*

Mr. McInnes stated that he has been employed in the Geological Survey since 1883. The regions in the west with which he is familiar, first the district between the Saskatchewan and Split lake on the Nelson, a country which he was over last summer, and secondly, the region lying between the west coast of Hudson's Bay and the northern part of Ontario, Lake Nipegon and the Lake of the Woods.

The whole region from Split lake to a line of about 40 miles north of the Saskatchewan is a clay covered country.

The witness passed through this country, went by the Burntwood river and came back by part of the Grassy river, and made a number of excursions inland between these two rivers. After leaving Split lake, ascending the river, this clay-covered country shows absolutely no boulders and no gravel. Even the shores of the lakes, until you reach a height of about 800 feet, show no gravel bars at all.

There is absolutely nothing to interfere with the cultivation of the soil there. It is a country that has been burnt over. Witness assumed that

---

* Reprinted from *Canada's Fertile Northland* edited by Capt. E. J. Chambers (Ottawa: Government Printing Bureau, 1907), pp. 65-67, 70. Reprinted by permission of Information Canada.

the Burntwood river got its name that way. It has been subject to repeated burns. At the present time it is covered by a very open forest. Grasses grow fairly luxuriantly. There are two species of this, blue joint grass and a wild rye, that are the prevailing grasses. He understood, though he is not very familiar with those grasses himself, from Professor Macoun, that these are very excellent meadow grasses and make excellent fodder.

Mr. McInnes left Norway House in the second week of June and made the circuit and came out at the Pas on September 6, so it was June, July and August he was there. He saw grass growing from eighteen inches to two feet high.

The witness computed the area of this country at about 10,000 square miles. He does not mean to say that all of that ten thousand square miles is good land, but the basin characterized by this deposit of clay has an area of about ten thousand square miles. It is bounded on the north by the Churchill river. The witness was at about the centre of the basin. The Indians told him it extended north to the basin of the Churchill river. Beyond that, northwards, instead of clay you get sand and gravel.

Starting at the Pas and proceeding towards Churchill the witness first passed through about 140 miles of country underlayed by the flat limestone of northern Manitoba. He walked for miles over hills of almost bare limestone with hardly any soil. Beyond that—that is above the contour he had spoken of where this clay was deposited, there is about 170 miles to Split lake, possibly in a straight line about as the railway is projected, that is characterized by these clay deposits.

As to the Flat Country

As to the flat country in Keewatin, beyond this clay area, it is a country of a different character. The witness proceeded from the Albany across country by the portage route to a large lake on the Agnooski river 100 miles, and then another 100 miles across to the Winisk lake and down the Winisk river to the sea, and he crossed through the country between Agnooski and the Winisk by three different routes,

perhaps 40 or 50 miles east and west between each route, and the country is very much the same character. It is a country that is very much denuded; that is to say, the original archaic rocks have been worn down to almost a plane. The elevations are very moderate. The only elevations to be seen are of glacial origin. They are old boulders and gravel. The country generally is characterized by these hills of boulder and gravel and intermediate valleys very largely muskeg. Except in the immediate valleys of the larger rivers there is very little land that would be suitable for agriculture, very little indeed, and that is a characteristic of the whole country Mr. McInnes was over; that is of the upper waters of the Agnooski and the Winisk rivers and down to about 150 miles from the sea. From the point specified, down to the sea, the country is of an entirely different character again; that is to say, it is country that is originally overlain by from a very few feet at its edge, to 100 feet or more, of boulder clay of a very tough impervious boulder clay, which holds up the water, and on which the drainage, up to the present time, is of a very imperfect character. The present drainage of that area is comparatively recent.

There is overlying this boulder a marine clay which holds very well defined marine fossils, some of the shells quite as well preserved as you will pick up on the seashore to-day. The witness picked up some of these species which showed that subsequent to glacial time that country up to the 450 foot limit was down in the sea.

A Great Keewatin River

The present drainage has only had since that time to work itself out, and has not yet become very perfect. An instance of that is seen in this Winisk river. There is a lake under the head of the Winisk from which the main river flows, and from which the west branch flows north. They come together at a point, following the main stream, 250 miles below, enclosing an island 250 miles long. There are two other islands of this character along the Winisk river, one 80 miles and the other about 50.

It is a good large river. Mr. McInnes estimated it in cubic feet per second, some 25,000 cubic feet per second. It runs in size somewhere between the Gatineau and the Ottawa, not as much as the Ottawa quite, but larger than the Gatineau. Over the whole of the country, the last 150 miles down to Hudson bay, granting the proper climate and granting proper drainage, this green clay would make an excellent soil. In fact it is quite the same as the clay in the vicinity of Ottawa, practically clay of the same soil. It is very impervious clay and the country is extremely fit, except for the moderate slope down towards the bay, and it occurs in east and west undulation, so that there is no drainage except by the larger rivers down to the bay. There are little streams running into the sides of the river, but they cut very sharp walled trenches, sometimes 80 feet, as steep as boulder clay will stand, and that means an angle of say 60 degrees, 80 to 90 feet high. You get on top of these banks and you have a mossy place, sometimes 6 feet of moss. It is never peat: never having turned into peat. It is simply a green moss which is pressed into layers of a couple of feet thickness at the bottoms of the 6 or 10 feet, but never apparently oxydized or never carbonized at all, practically unchanged. The growth is going on still. It is merely the successive layers which are pressed down by subsequent layers on top of them, so that in places the thickness is quite ten feet. The first week in August Mr. McInnes got down to the sea coast and spent a month there. There was an ice barrier when he reached Hudson bay, off the mouth of the Winisk river. It had grounded about five miles out. It is very shallow water. It extends out four or five miles exceedingly shallow, the large boulders sticking out in high water. In low water there are extensive mud flats running out four or five miles from the shore, and the company's fishing boat had to make a circle of eight miles out of the bay before they could run up the coast.

Agriculture

There are no grasses in that mossy district in the valley of the Winisk. A river of that size in places has some shores, perhaps a quarter of a mile, here and there, beyond the actual shore of the river, and it is grassy there. That is, there are occasional bottom lands, but there is no extent of them. The witness did not think there is an agricultural country in that eastern district. It is entirely different from the country he had been previously speaking of.

Upon the Nelson river wheat has been grown successfully at Norway House, and also at Cross lake. Of course, he could see that they grow no grain at any of their posts nowadays. In the old days they grew it and ground it in hand mills. Witness saw potatoes that were grown about 50 miles north of the Pas. There were quite showy potatoes, great large fellows like those you see exhibited in fairs—tremendously large, grown on practically new land, and they had a very large crop of them. Mr. McInnes did not eat any of them. The Nelson has its source within forty miles of the Rockies.

There are no settlers in the Nelson district. The Indians, however, grow potatoes at several points, even in the northern part of it, as far north as Nelson House, about latitude 55. On July 11, when the witness arrived at Nelson House, the Indian potatoes had vines about eleven inches high, and were almost ready to flower. When he got out on September 6 to the Saskatchewan, at the Hudson bay post there, at the Pas, Indian corn was very well headed out, with very large fine ears quite ready for table use, and there was no frost until September 29. He knew that because he stayed there until then.

With eighteen hours of the day light, and no frost in the summer, vegetation is rapid. In a country where you can ripen Indian corn you can grow practically anything.

Mr. McInnes drew the attention of the committee to the fact that there is a very large area immediately adjoining the Saskatchewan river from a little this side of Prince Albert, clear down to the mouth of the river of very swampy land. In fact for a long time they thought they could not build a railway in to the Pas on that account. It occurred to Mr. McInnes going down that stretch of country that the only thing that has prevented the Saskatchewan draining this area is the occurrence at the mouth of the Saskatchewan of what is known as 'the Grand Rapids,' with a fall of 100 feet. This fall is in length a distance of about 3½ miles or there-abouts, and Mr. McInnes suggested that there is a possibility that these marshes might be done away with by blasting out the rock, thus increasing the speed of the river and lowering the basin of the Saskatchewan and draining that swampy country. It would bring into culti-vation a great many thousand square miles of as fine land as could possibly be found. It is all alluvial land of the best possible character.

Climate

Mr. McInnes said he could not very closely indicate the isothermal line on the part of the country he had explored last year, but he could say that that country averaged from four to five degrees in the summer months higher temperature than the same latitude further west. He thought that the isothermal line which would go past the north end of the country he had been speaking of would come down as far as the north shore of Lake Superior, which would be a very long distance south. He had records kept during all summer of the temper-atures through that western country, and he had a summary of the record kept in the pre-ceding summers.

He was rather surprised at the warmth of that western country in summer. He was sur-prised at the way heat kept up in the evenings. He kept the thermometer readings morning,

noon and 6 o'clock in the evening, and found at 6 o'clock temperatures were almost as warm as the noon temperatures. That country has a very long day in summer. The day in those high latitudes is very much longer, and the growing time proportionately longer. In June they have about eighteen hours of daylight.

As to the district where he found the 170 miles of agricultural land he had described, he only reached there about the middle of June. There was no frost in the balance of June or in July, and no frost in August, excepting once, on, he thought, the 29th, when the thermom-eter dropped just to freezing point. There was not enough frost to touch vegetation at all in the valley of the river where he was. He noticed when he got out to the Saskatchewan there was rather a high ridge on which there were a lot of half-breed settlers. He got there on September 6 and noticed on top of the hills where they had potatoes that they had been touched just on the tops, but down in the valleys the potatoes in the garden of the Hud-son bay post had not been touched at all. He presumed that frost was on August 29.

The witness had often been over the Cana-dian Pacific Railway between Lake Nipissing and Port Arthur, and the country he had tra-versed from the Pas eastwards as compared with the country north of Lake Superior was much superior.

The evidence reproduced above is only a portion of material available. It does, however, return us forcefully to the initial discussion of Canada as an ana-logue of an introductory physical geography course. Early explorers lived by their knowledge of navigation and the nature of the globe, the sequence of seasons, and the positioning of latitude and longitude on a chart. As we saw from Champlain's work, he records essentially the nature of the land, its hills and mountains. Once the explorers have found the land, its geomorphology is recorded next: sites for fortification and settlement, rapids which are impassable, rivers providing good access routes, and lakes providing freedom of travel are all items of importance.

Settlement is the location of people amongst or upon these topographic forms and, as the settlers attempt to cultivate crops over a period of time, they build up a picture of the climate at that point. We may assert that initially they have a record of the weather and as time passes this becomes a record of climate. Thus we began to know in detail the climate of settled areas but we had much less reliable data for other areas until the weather satellites gave us a global picture of the behaviour of our atmosphere.

Vegetation and soils may be viewed in a similar way. Early settlement under government supervision was on lands which had been surveyed. The surveyors' reports frequently contain notes on the vegetation associations at each site and the nineteenth century Ontario settlers (see Kelly above) associated these with soil type. There was little cognizance of the fact that soils could be chemically depleted or of the fact that soils could be washed away. Only in the last fifty years were soils seriously studied as a resource.

We propose then that in the growth of Canada, man was at first aware of the position and shape of the land mass. Slowly he learnt the morphology of it and, settling amongst the landforms, he came to a knowledge of the climate. Finally these items have been studied in the recent past and evaluated together as they are reflected in soils, soil-forming processes, and vegetational patterns. It is this sequence we follow below in our discussion of the component parts of physical geography. The sequence is also the one adopted in the published volumes of papers for the twenty-second International Geographical Congress held in Montreal in 1972 (see Adams 1972), and it is fitting that such a sequence should be preserved.

Chapter 4

The Component Parts of Contemporary Physical Geography

In the preceding chapters, the component parts of physical geography were indicated. It is a recurrent theme in those chapters that man's perception and understanding of his physical or natural environment has increased when the environment has been exploited. In history, this exploitation has been for minerals, animal products, and vegetable products, the latter two including wild and domesticated plants and animals. Physical geography is surprisingly oblique to these items, and the traditional content of introductory courses rarely includes the study of distribution of such products.

Thus introductory physical geography rarely includes material about the location and distribution of precious metals or industrial minerals. Similarly, it rarely discusses in useful detail the limits of the zones in which certain plants of economic importance are grown. We might expect that these items would legitimately be a part of physical geography because they exist irrespective of man's activity. The lack of information about the domain of animals of economic significance is also surprising and one could argue that Homo sapiens was an animal to be included in such a study. The domain of animals has emerged as an ingredient of cultural geography (Wagner and Mikesell 1962) and it is a reflection on the calibre of physical geography that Vavilov's (1951) work should not emerge as part of physical geography.

Physical geography therefore has a poorly defined content. It is usual to find geomorphology, climate, soils and vegetation included as ingredients under the name of physical geography and it is useful to ask why this should be. In certain respects, this is a reflection of the rapid growth in knowledge during the twentieth century, for during that time we have moved from a horse and buggy era to an era of space travel and satellites which record weather and earth resources.

As we have seen above, the early aim of explorers and geographers was to visit exotic places and report back with a description of these places. The number of explorers increased and so did the number of their reports. Standardization began and from this emerged the convention of describing the landscape and the peoples in separate sections. As knowledge increased still further the number of these specialist sub-headings increased, and a clear example of this is provided by the divisions of the evidence given to the select committee of the Senate in 1907 (see Chapter 3). This accepted format of reporting exploration invariably included sections on the climate, landscape, and vegetation although the latter may appear as forestry, grassland, and agricultural potential.

Obviously the process of comprehensively describing a place became more and more elaborate as the knowledge of landforms, climate, and vegetation increased. Things which came to be directly controlled by man, e.g., agriculture and forestry, were separated from the physical geography complex and these in turn became studies in their own right. The fragmentation of knowledge into specializations has continued to the present day. Pleistocene geology developed as a distinct sub-field of geology and solid geology developed an off-shoot, economic geology. In Europe geomorphology was considered the pursuit of geographers and the geologists in general focussed their attention on hard rock geology. This influence carried to Canada. In the United States geomorphology developed in both the geography and geology discipline areas with geology becoming dominant. Geomorphology thus became a hybrid in North America, belonging in some cases within geological studies, in others in a group of geographical studies.

Climate studies became based less on intuition and folklore and more on factual observations. The growing network of weather stations and the emergence of atmospheric science led to the specialist study of meteorology. The development of telegraph systems meant that for the first time weather reports could be received from remote points more rapidly than the speed of movement of the weather systems themselves. Speed of movement also increased for man himself and the trans-continental railroad had a telegraph system which provided weather reports pertinent to the operation of the system.

The growth of travel facilities and the increasing speed of travel increased the need for weather reporting. The use of aircraft made this need very great. The strategic value of aircraft during the early stages of the cold war made weather reporting a significant military activity and polar meteorology as a specialist science boomed.

If we look at developments in the study of the environment during the past half century we can indeed see a clear demonstration of the old adage that specialists strive to know more and more about less and less until they know everything about nothing! The components of physical geography, or physiography as perceived in the 1920 era, subsequently grew into detailed studies of micro-meteorology, micro-morphology, local variability of soils, and plant species counts. Literature on these topics is available in considerable quantities. In spite of this, physical geographers are not equipped to advise the Canadian government on environmental matters such as the impact of pipelines on the physiography of the Arctic or the causes of the landslides in Eastern Canada (Fyles 1972). In total our knowledge of specialist topics has increased but our ability to comprehend the complex workings of the physical environment has not increased in like manner.

Perhaps the response to this situation is emerging in the resurgence of environmental studies which has been noticeable in the past decade. Many of these studies, however, regard environment as a narrowly defined term for use within a specialist jargon apparently created to restrict the use of such studies in any interdisciplinary study of environment.

There is, however, a growing trend towards interdisciplinary study and this perhaps is a further consequence of the specialist approach of recent decades. In this area, physical geography is firmly established. The study of physical environment has always been possible and sub-dividing such studies into other subject areas has been a recent fashion. It has created the impression that physical geography "borrows" techniques and results from other subject areas. In actual fact, the other subject areas have developed as narrow specialist studies, each concerned with one aspect of the physical environment, and each should automatically be integrated with the other specialist studies when the results have been determined.

The failure to integrate the components after they have been artificially separated for analytical study has implanted the concept of "pure study" where an abstraction can be studied for its own sake. We thus create a sterile type of physical geography by failing to integrate the segments we understand. Such integration would produce a greater understanding of the complex interaction which is the physical environment. Indeed such integration is frequently discussed as desirable and dismissed as too complex to attempt.

Physical Geography

So far physical geography has been only briefly defined. It is in some cases equated with physiography, in other cases it is specifically identified as the study of geomorphology, climate, and soils or generally considered to be the study of the physical or natural environment. The most interesting fact is that physical geography exists. It exists as the specific morphology of the earth in every place, the climate of that place, the interaction of these two to modify each other and to influence the soils and vegetation of the place.

There is no ready definition of this complex, since the complex itself is not static, nor are its individual components. The complex which gives a place its physical geography is continuous over the surface of the globe. In the oceans a layer of saline water exists between the crust of the earth and the atmosphere but interaction still continues. Sediments are deposited, transformed, modified, and sorted; vegetation exists and animal life is also present.

The whole complex which is physical geography is dynamic through time and space and the very concept of seasons is a time-space continuum. The landforming processes seem to have an impact on the land which is conditioned by the space they occupy and the ways in which this changes with time.

The glaciations of the quaternary offer a very clear insight into this. Ice encroached on the North American continent from the polar regions and, although the reasons for this are not clearly established, several theories exist. This ice mass has important meteorological and climatic implications since the climatic conditions changed as the ice advanced. In consequence, the spatial systems of plants and animals also changed and the nature of the soil must have been similarly affected.

The ice moved further south and vegetation became much reduced in quantity; the region adjacent to the advancing ice front was overwhelmed as the ice mass progressed. As the ice moved across the land, climate and weather patterns changed; vegetation and animal life patterns changed. The ice itself removed some landforms and created others. The consequences of the presence of ice in our country were so great that the whole of the pre-existing physical geography was transformed.

The ice mass did not recognize political boundaries—it had no concept of "Canada"—but rather it was part of the global time-space continuum. A similar series of ice-masses moved across Europe and similar displacement of vegetation patterns occurred there. In both areas the complex which is physical geography was much affected. It is that complex which becomes sub-divided into specialist subject areas for study. This sub-division is not because we cannot comprehend such a complex of factors but rather because we have no means of measuring the total complex. Our sophisticated scientific methods and systems of measurement cannot convey in the currently fashionable quantitative sense what the nature of the complex is and we cannot therefore *measure* changes in the total complex.

The physical geography of all places continues to exist as the total environment regardless of the political activity of man. National boundaries have relatively little significance in the climate-landforms-vegetation-soils complex through space and time on the surface of the globe. Similarly, academic dispute about the "subjects" which should be a part of physical geography does not in any way change the environment.

Man however does have an impact on the physical geography complex. During the four centuries since Western man invaded the North American continent, he has had an effect which has many parallels to that of the Pleistocene ice mass.

Man moved on to the continent, and initially the animal population was changed with the introduction of the horse and the virtual elimination of the bison. As the influx of man increased, the vegetation patterns were changed also.

On a continental scale man has not yet effected much change in the landforms but metropolitan areas may be thought of as extensive karst-like regions. The morphology of the city (in the terms of a geomorphologist) is a deeply dissected peak with underground drainage approximating, in many aspects, karst topography. The vegetation of cities is much affected by the karstic environment, and the soils manipulated to man's specifications have lost many of their plant nutrients and their inherent fertility. The hydrology of urban areas is also affected and much of the precipitation is rapidly carried into the sewer system and out of the region instead of percolating into the soil to recharge the ground water resource.

Studies of urban climate indicate that man has changed the climate over metropolitan areas and the full impact of these changes may not be understood for some considerable time.

In four centuries man has profoundly altered the physical geography of Canada in certain places. If we speculate on the amount of change which would result from man's activity over several millenia we may be witnessing a phenomenon comparable to the Ice Age in its impact on physical geography.

On a mundane level, we shall restrict these far-ranging thoughts and speculations to the compartmentalized thought patterns developed in our systems of education. These compartmentalized thought patterns should not be permitted to dominate our personal thinking because the fully integrated natural environment will continue to exist. It is undeniably convenient to exclude important items from our consideration of physical geography and to consider only certain aspects of it. Add to this the fashionable image created by "specializing" in only one aspect of physical geography and the convenience, together with the apparent prestige, leads to a powerfully entrenched system of studies whereby we are led to consider the landform but not its soil or vegetation. We are also led to consider climate as a mathematic expression of process and to consider soils and vegetation as items to be presented on a two-dimensional map in such a way that their delicate adjustment to topography and microclimate are obscured.

In this framework there is currently some sign of soul searching amongst the group of scientists. Geomorphologists recently have begun to ask the questions, what do we study, why do we study it, and how do we study it. These questions were posed by several writers, and in the Canadian academic arena there is perhaps the most realistic evaluation of the component parts of physical geography.

Components

If we attempt to define the components of physical geography, we may do so in several ways. Definition can be by direct reference to the commonly listed subdivisions, namely, climate, geomorphology, and soils and vegetation. This definition does nothing more than echo the convenient sub-divisions of regional description exemplified by the evidence to the Senate committee which was reprinted above. It provides a working frame of reference which seems to emphasize the fragmentary nature of physical geography and it highlights the links between geomorphology and geology, between climate and meteorological physics, and among soils and vegetation and agricultural studies, botanical studies, forestry, pedology, and ecology.

Such a division is commonly used. It is used without a full realization that within it is embodied a further sub-division of physical geography into the inventory and the process studies.

Inventory Studies

Inventory studies are those which record features and their distribution and are frequently synthesized to produce a regional picture. This type of information is recorded and published as maps of Canadian climate, vegetation, soils, landforms, etc. Only within the last year has a descriptive analytical account of Canada's landforms been published (Bird 1972). This volume, which is very complete in its coverage of the country's landscape, provides an important addition to the available physical geography literature for Canada. Similar volumes on climate and vegetation and soils would be invaluable companions to it; together these would constitute a study of individual elements of physical geography as each varies in time and space over the face of this country.

These elements do vary in time and space, a fact which deserves more attention than is usually accorded to it. Even the present configuration of landforms is new, because, on the geological timescale, the landscape we now observe was fashioned less than 12,000 years ago. Indeed changes are still taking place although, at the present-time, these changes on a national scale are slow. Climate is by its nature a record of changes from day-to-day and season-to-season and even these changes can be viewed over the millennia so that major modifications in climate can be recorded from palynological evidence. Vegetation as recorded at the present day has been modified greatly in the past two centuries by the activities of man. On the larger time scale the succession of vegetation, which established itself on the landscape left by the retreating Wisconsin ice-sheet, has shown major change through time. The total picture of the present physical geography of Canada as an event in time is not usually stressed. It is, however, an ingredient in our perception of physical geography which may be reinforced by the data available from the Earth Resources Technology Satellite (ERTS) which is recording data about the whole of Canada once every eighteen days. These data (subject to problems with cloud cover) provide a permanent record of change throughout the seasons and also indicate changes resulting from transient events such as snow-storms, drought, and floods. This new dimension of time in our perception of physical geography is one which is easily transferred to other geographical studies. The rapid rates of change in the patterns of human geography have been brought to our attention as cities have spread over many square miles. Perhaps too little attention is paid to the rates of change in vegetation, animal populations, or even landforms which result from man's occupancy of the land.

Studies of Process

Studies of the process of change within each of the component parts of physical geography have attracted a great deal of attention and effort. This may be attributed to the specialized nature of such studies and the "scientific aura" associated with the quantitative nature of the laboratory data which has been gathered. Process studies have an appeal because controls established in the laboratory make a process "manageable."

This is in effect the ultimate in specialization where attention is focussed on one aspect of one sub-division of physical geography. Studies of this type have a considerable importance because the precise relationships between the measured parameters can be determined. Two types of process studies have resulted: those which are based on physical laws and which examine relationships and change at the microscopic level; and those which imply relationships between different attributes of a system. The latter studies can be more easily applied in the field where relationships between components can be determined and used as indicators to establish the framework for further research. This is an imprecise approach but it does analyze material and situations which have developed under the operation of the full complex of factors controlling physical geography. The laboratory mode alone provides precise data but this is usually unrelated to the realities of the field situation because it is difficult to determine the relative importance of the individual processes studied. Scale also becomes a problem in extrapolating the results of a small-scale laboratory study into the large-scale complex which is the physical geography of an area.

Some of the concepts of a systems approach attempt the integration of the specialist studies of geomorphology, climate, soils, and vegetation in order to create a stylized concept of physical geography. The relative importance of the links between different parts of the system is, however, difficult to determine.

These difficulties exist and the study of physical geography has, for the past two decades, adopted the specialist approach with broadly defined areas of concentration. These include geomorphology, climatology, soils and vegetation; we produce in the following sections a review of the present state of these traditional areas in the Canadian context. This is done to make the volume comprehensible and to make it easy to use alongside existing texts. The important message for physical geographers, however, appears to be "get it together." The resurgence of ecology, environmental studies, and other such integrated studies of environment indicates a current concern with the total environment. This is the real subject matter of a fully integrated physical geography, but too frequently in present-day studies the essential geographic components are omitted, and we see studies of environment concerned only with atypical plants and their associations or with the plant associations limited to a single exotic soil type. A fully integrated study of landforms, climate, soils, and vegetation is not yet regarded as the necessary aim of environmental studies.

It is particularly interesting to note that physical geography has not been readily accepted as a full member of the pure sciences. The emphasis on the lack of rigour in many studies and an undue stress on the material "borrowed" from allied sciences has produced a tendency for physical geography to be under-rated. This is in direct contrast to the "new" studies emerging from integrated interdisciplinary work between small groups of the pure sciences. The partial understanding resulting from such activities must then be integrated with similar studies from other combinations of relevant disciplines and ultimately we may see the "pure sciences" integrating their efforts to produce the background necessary for a full study of the natural environment. It is a moot point whether members of the "pure" sciences will realize that together they will finally achieve their nadir by becoming physical geographers!

Following such a whimsical view of the future, it is necessary to return to the present realities of Canadian physical geography. Studies during the academic life of physical geography have tended towards specialization and the following sections reflect this. It is somewhat paradoxical to move from a plea for integrated physical geography into a fragmentation of the subject matter but, at the present time, this is the context in which much of the work has been undertaken.

BIBLIOGRAPHY

Adams, W. P. and F. M. Helleiner (Eds.): 1972, *International geography, 1972*, 2 vols., University of Toronto Press, Toronto, 1354 pp.

Bird, J. B.: 1972, *The natural landscapes of Canada*, Wiley, New York, 191 pp.

Bishop, M.: 1948, *Champlain: the life of fortitude*, Knopf, New York, 364 pp.

Brunger, A. G.: 1972, "Analysis of site factors in nineteenth century Ontario settlement" in W. P. Adams and F. M. Helleiner (Eds.), *International geography 1972*, University of Toronto Press, Toronto, pp. 400-402.

Careless, J. M. S. and R. C. Brown (Eds.): 1967, *The Canadians 1867-1967*, Macmillan, Toronto, 856 pp.

Chapman, L. J. and D. F. Putnam: 1951, *The physiography of Southern Ontario*, University of Toronto Press, Toronto, 284 pp.

Chorley, R. J., A. J. Dunn and R. P. Beckinsale: 1964, *The history of the study of landforms*, Vol. 1, Methuen, London, 678 pp.

Chorley, R. J. and B. A. Kennedy: 1971, *Physical geography, a systems approach*, Prentice-Hall, London, 370 pp.

Farb, P.: 1963, *Face of North America*, Harper and Row, New York, 316 pp.

Fyles, J. G.: 1972, Remarks reported in "Discussion: Quaternary Geomorphology," in E. Yatsu and A. Falconer (eds.) *Research methods in pleistocene geomorphology*, Geo Abstracts, Norwich, pp. 235-239.

Head, C. G.: 1971, "The changing geography of Newfoundland in the eighteenth century," Ph.D. thesis, University of Wisconsin.

Head, C. G.: 1972, "The establishment of year-round settlement at Newfoundland," in W. P. Adams and F. M. Helleiner (Eds.) *International geography 1972*, University of Toronto Press, Toronto, pp. 425-427.

Morison, S. E.: 1972, *Samuel de Champlain, father of New France*, Little-Brown, Boston, 299 pp.

Strahler, A. N.: 1960, *Physical geography*, (2nd. edition), Wiley, New York, 534 pp.

Strahler, A. N.: 1972, *Physical geography*, (3rd. edition), Wiley, New York, 733 pp.

Vavilov, N. I.: 1951, "The origin, variation, immunity and breeding of cultivated plants: selected writings of N. I. Vavilov," translated from the Russian by K. Starr and cited in P. L. Wagner and M. W. Mikesell, *Readings in cultural geography*, University of Chicago Press, Chicago, 1962, 584 pp.

Wagner, P. L. and M. W. Mikesell: 1962, *Readings in cultural geography*, University of Chicago Press, Chicago, 584 pp.

Warkentin, J.: 1968, *Canada, a geographical interpretation*, Methuen, Toronto, 608 pp.

Yatsu, E., F. A. Dahms, A. Falconer, A. J. Ward and J. S. Wolfe (Eds.): 1971, *Research methods in geomorphology*, Geo Abstracts, Norwich, 140 pp.

Yatsu, E. and A. Falconer (Eds.): 1972, *Research methods in pleistocence geomorphology*, Geo Abstracts, Norwich, 285 pp.

Section 2
Geomorphology

Chapter 5

Geomorphology in Canada*

J. T. Parry

In the Beginning

The beginnings of geomorphology in Canada are difficult to discover, but as in other countries the roots are intergrown with those of geology and scientific exploration. Undoubtedly the first writer to demonstrate a clear appreciation of the variations in regional character from one part of Canada to another was David Thompson. It is apparent from his journals that Thompson frequently prepared a complete geographical synthesis of the areas traversed, in which was included a detailed description of the topography.[1] It is in these descriptions that the first contributions to Canadian geomorphology are to be found.

In the period 1850-1900, Canada, like the United States, had its scientific frontiersmen. It was the period of safari-like reconnaissances in which a few scientists, usually assisted by local guides, prepared field notes on what they could observe along the rivers and portages and from the vantage points of high summits. The reports of William Logan and his small group in the Geological Survey provided the first reliable statements about the geology and physiography of the settled parts of Canada, while the reports and maps of the Palliser expedition, 1857-59, which was sponsored by the Royal Geographical Society, and the Hind-Dawson expeditions, 1857-58, supported by the Canadian government, made available the first accurate information about the western Prairies and the adjacent cordilleras. A very significant contribution to regional physiography was made by James Hector, chief scientist of the Palliser expedition, in identifying the three prairie steps and in distinguishing the major vegetation zones of the area,[2] while to H. Y. Hind must be given the credit of reviving the land ice theory and proposing that an ice sheet of continental proportions had covered the whole interior of Canada.[3]

In 1870, the responsibilities of the Geological Survey of Canada were increased to include the western territories. During the next thirty years the members of the Survey carried out a series of magnificent exploratory traverses, and almost everything that was known of the remote areas of the Rocky Mountains and the Shield was derived from their journals. The reports of such men as G. M. Dawson, R. Bell, R. G. McConnell, J. Macoun, J. B. Tyrrell, J. Richardson, R. W. Ells, J. W. Spencer, and A. P. Low may be less well known than those of J. W. Powell, G. K. Gilbert, C. E. Dutton, A. R. Marvine, and F. V. Hayden, but their writing shows the same freshness of approach and freedom of opinion and the same appreciation of the significance of landforms in interpreting the geological history of an area.

The first major achievement of the Survey was the publication of the *Geology of Canada* which appeared appropriately enough in 1863,[4] twenty-one years after the establishment of the Survey, and provided the first comprehensive treatment of Canada's structure and scenery. In later years, the explorations of members of the

* © J. T. Parry, Department of Geography, McGill University, Montreal, 1973. Reprinted by permission of the author.

Survey in different parts of Canada led to the investigation of several basic problems in geomorphology, for example, in the writings of G. M. Dawson one finds a review of the relative merits of the land ice and iceberg hypotheses for the origin of glacial drift.[5] R. G. McConnell provided the first pieces of evidence for estimating the size and depth of the continental ice mass.[6] A. P. Low correctly identified abandoned shorelines and high level deltas in Labrador and around James Bay and demonstrated their significance in the study of postglacial isostatic response.[7] W. Upham outlined the successive stages of Lake Agassiz and provided evidence of its proglacial character.[8]

It must be remembered that the treatment of land forms in the early geological survey reports was more or less incidental. With the turn of the century, Canadian geology followed the European lead and gave little encouragement to geomorphology, in spite of the successful debut of this offspring in the United States. The studies of regional physiography, inspired by W. M. Davis, which form such an impressive part of the American geological literature, are almost completely lacking from the Canadian; the only significant contribution is the *Physiography of Nova Scotia* by J. W. Goldthwait,[9] which deserves recognition as the first Canadian monograph to provide a geomorphological approach to landscape study.

The Founding Fathers

As in the nineteenth century, it was external stimuli which led to further progress in Canada. The new wave of European teaching after the First World War was introduced to Canada very largely through the endeavours of Raoul Blanchard, who made Canada his country of adoption. In 1927 Blanchard, who was then professor of geography at Grenoble and already well known for his research work on many Alpine topics, was invited to the University of Chicago to give a course of lectures. This provided the occasion for his first visit to Canada and so commenced a long and close association with French Canada, the true significance of which can be judged from the deep personal affection for Blanchard and the high opinion of his work which is expressed by so many of the contributors to the *Mélanges géographiques canadiens offerts à Raoul Blanchard*.[10]

The manifesto of geography which Blanchard presented in Canada was based on the teachings of Paul Vidal de la Blache, the founder of the French school, and so it is not surprising to find that regional geography is given great emphasis. However, Blanchard had a proper appreciation of the significance of the physical setting. The importance of geology and structure as a basis for regional differentiation is frequently underlined, and his writings contain many penetrating observations on landforms and the sculpturing processes.

Blanchard's achievement in Quebec had its counterpart in Ontario, where Griffith Taylor assumed a pioneer role in accepting the invitation of President Cody to establish a department of geography at the University of Toronto in 1935. Taylor found in Canada the same sort of challenge he had met in Australia and he devoted himself, over a period of eighteen years, to the building of the first independent geography department.

Taylor's undergraduate training was in geology with a strong emphasis on mining geology, and his first university appointment was as a demonstrator in geology. However, he also developed a keen interest in physiography and had the unique opportunity, while still a research student at Cambridge, of accompanying W. M. Davis on a field excursion through the Alps. In 1910, he returned to Australia to take up an appointment with the Weather Service, and this led to his being attached to Scott's expedition to the Antarctic as meteorological observer.

It is not surprising that Taylor placed great emphasis on the physical aspects of geography. He was more convinced than Blanchard of the value of geomorphology for its own sake and not simply as the preliminary to regional synthesis. In his second year at Toronto, Taylor instituted a laboratory period to demonstrate the physical basis of geography and to teach students cartographic techniques and the use of surveying and draughting instruments. Much of the laboratory work was based on the local area so that students could benefit from field studies, and when the honours program was established in 1940, provision was made for more intensive field work.

Blanchard and Taylor, the first two honorary presidents of the Canadian Association of Geographers, founded geography in Canadian universities and, by the same token, they established geomorphology as a part of the geographical curriculum. Progress was slow but, at the outbreak of the Second World War, geography, with geomorphology as an intrinsic part, was being taught at five universities. The number of professional geographers could be counted on one hand, but already some research in geomorphology was being done; for example, Taylor's first publications on Canadian topics dealt with topographic control in the Toronto area[11] and the structural basis of Canadian geography[12] while Blanchard's *études canadiennes*, which appeared in eight parts between 1930 and 1939,[13] provided the first physiographic treatment of southern and eastern Quebec.

The New Frontier

The decade 1940-50 was very significant, both for geography and geomorphology in Canada. A. H. Clark, in commenting on the period, noted that Canadian geographers had been engaged in a serious struggle for recognition, indeed actual survival.[14] However, by 1946 the battle was won, geography had emerged in six universities,[15] and the federal government, having been convinced of the value of geography in developing national resources, had established a Geographical Bureau (1947), which was elevated to the status of a Branch in 1950. It is interesting to note that this same battle for recognition had been fought by European geographers more than half a century before. Once accepted the results were the same; geography became respectable, and by the same token, geomorphology was also accepted into the halls of learning as a suitable mentor in teaching and research. Although it is important to recognize that geographers have not been the only contributors in the development of geomorphology, nevertheless it is apparent that it was only with the establishment of geography that the way became open for further progress.

An analysis of the publications in geomorphology according to topic indicates that in the 1940s studies of glaciers and regional physiography were most common. However, the most significant aspect of the geomorphological research work of the 1940s is the seminal nature of several of the studies undertaken at this time, studies which were continued in the next decade and which provided the signposts for the future development of Canadian geomorphology. Among the published works of this type which could be cited are the reports on glaciology and glacial geomorphology resulting from P. D. Baird's expedition to Baffin Island,[16] R. F. Flint's analysis of the growth of the Labrador and Keewatin ice sheets,[17] J. L. Jeness' review of permafrost in Canada,[18] Putnam and Chapman's series of articles which were expanded into a book on the physiography of southern Ontario,[19] the investigations of patterned ground in the St. Elias Range, Yukon Territory, and in Victoria Island by R. P. Sharp[20] and A. L. Washburn,[21] and V. Tanner's study of Labrador.[22]

The great significance of the work of Tanner and Putnam and Chapman can be judged from the fact that every subsequent publication dealing with either Labrador or southern Ontario draws upon their material. *The Physiography of Southern Ontario*, now revised in its second edition, is the result of painstaking and detailed field work over an area of nearly 30,000 square miles, involving the mapping of landforms and surface materials at a scale of one inch to the mile.

In much the same way, Tanner's work has become a standard reference on Labrador. It was based on observations made during the Finland-Labrador expedition, 1937, and the Tanner-Labrador expedition, 1939, and so the treatment is less detailed than that of southern Ontario. However, a complete regional physiography is attempted, with the systematic examination of the geology and structure of Labrador, the denudation chronology in late Cenozoic times, and the effects of the Pleistocene glaciation with its attendant isostatic response.

One of the noteworthy features of the periodical literature of the 1940s is the increasing proportion of articles devoted to northern Canada. An important stimulus to northern research was provided in 1945 with the establishment of the Arctic Institute of North America, an international agency with American, Canadian, and Greenland representation on its board of governors. The Institute has promoted research in the Arctic and Subarctic through its library facilities, its publication program, and its generous grants for field work. In its first ten years of operation the institute helped to finance 177 projects, of which twenty-one were of a geomorphological nature.[23]

The interest in the Arctic and Subarctic reflected in the geomorphological literature of the 1940s is indicative of the very real concern of the governments of Canada and the United States in obtaining information about all aspects of the environment of the North. This sudden awareness of the northland is understandable in view of the strategic significance of the area in the days of the "cold war," before the development of intercontinental missiles, when the Soviet Union and the Western world faced each other over the pole, and it appeared that the most likely theatre of operations would be the Arctic. It was this situation which led to the construction of northern airfields and the aerial photography of vast stretches of Labrador-Ungava and Keewatin.

The possibilities of using this unrivalled source of information in physiographic investigations were realised almost immediately. Indeed, some pioneer work using air photographs (mainly high and low level obliques) had already taken place in the 1930s; for example, J. T. Wilson's examination of the drumlins of Nova Scotia [24] and the eskers northeast of Great Slave Lake,[25] and N. E. Odell's study of the regional geomorphology of northern Labrador.[26] The availability of an extensive air photo coverage produced a revolution in geomorphological thinking in Canada, which marked the departure from the European tradition, with its emphasis on detailed studies of small areas and its overtones of nineteenth century positivism—an outlook quite unsuited to the Canadian situation, where the vast areas to be studied, the difficult field conditions, and the small numbers of researchers made a mockery of such a minuscule approach. Air photo interpretation provided a new perspective, ideally suited to regional physiography in general, and glacial geomorphology in particular. The spectacular results that could be achieved with "aerogeomorphology" were demonstrated in the late forties and early fifties.

Go North, Young Man

There are several commentaries on the status of geography in Canada in the early fifties, and it is apparent that there was an increasing interest in physical geography at both the university and the government level. M. R. Dobson's analysis of geography in Canadian universities indicates that in 1950, general courses in physical geography were offered at seven universities[27]: British Columbia, Laval, McGill, McMaster, Montreal, Toronto, and Western Ontario, as part of an honours or concentrated program in geography. However, specialization in geomorphology at the graduate level was only possible at four universities: British Columbia, McGill, Montreal, and Toronto. It is interesting to note that four of these universities also offered specialized courses on northern Canada or the Arctic.

McGill University was able to take the lead in this development as the result of two important circumstances, firstly, the allocation of funds for research on the surface characteristics of northern Canada by the Defence Research Board which permitted the organization of the McGill University Research Group, and secondly, the establishment of the Subarctic Research Laboratory at Knob Lake.

The work of the McGill Research Group was the direct outcome of the new scale of operations permitted by air photo coverage. Their task was the investigation of the surface characteristics of Ungava-Labrador, an area of 500,000 square miles. This involved field reconnaissance to establish appropriate classification systems for vegetation and landforms and the elaboration of air photo interpretation keys.[28] A parallel project, sponsored by the Arctic Institute of North America, involving some members of the McGill Group, was the preparation of reconnaissance maps of surficial deposits and structural features for the whole of Ungava-Labrador.[29]

The McGill Subarctic Research Laboratory was constructed in the summer of 1954, with the assistance of mining and construction companies involved in the development of the Knob Lake iron ore deposits. The Laboratory has served as a centre for the field study of a wide range of Subarctic topics. Even in the early years, the research program covered eight different fields ranging from limnology to ionospheric observations, but it was apparent from the first that the Station provided magnificent opportunities for geomorphological research, particularly glacial geomorphology.[30]

McGill University was not alone in its emphasis on northern research. Although none of the research projects at other universities approached the size of the McGill program, there were several similar applications of aerogeomorphological techniques; for example, at Toronto University, W. G. Dean's investigations of the drumlinoid landforms of the "Barren Grounds" and J. T. Wilson's study of the glacial features between the Mackenzie River and Hudson Bay.[31]

Besides encouraging independent research on the geomorphology of northern Canada of the type discussed above, several departments of the federal government were responsible for major programs dealing with particular aspects of the Arctic and Subarctic environment. At the Geological Survey of Canada, which had increased the scale of its operations in the north after the Second World War, it was found that conventional methods were quite inadequate for even exploratory level mapping over the vast areas involved, and so a new approach was tried in 1952. Two helicopters were used by a party of geologists for aerial traverses and spot landings, with the result that 57,000 square miles of southern Keewatin were covered at the exploratory level (allowing mapping at the scale of 1:500,000) in a single summer.[32] "Operation Keewatin" was so successful that the same methods were adopted in "Operation Franklin" in 1955. Although not specifically geomorphological in intent, much valuable information on surficial deposits and landforms was obtained from these surveys.

The Division of Building Research, National Research Council, became involved in northern research because of the peculiar problems associated with construction in the Arctic and Subarctic. The wartime Canol project and the activities of the U.S. Army Corps of Engineers in the late forties provided the first information about permafrost, and after a general survey of permafrost conditions in northern Canada, the National Research Council decided to establish a northern research station at Norman Wells in 1952 to serve as a base for field and laboratory studies. Particular attention was given to the distribution of permafrost and the location of sporadic and discontinuous permafrost bodies in the southern fringe area. In addition, the factors influencing thermal conditions in the ground were studied.[33]

During the same period the Geographical Branch, which had become the chief employer of geographers in Canada, was also involved in the study of ice in northern Canadian waters.[34] The survey was concerned with all aspects of floating ice, the different types of ice, the distribution and movement, the factors governing formation and decay, and the associated navigational problems. All the existing documentary information on floating ice was collated, and the distribution of sea-ice in the Canadian Arctic was examined on all available air photographs taken between 1947 and 1952. Initially only Canadian Arctic waters were included, but the survey was later extended to include the Gulf of St. Lawrence and eastern Canadian waters.

In addition, the Branch also became involved in terrain studies. Commencing in 1952, a series of reports on surface conditions in northern Canada was prepared. Existing information was supplemented by air photo interpretation, the aim being to describe all aspects of the surface: deposits and landforms, vegetation, and water bodies including their freeze and break-up characteristics.

The North became the new frontier in Canada, and the acceptance of this challenge by Canadian geomorphologists obviously provides one of the highlights in any review of the development of geomorphology in the twentieth century, in much the same way that the survey of the western frontier created the most dramatic episodes in the nineteenth century.

A Decade of Research at the Universities and the Institutes

Many new graduate schools of geography were established during the late fifties and early sixties. In 1955, there were only seven Canadian universities offering advanced degrees in geography, but this total increased to eleven in 1960 and twenty in 1966. Inevitably, this development allowed increasing scope for research in geomorphology and prompted the elaboration of new methods and the investigation of a wider range of topics.

McGill University confirmed its position as a centre for geomorphological work with continued emphasis on its Arctic program. At the Subarctic Research Laboratory, studies of glacial geomorphology soon became predominant. Of the thirty-two theses which have been prepared by graduate students working at the laboratory prior to 1966, exactly a half deal with topics in glacial geomorphology, and of the ninety or so papers published in different journals as the result of work carried out at the laboratory in the same period, nearly a half deal with some aspect of glacial geomorphology.[35] The work of J. D. Ives, who was director from

1957 to 1960, had a considerable influence on the pattern of research. The basic problems of the deglaciation (including the location of the last ice centre), which had been studied initially in the heart of the peninsula, were investigated farther afield, and related problems, such as the late-glacial marine transgression, the proglacial lakes, and the isostatic response to deglaciation were also examined. In addition, there have been studies of organic terrain, periglacial features, fluvioglacial morphology and mass wasting, as well as snow and ice surveys in the immediate vicinity of Schefferville. At the present time, there is increasing emphasis on more detailed investigations, and it is very likely that the next twelve years in the history of the laboratory may be even more productive than the last.

Continued interest in the Arctic is reflected in the two major research programs undertaken at McGill University Geography Department in the late fifties. The first of these, directed by J. B. Bird, was supported by the R.A.N.D. Corporation, a private American agency undertaking research for the United States Air Force. A research group was formed in 1955, and their task was the production of a series of reports and maps on the physiography of the southern Canadian Arctic between Baffin Island and Banks Island.[36]

The second project, the Jacobsen-McGill expedition to Axel Heiberg Island (1959-1972), led by F. Muller, received support from the National Research Council and from private sources. All aspects of the physical geography of the central part of Axel Heiberg Island were investigated,[37] and special attention was given to glaciology and the techniques of glacier mapping.[38]

During the 1960s an increasing interest in quantitative studies is apparent at McGill as evidenced by the research of M. A. Carson on the mechanics of erosion[39] and the processes of slope development.[40] Increasingly there is a need for geomorphologists to master the principles of fluid and solid mechanics and to relate these to their studies of exogenic processes. Carson has successfully achieved this aim with regard to slope processes, and in addition, has demonstrated the value of a systems approach in geomorphology and the significance of inductive-deductive models in the development of geomorphological theory.

All of the French-speaking universities in Canada now offer specialized courses in geomorphology, but only two of them have developed strong research programs in this field. At l'Université de Montréal, the main emphasis has been on regional physiography and morphological mapping. In the latter field, G. Ritchot has been one of the first to apply the techniques developed in France to sample areas in southern Quebec. At l'Université Laval, where the Institut de Géographie became independent in 1958, considerable attention has been given to regional physiography, but far more significant has been the contribution to *la geomorphologie froide* (*glaciare et periglaciare*) inspired by L. E. Hamelin[41] One of the first pieces of research was the collection of information for the *Carte préliminaire de phénomènes periglaciares du Canada,* first published in 1960.[42] A comprehensive study of periglacial processes and features in Canada was published in the following year,[43] and a few years later this was supplemented by an illustrated glossary of periglacial phenomena prepared by L. E. Hamelin and the late F. A. Cook.[44]

The establishment of the Centre d'Etudes Nordiques in 1961 gave an additional stimulus to the geomorphology of cold regions, which was felt both at Laval and farther afield. Twenty-five percent of the seventy or so projects supported by the C.E.N. during the past five years have been in geomorphology, and of these, fifteen have been presented as theses for higher degrees at Laval and other universities. Nearly a quarter of the twenty-three publications of the C.E.N. deal with geomorphology, particular attention being given to regional physiography. In 1962 facilities at Fort Chimo were made available by the Ministère des Richesses naturelles, and this station has undoubtedly stimulated interest in the Ungava Bay area.

The colloquium on the Quaternary of Quebec held at Chicoutimi in 1968 provided an occasion for the discussion of a wide range of problems by francophone and anglophone geomorphologists, and the resulting publication[45] is a useful resumé of the work of French-Canadian geomorphologists in the late sixties.

At the Ontario universities during the last ten years, research in geomorphology has been comparatively limited; however, mention must be made of some recent developments which augur well for the future. The University of Western Ontario has recently established a field station on the shores of Lake Erie, near Port Bruce, which will serve as a centre for the study of the shoreline and mass wasting processes which affect the high bluffs along the lake margin. At the University of Guelph, E. Yatsu has produced one of the few quantitative treatments of the factors controlling rock resistance to weathering and erosion that has appeared in the geomorphological literature.[46] At McMaster University, D. C. Ford has specialized in the study of solution processes in Ontario and British Columbia, with particular emphasis on cavern genesis and sedimentology. An inventory of Ontario caverns has been prepared, and research is continuing on karren development along the Niagara escarpment,[47] doline formation in Ontario, cavern systems in the Selkirk Range, B.C.,[48] and the isotope dating of drip-stone features.[49]

In the west of Canada, the University of British Columbia has offered graduate work in geomorphology since 1948, and although there have been few theses presented in this field, there has been no lack of research. The studies of the Mackenzie delta initiated by J. R. Mackay and J. K. Stager in the fifties, and continued through to the present, have been particularly outstanding because of their comprehensive nature. Over the years, almost every aspect of the geomorphology and hydrology of the delta has been investigated: the landforms as such, oriented lakes, river channels and pingos; the geomorphic history, with particular emphasis on the late-glacial episode; and finally the contemporary processes, solifluction, shoreline recession, the break-up pattern of river ice, and freezing and thawing in the active layer.[50]

The expansion in geomorphological research at the universities during the last decade has been paralleled, and indeed, in several instances, complemented by research supported by private institutions. The Arctic Institute of North America has continued its support of Arctic and Subarctic projects and between 1956 and 1966, thirty-one awards representing ten percent of the total were made for geomorphological research. Another organization, the Geological Association of Canada, made an important contribution to geomorphology with the publication of the Glacial Map of Canada in 1958.[51] This map is, in a sense, the outcome of all previous research on the glacial geomorphology of Canada, but it owes most to the studies in aerogeomorphology undertaken in the late forties and fifties, which have been mentioned above.

A Decade of Research in the Federal and Provincial Government

In several government departments there has been an increasing interest in both pure and applied geomorphology in recent years. The National Research Council has supported several pioneer programs of a multi-disciplinary type including various aspects of geomorphology. For example, the Photogrammetric Research Section provided the ground control and produced the maps of Salmon Glacier for the University of Toronto Expedition (1956-57) and was responsible for the glacier maps of the Jacobsen-McGill Axel Heiberg expedition, which was discussed above. In the Division of Building Research, investigations into the nature and distribution of permafrost have continued. Field studies of permafrost and the site indicators, such as vegetation, drainage, and soil type, have been made in several parts of northern Canada, notably by J. A. Pihlainen and R. J. E. Brown. In 1960, Brown prepared a provisional map showing the southern limits of the different types of permafrost and, recently, he completed a book which is based on the results of detailed field work in the Mackenzie District and the Prairie provinces.[52] The National Research Council has also supported research on organic terrain and mass wasting through its Committee on Soil and Snow Mechanics and its sponsorship of the Muskeg Research Conferences.

The Defence Research Board continued to support research on northern Canada during the late fifties and early sixties, with particular emphasis on air photo interpretation techniques and ground studies of the polar continental ice shelves. In the former field, the main contributions to geomorphology were the collection and interpretation of a representative sample of air photographs of Arctic Canada by M. Dunbar and K. R. Greenaway,[53] and the preparation of a series of air photo interpretation keys for organic terrain by N. W. Radforth.[54]

The Geological Survey of Canada has paid considerably more attention to the study of Pleistocene stratigraphy and landforms during the last decade. Facilities for radiocarbon dating and palynological work have been established, and investigations of unusual structures such as cryoturbation features and glacio-tectonic forms have been included in an expanded program for the mapping of surficial deposits in the settled parts of Canada. Unfortunately these studies are too numerous to discuss here.

The most substantial contribution to geomorphological research in the last decade has been made by the Geographical Branch. This is not surprising in view of the size of the staff, which at one stage included the largest group of geomorphologists in Canada, and the research interests of the last director, J. D. Ives, who after specializing in geomorphology, became director of the McGill Subarctic Research Laboratory and devoted his attention to the study of the glacial and post-glacial history of Labrador-Ungava. The Branch was able to develop several major research programs which were beyond the scope of universities or private organizations; some of these were continued from the early fifties,[55] others were begun more recently.[56]

Several new research programs, such as the study of periglacial phenomena were begun in the late fifties. The most important contribution to this program was made by F. A. Cook. His early work included studies of the types of patterned ground on Cornwallis Island and an investigation of the thermal regime of the active layer in the same area.[57] A three-stage program was started in 1958, involving the compilation of information on periglacial phenomena from printed sources,[58] the preparation of a glossary of periglacial terms, and finally the elaboration of techniques for the continuing study of periglacial processes. In addition, several experimental maps of periglacial phenomena were prepared, for example, B. Robitaille's map of the Mould Bay area, Prince Patrick Island (1:72,000),[59] D. A. St. Onge's maps of Ellef Ringness Island (1:250,000—seven colours), and Isachsen (1:30,000, also in seven colours) which provides sets of symbols and shading for structural, fluvial, marine, nivial, glacial, solifluction, and cryoturbation features.[60]

In 1961, the Branch embarked upon an even more specialized program in the eastern Arctic. After a preliminary reconnaissance, the Penny and Barnes ice caps in central Baffin Island were selected for long-term studies of glaciology, glaciometeorology, and glaciofluvial processes. Pit studies, gravity and seismic traverses, and mass balance studies have been carried out on the ice caps, and test studies of some ice cored moraines have been completed. The glacial landforms adjacent to the Barnes Ice Cap have been examined in detail, and till fabric and lichenometric techniques have been used to unravel the recent geomorphological history. In addition, sections of both the east and west coast have been studied, partly for the evidence they provide on the postglacial isostatic readjustment,[61] and partly in search of information on present day coastal processes. The Baffin Island program was conceived as part of a long range scheme for the detailed study of a representative cross section of the eastern Arctic.[62] The project continues, although on a smaller scale, as part of the research program of the Institute of Arctic and Alpine Research at the University of Colorado, where several of the former members of the Geographical Branch are now based.

It is apparent from this review that at both government and university levels the amount of geomorphological research has increased rapidly in the last decade. This is reflected in the proportion of papers on geomorphological topics presented at meetings of the Canadian Association of Geographers which has increased from a quarter of the total in the first seven years of the Association's history to a third in the last seven years. It is also reflected in the steady increase in the total volume of geomorphological literature, two significant aspects of which are worth underlining. Firstly, there has been a relative decline in the numbers of reconnaissance-type studies and an increase in the numbers of detailed field investigations. Secondly, there are indications of a move toward the inclusion of ancillary techniques such as lichenometry, palynology, fabric analysis, and geochronology (particularly radiocarbon dating methods) with standard geomorphological procedures such as the study of morphology and stratigraphy. In a few instances, the projects have been truly interdisciplinary involving scientists in cognate fields.

Canadian Geomorphology Perspective

Geomorphology in Canada has come of age, and although the number of geomorphologists in the country is still less than one hundred, (representing perhaps ten percent of the geographical profession) this group has made a very significant contribution at the national and the international level. It is difficult to assess the contribution at the international level because there are no adequate criteria for a truly objective analysis. In order to make an assessment of the Canadian contribution to the geomorphological literature, it was decided to use *Geo Abstracts*[63] and make counts of the articles appearing in the different sections. In its early years, *Geo Abstracts* did not provide a comprehensive coverage of the geomorphological literature; nevertheless, it is the only review of its type, and it is a better index of research activity than the papers presented at international meetings or the contents of individual journals.

The graphs in Fig. 1 are derived from the analysis of the 14,370 articles which appeared in *Geo Abstracts* over the twelve-year period 1960-1971. In the first graph it can be seen that there is an increase in the total numbers of Canadian

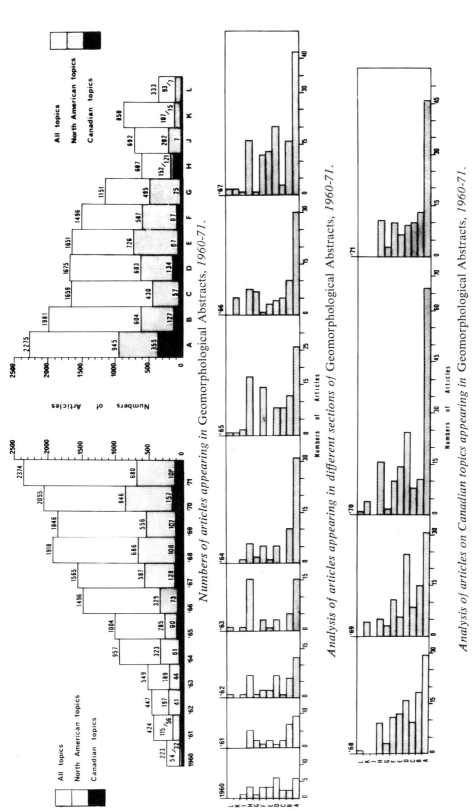

Numbers of articles appearing in Geomorphological Abstracts, 1960-71.

Analysis of articles appearing in different sections of Geomorphological Abstracts, 1960-71.

Analysis of articles on Canadian topics appearing in Geomorphological Abstracts, 1960-71.

Figure 1. The Canadian contribution to Geomorphological Abstracts, 1960-71 A:
glacial geomorphology and glaciology; B: regional physiography; C:
weathering and mass wasting; D:
non-glacial aspects of the Pleistocene; E: coastal geomorphology and oceanography; F: miscellaneous; G: fluvial geomorphology; H: periglacial studies; J: karst studies;
K: structural geomorphology; L: aeolian studies
and the geomorphology of arid areas.

publications each year until 1966, but this appears as a relative decline when considered as a percentage of the overall total (Table 1). In all probability, the Canadian contribution has remained fairly constant at about 6 percent of the total; the apparent decline is simply due to the fact that *Geo Abstracts* has progressively widened its scope, and the early issues, while including all the Canadian material, were far from providing a complete world coverage. When considered as a percentage of the North American total, the Canadian contribution has again shown a relative decline since 1960, apparently for the same reason. The present figure of about 17 percent would seem to be representative since, with the exception of 1965, it has remained fairly steady at this level since 1962.

In the upper right hand graph (Fig. 1), the total numbers of papers appearing in the different sections of *Geo Abstracts* are shown with an indication of the North American and the Canadian contributions. It can be seen that glacial studies, including glacial geomorphology and glaciology (column A), have attracted the most attention representing 15.8 percent of the total literature (Table 2). Canadian geomorphology has followed this general trend, and the Canadian contribution to glacial studies over the last twelve years has been very significant: 355 publications, accounting for 15.6 percent of the total.

Regional physiography, perhaps for historical reasons, has been a popular field of research (13.8 percent of the total literature), and here again Canadian workers have made a distinct contribution (6.4 percent of the literature). The proportion of studies of this type in a nation's scientific literature is often a good index of both the stage of economic development and the extent of geomorphological coverage.

The next three categories each account for 11 percent of the total literature (columns C, D, E, Fig. 1), but as can be seen the Canadian contribution varies considerably (Table 2). The Canadian interest in the non-glacial aspects of the Pleistocene epoch (4.7 percent of the literature) is understandable, and the comparatively large number of articles on solifluction account for a similar Canadian contribution to weathering and mass wasting (4.0 percent). The involvement of Canadians in coastal geomorphology and oceanography (9.3 percent of the literature) is hardly surprising in view of the extent of Canada's coastline and the area of inshore waters.

The remaining five categories account for only a small proportion of the total world literature, which is explained by the fact that less emphasis is placed on them in teaching, and there is less opportunity for their study. The Canadian contribution is very restricted except in periglacial geomorphology, where Canada competes with Poland for dominance of the field. It is interesting to note that this is the only topic where the Canadian contribution (20 percent of the world total) is greater than the American. The paucity of Canadian studies in the other categories is difficult to explain, particularly in fluvial and structural geomorphology, because the Canadian landscape provides almost unrivalled opportunities for work in these fields.

The variation in the Canadian contribution to the different fields of geomorphology is very apparent from the lower graph (Fig. 1), where the numbers of publications are shown according to topic in each of the years 1960-71. Over such a short time period it is difficult to establish trends, but it seems safe to conclude that there is a continuing interest in glacial and periglacial studies, with only a limited and sporadic interest in other topics. Canadian geomorphology is still essentially European in its outlook: its working methods are traditional. Although

TABLE 1
The Canadian Contribution to the Geomorphological Literature 1960-71

	1960	1961	1962	1963	1964	1965	1966	1967	1968	1969	1970	1971	Av.
Canadian contribution as a % of the world total	14.3	13.2	9.2	8.0	6.4	8.9	4.8	8.2	5.6	5.8	7.6	4.6	6.7
Canadian contribution expressed as a % of the North American total	59.2	48.7	20.8	23.2	18.8	31.5	21.9	21.8	15.7	19.2	18.5	16.0	20.7

TABLE 2
The Canadian Contribution to Various Geomorphological Topics 1960-71

		Contributions of all countries to each topic expressed as a % of the total	Canadian contribution to each topic expressed as a %
A	Glacial geomorphology and glaciology	15.8	15.6
B	Regional physiography	13.8	8.8
C	Weathering and mass wasting	11.5	4.0
D	Coastal geomorphology and oceanography	11.6	9.3
E	Non-glacial aspects of the Pleistocene	11.5	4.7
F	Miscellaneous	10.4	6.0
G	Fluvial geomorphology	8.0	1.7
H	Periglacial studies	4.2	8.4
J	Structural geomorphology	4.8	0.5
K	Karst studies	5.9	1.0
L	Eolian studies	2.3	0.5
		Total 99.8	Average 5.5

glacial geomorphology and regional physiography have received the most attention, the results have been significant at the local rather than the universal level, and the important Canadian contributions to the progress of geomorphological science as a whole have been rather in the realm of aerogeomorphology and periglacial studies. Quantitative geomorphology has been largely ignored except in glaciological investigations where the stimulus has come from physics and meteorology and slope studies where the impetus has come from engineering.

Canadian Geomorphology—The Prospect

From this inventory of studies in Canadian geomorphology several interesting features have emerged, and it is valuable to have seen these in their historical context because it provides the proper perspective for the future. If progress in geomorphology is to continue at a steady pace, it is important that some attention be given at the present time to the direction of future development. This is obviously a complex question, but an approach can be made by considering some of the component parts, such as the choice of topics for investigation, the methods of data collection and storage, and the procedures of interpretation and analysis.

The first of these factors, the choice of topic, in fact involves several choices, including the field of study and the study area. It is apparent that Canada with its range of conditions from the muskeg lowlands of the Shield to the high mountain chains of the Rockies, and from the ice-clad islands of the Arctic Archipelago to the sage brush flats of the dry prairie offers a tremendous variety of terrain types and geomorphic processes.[64] Canada's natural laboratory has been richly furnished. The review of investigations that has been presented in the last few pages indicates that some parts of the laboratory have been most effectively used, resulting in important contributions to glacial and periglacial geomorphology; however, into other parts of the laboratory we have scarcely ventured. Many branches of geomorphology are practically untouched in Canada, and remarkable opportunities exist for both landform and process studies in weathering and mass wasting, coastal, structural and fluvial geomorphology. It should be noted that there is a critical need for sound geomorphological work in the settled parts of Canada as a basis for economic and regional studies, and if one accepts the view that the researcher should be concerned to some extent with aiding national development, there is good reason to advocate the rerouting of some of our research effort from northern to southern Canada.

The methods of data collection and the procedures of interpretation and analysis hinge directly one upon the other, since the type of program determines the form in which field evidence has to be presented. During the last decade, increased emphasis has been placed on experimental and analytic techniques demanding precise information about the amounts of material and energy involved in geomorphic processes and the expression of surface form in suitably stratified arrays of morphometric data. Developments in other fields such as engineering and physics are directly responsible for this trend in geomorphology, and it is now apparent that it is more than a passing fancy. Quantification in geomorphology has two aspects which it is well to keep distinct. Firstly, there is precise measurement: the monitoring of geomorphic processes and the morphometry of landforms; and secondly, model building: the establishment of geomorphological relationships by statistical and mathematical methods.

Careful observation and attempts at precise measurement have been part of geomorphology since its beginning in the seventeenth century, and additional efforts in this direction are to be welcomed. Until quite recent times the problem of land-form quantification presented a serious obstacle to further progress. Accurate contoured maps provided the first comprehensive quantitative statements about landforms, and geomorphologists were quick to use them for simpler procedures such as profiling and altimetric frequency analysis, but it is only in the last few years that the possibility of translating cartographic data into morphometric data has been demonstrated. The full potential of this approach has not yet been realized, perhaps because of a distrust for data derived from topographic maps. However, there is no reason why morphometric data cannot be obtained in other ways, for example, from multiple profiles provided by airborne radar altimeters or gas lasers, or from measurements made on a stereo-model with a sophisticated plotting instrument, such as the Wild orthograph linked to a digital read-out. All discreet landforms, whether positive or negative, can be expressed in morpho-metric terms such as length, breadth, elongation, peakedness, profile area, and volume, while assemblages can be expressed in terms of spacing, density, parallel-ism, local relief, etc., thus allowing direct comparison of one individual in a group with another, or the group as a whole with another group.[65] Once a sufficient population has been sampled, it should prove possible to devise suitable class intervals on a naturalistic basis.

In the same manner, possibilities exist for improving our knowledge of geomorphic processes by the collection and analysis of numerical data. The current trend in geomorphology involves increasing use of basic physical and chemical approaches to investigate the inner workings of particular processes and it is ap-parent that such problems demand a much greater use of mathematics, statistics, and computer science than has been customary in the past. Some geomorphological processes are more amenable to careful monitoring than others, and Canadian geomorphologists are fortunate in having within their territory a variety of sub-aerial processes, many of which are sufficiently rapid to yield significant results after a few years of study. The prime opportunities lie in the study of fluvial, glacial, coastal, and certain of the more rapid mass wasting processes. In many cases, study sites lie close to universities or within reach of moderately financed expeditions.

The final aspect of field data collection is that of storage. So far, the only way in which geomorphic data has been stored is in publication, manuscript, and map form, which has meant that retrieval has been more or less determined by the thoroughness of the bibliographic search. The literature becomes more voluminous every year, and it is apparent that Canadian geomorphology would benefit from the establishment of a national data storage system. The advantages of such a system are obvious enough.[66] It would allow for the automatic storage of current data, thus providing a rapid retrieval system. It would permit the immediate comparison of different sets of data, and by accumulating information it should lead to the discovery of the actual frequency distributions of particular geomorphic attributes, and allow the recognition of trends, associations, and groupings.

The possibilities for the development and testing of theoretical models are very considerable; however, it must be remembered that we are dealing with a variable mixture of solids, liquids and gases, and the parameters in the equations will be extremely difficult to establish. The discovery of cause-effect relationships between process and landform is complicated by interactions between processes and by human activity. Once the time factor is added, it becomes even more complex, because of the difficulty of determining how much each landform owes to present processes and how much to processes operating in the past, when climatic conditions may have been very different. Another complicating factor is introduced by the diastrophic processes.·

Perhaps the best prospects at the present time are for studies of scale models, restricted to a particular situation where the conditions can be set up as required, thus eliminating the complex variables such as diastrophic and climatic history, and allowing the systematic investigation of the process and response factors. In many situations, the scaling laws still have to be worked out, and it is difficult to maintain similarity between the model and the actual situation, particularly with regard to the dynamic dimensions. The establishment of a geomorphological laboratory for research on scale models is well within the capabilities of many universities in Canada; however, for long term projects it might be more effective for such a facility to be sponsored by a government organization such as the National Research Council.

A well-argued case for an autonomous geomorphology has been presented quite recently in Canada,[67] and it has been apparent for some time that geomorphology cannot develop to its full extent if it is obliged to relate its work at all points to the general field of geography. Most of the recent demands for a geographical geomorphology have come from the United States, although some echoes have been heard in Canada, and it becomes increasingly apparent that to accede to such demands would reduce geomorphology to the status of a branch of applied geography, its field restricted to the provision of landform description suitably predigested to allow for the socio-economic treatment of one area after another. Such a future is quite unacceptable.

Canadian geomorphology has come of age, and it must be independent of geographical control to achieve its full potential. However, a self-conscious and defensive isolation would be equally damaging. Modern science is becoming increasingly synoptic, and geomorphology can both contribute to, and benefit from, an interdisciplinary dialogue among the natural sciences. By accepting rigorous scientific standards in its procedures and encouraging collaboration with researchers in allied fields, Canadian geomorphology can make an even more effective contribution to earth science in the future.

REFERENCES

1. J. Warkentin (ed.), *The western interior of Canada — a record of geographical discovery, 1612-1917*, McClelland and Stewart, Toronto, 1964, pp. 91-105.
2. J. Hector, "Physical features of the central part of British North America, with special reference to the botanical physiognomy," *Edinburgh New Philosophical Journal, New Series*, XIV, No. II, 1861, pp. 263-68.
3. H. Y. Hind, "Observations on supposed glacial drift in the Labrador Peninsula, Western Canada, and on the south branch of the Saskatchewan," *Quarterly Journal of the Geological Society of London (Q. Jl. Geol. Soc. London)* XX, 1864, pp. 122-30.
4. *Geology of Canada, 1863*, with a preface by W. E. Logan. Geological Survey of Canada, 1863, (Geol. Surv. Can.).
5. G. M. Dawson, "Report on the country in the vicinity of the Bow and Belly Rivers, North West Territory," *Geological and Natural History Survey of Canada, Report of Progress, 1882-83-84*, 1885, Part C, pp. 149-152.
6. R. G. McConnel, "Report on an exploration in the Yukon and Mackenzie Basins, N. W. T.," *Geol. Surv. Can. Annual Report, 1888-89*, New Series, IV, 1890, Part D, pp. 24-29.
7. A. P. Low, "On explorations in James Bay and the country east of Hudson Bay drained

by the Big, Great Whale, and Clearwater Rivers," *Geological and Natural History Survey of Canada, Annual Report, 1887-88*, New Series IV, 1889, Part J, pp. 26-33 and 61-62.

8. W. Upham, "Report of exploration of the glacial Lake Agassiz in Manitoba," *Geological and Natural History Survey of Canada, Annual Report, 1888-89*, New Series IV, 1890, Part E, pp. 1-110.

9. J. W. Goldthwait, *Physiography of Nova Scotia*, Canada Dept. of Mines, Geol. Surv. Memoir, 140, 1924.

10. L. E. Hamelin, (ed.), *Mélanges géographiques canadiens offerts à Raoul Blanchard*, Les Presses Universitaires Laval, 1959.

11. G. T. Taylor, "Topographic control in the Toronto region," *Canadian Journal of Economics and Political Science*, IV:4, 1936, pp. 493-511.

12. G. T. Taylor, "The structural basis of Canadian geography," *Canadian Geographic Journal (Can. Geog. Jl.)*, XIX:5, 1937, pp. 297-303.

13. A complete list of Blanchard's works is given in L. E. Hamelin (ed.), *Mélanges géographiques canadiens offerts à Raoul Blanchard*, op. cit., pp. 35-45.

14. A. H. Clark, "Contributions to geographical knowledge of Canada since 1945," *Geographical Review (Geog. Rev.)* XL, 1950, pp. 285-312.

15. G. H. T. Kimble, "Geography in Canadian universities," *Can. Geog. Jl.* 108, 1946, pp. 114-115.

16. P. D. Baird et al., "The glaciological studies of the Baffin Island expedition 1950," *Journal of Glaciology (Jl. Glac.)* 11:2, 1952, pp. 2-9.

17. R. F. Flint, "Growth of the North American ice sheet during the Wisconsin age," *Bulletin of the American Geological Society (Bull. Am. Geol. Soc.)* 54, 1943, pp. 325-362.

18. J. L. Jeness, "Permafrost in Canada," *Arctic* 2, 1949, pp. 13-27.

19. The material in these articles was finally published in book form; L. J. Chapman and D. F. Putnam, *The Physiography of Southern Ontario*, University of Toronto Press, 1951.

20. R. P. Sharp, "Soil structures in the St. Elias Range, Yukon Territory," *Journal of Geomorphology (Jl. Geom.)* V:4, 1942, pp. 274-301.

21. A. L. Washburn, "Patterned ground," *Revue de Géographie Canadien* 4, 1950, pp. 5-59.

22. V. Tanner, "Outline of geography, life and customs of Newfoundland-Labrador," *Acta Geografica* 8, 1944, pp. 1-906.

23. A full list of the field research projects sponsored by the Arctic Institute of North America between 1945 and 1955 is given in *Arctic* 7:3-4, 1955, pp. 354-366.

24. J. T. Wilson, "Drumlins of Southwest Nova Scotia," *Trans. of the Royal Society of Canada (R. Soc. Can.)* 32, Series 3, Sect. IV, 1938, pp. 41-47.

25. J. T. Wilson, "Eskers north-east of Great Slave Lake," *Trans. R. Soc. Can.* 33, Series 3, Sect. IV, 1939, pp. 119-129.

26. N. E. Odell, in A. Forbes, *Northernmost Labrador mapped from the air*, American Geographical Society (Am. Geog. Soc.), 1938.

27. M. R. Dobson, *Geography in Canadian universities*, Misc. Paper No. 2, Geography Branch Dept. of Mines and Technical Surveys, 1950.

28. F. K. Hare, *A photo-reconnaissance survey of Labrador-Ungava*, Geog. Branch Memoir 6, 1959.

29. M. C. V. Douglas and R. N. Drummond, *Air photograph interpretation of Quebec-Labrador*. 35 volumes, 1953. (Unpublished reports arranged by N. T. S. map areas prepared for the Defence Research Board of Canada.)

30. R. N. Drummond, "Research at the McGill Sub-arctic research laboratory," *The Canadian Geographer (Can. Geog)* 11, 1958, pp. 45-46.

31. Much of this research was published and a complete reference list is given in J. B. Bird, *The physiography of Arctic Canada*, Johns Hopkins University Press, 1967.

32. C. S. Lord, "Operation Keewatin, 1952, a geological reconnaissance by helicopter," *Precambrian*, 26:4, 1953, pp. 26-30.

33. A general review of research prior to 1955 is provided in R. F. Legget, "Permafrost research," *Arctic* 7:3 and 4, 1955, pp. 153-158. The National Research Council maintains

a complete bibliography — *List of Publications on Permafrost and Building in the North,* 1970.

34. J. K. Fraser, "Canadian ice distribution survey," *Arctic Circular* 5:5, 1952, p. 56.

35. P. W. Adams, "The laboratory in 1964," *McGill Subarctic Research Papers* 22, pp. 9-13 1972.

36. RAND Corp., *A report on the physical environment of the Great Bear River area, N. W. T.* RM-1222-1-PR, 1963. The other reports have the same general title and cover northern, central and southern Baffin Island, the Quoich River area, the Thelon River area and Victoria Island respectively.

37. Reports on the activities over the first four years were published in the *Axel Heiberg Island Research Reports, McGill University—Preliminary report 1959-1960* and *Preliminary Report 1961-1962.*

38. F. Muller, "Large scale maps for glaciological research in the Canadian High Arctic," *Revista Cartografica.* Anno 12:12, 1963, pp. 315-324.

39. M. A. Carson, *The mechanics of erosion,* Pion Press Ltd., 1971.

40. M. A. Carson and M. J. Kirby, *Hillslope form and processes,* Cambridge University Press, 1972.

41. L. E. Hamelin, "Bilan vicennal de géomorphologie a l'Institut de Géographie de Québec," *Bulletin de l'Association des géographes de l'Amérique francaise,* 10, 1966.

42. J. C. Dubé and L. E. Hamelin, *Carte preliminaire de phénomènes périglaciaires du Canada,* 1960.

43. L. E. Hamelin, "Périglaciaire du Canada: idées nouvelles et perspectives globales," *Cahiers de Géographie de Québec* 10, 1961, pp. 141-203.

44. L. E. Hamelin and F. A. Cook, *Illustrated glossary of periglacial phenomena,* Les Presses de l'Université Laval, 1967.

45. "Le quaternaire du Québec, numero special," *La Revue de Géographie de Montréal,* XXIII:3, 1969.

46. E. Yatsu, *Rock control in geomorphology,* Sozosha Publishing, 1966.

47. D. C. Ford, "Research methods in karst geomorphology," *Research Papers in Geomorphology, First Guelph Symposium 1969,* 1971, pp. 23-47.

48. A. Pluhar and D. C. Ford, "Dolomitic karren of the Niagara escarpment, Ontario," *Zeitschaft fur Geomorphologie* 14, 1970, pp. 392-410.

49. D. C. Ford, "Geological structure and a new explanation of limestone cavern genesis," *Trans. of the Cave Research Group* 13:2, 1971, pp. 81-94.

50. J. R. Mackay has published many articles on the Mackenzie delta, but his most comprehensive treatment of the area is "The Mackenzie Delta Area," *Geog. Branch Memoir 8,* 1963.

51. *Glacial map of Canada,* Geological Association of Canada, 1958.

52. R. J. E. Brown, *Permafrost in Canada — its influence on northern development,* University of Toronto Press, 1970.

53. M. Dunbar and K. R. Greenaway, *Arctic Canada from the air,* Ottawa, 1956.

54. N. W. Radforth, *Organic terrain organization from the air (altitudes 1,000-5,000 feet).* Handbook No. 2, D. R. 124, D. R. B. Ottawa, 1958.

55. N. L. Nicholson, "The Geographical Branch, 1947-57," *Can. Geog.* 10, 1957, pp. 61-68.

56. A useful summary of the activities of the Geographical Branch has appeared in the *Newsletter* of the Canadian Association of Geographers from 1960 onwards.

57. F. A. Cook, "Geographical Branch studies in periglacial geomorphology," *Cahiers de Géographie* 7, 1959, pp. 209-10.

58. F. A. Cook, "A selected bibliography on periglacial phenomena in Canada," *Geog. Branch, Bibliographical Series,* No. 24, 1959.

59. B. Robitaille, "Présentation d'une carte géomorphologique de la région de Mould Bay, Ile-du-Prince-Patrick, Territories du Nord-ouest," *Can. Geog.* 15, 1960, pp. 39-43.

60. D. A. St. Onge, "Géomorphologie de l'Ile Ellef Ringness, T. du Nord-ouest," *Geog. Branch Memoir 11,* 1964.

61. J. T. Andrews, "Pattern and cause of variability of postglacial uplift and rate of uplift in Arctic Canada," *Journal of Geology (Jl. of Geol.)* 76, 1968, pp. 404-425.

62. O. H. Loken and J. T. Andrews, "Glaciology and chronology of fluctuations of the ice margin at the south end of the Baffin ice cap, Baffin Island, N. W. T.," *Geographical Bulletin (Geog. Bull.)* 8:4, 1966, pp. 341-359.

63. Geo Abstracts is edited and published by K. M. Clayton, University of East Anglia, Norwich, U. K. Series A deals with the geomorphological literature.

64. J. B. Bird, *The natural landscapes of Canada: a study in regional earth science*, Wiley of Canada, 1972.

65. J. T. Parry and J. A. Beswick, "The application of two morphometric terrain-classification systems using air-photo interpretation methods," *Photogrammetria* 29:5, 1973, pp. 153-185.

66. A convincing case for a national storage system for geological data has recently been produced. *Interim Report of the Committee on Storage and Retrieval of Geological Data in Canada*, Geol. Surv. Can., Paper 66-43, 1966.

67. L. E. Hamelin, "Géomorphologie — géographie globale — géographie totale associations internationales," *Cahiers de Geographie de Quebec* 16, 1964, pp. 199-218.

Chapter 6

The Changing Interpretation of Landforms

In his opening remarks Parry notes that the origins of geomorphology in Canada are difficult to discover. The fact is equally true if we attempt to discover the origins of geomorphology in a global context; the most valuable compendium of material for these purposes is the history of geomorphology by Chorley et al (1964), which provides a wealth of historical material.

In his earlier paper Parry (1967) summarizes the events of geomorphic significance in the latter part of the nineteenth century. He says,

> In Europe, in the 1840s, there was a shift in the climate of opinion, so that uniformitarianism penetrated where catastrophism has previously prevailed. This change was largely the result of the work of Charles Lyell, and his visits to North America in 1841 and 1845 ensured successful transplantation of the new doctrine. Unfortunately, Lyell's uniformitarianism was marred by his overemphasis of the role of marine processes in landscape sculpture, and his insistence on the iceberg-drift origin of glacial deposits. Thus, while he accepted a fluvial origin for the Niagara gorge, and correctly ascribed the features on Mount Royal and the Scarborough bluffs to wave action, he also believed that intensive erosion could result from the mass movement of ocean water, and so the overdeepened basins of the Great Lakes were considered to be the result of the swirling action of estuarine currents. The reader of Lyell's *Travels in North America* is presented with a vision of a land half-sunk beneath Arctic seas, in which icebergs ground against one another, and became stranded along the coastlines. The simpler land-ice theory of Louis Agassiz was eclipsed for nearly a quarter of a century.

> One of Lyell's converts, William Logan, was appointed to the newly created post of provincial geologist in 1842. This appointment marked the beginning of the work of the Geological Survey in Canada, an important event in the history of landform study because, although the immediate concern of Logan and his assistants was the examination of Canada's coal and mineral resources, their long-term project was nothing less than a field survey of all the provinces. Involved in this was not only geological mapping, but topographic survey as well, since virtually no hypsometric data were available at that time. (Parry, 1967).

Lyell's paper of 1842 which describes the landforms and geomorphic history of the Great Lakes-St. Lawrence area is reprinted here as a clear example of his reasoning processes and methods of investigation.

ON THE RIDGES, ELEVATED BEACHES, INLAND CLIFFS AND BOULDER FORMATIONS OF THE CANADIAN LAKES AND VALLEY OF ST. LAWRENCE; BY CHARLES LYELL, ESQ., F.G.S., F.R.S., &c.*

After adverting to a former paper on the recession of the Falls of Niagara, and the observations which he made jointly with Mr. Hall in the autumn of 1841, (see Proceed. Geol.

* Reprinted from *The Proceedings of the Geological Society of London 1842-3*, Vol. 4, pp. 19-22.

Soc. Vol. III, p. 595,) Mr. Lyell gives an account of additional investigations made by him in June, 1842; in the course of which he found a fluviatile deposit similar to that of Goat Island, on the right bank of the Niagara, nearly four miles lower down than the great Falls. The fresh-water strata of sand and gravel here alluded to occur at the Whirlpool. They are horizontal, about forty feet thick, plentifully charged with shells of recent species, and are placed on the verge of the precipice overhanging the river. They are bounded on their inland side by a steep bank of boulder clay, which runs parallel to the course of the Niagara, marking the limit of the original channel of the river before the excavation of the great ravine. Another patch of sand, with fresh-water shells, was detected on the opposite or western side of the river, where the Muddy Run flows in, about one mile and a half above the Whirlpool. From the position of these strata it is inferred that the ancient bed of the river, somewhere below the Whirlpool, must have been three hundred feet higher than the present bed, so as to form a barrier to that body of fresh water, in which the various beds of fluviatile sand and gravel above-mentioned were accumulated. This barrier was removed when the cataract cut its way back to a point further south. The author also remarks, that the manner in which the fresh-water beds of the Whirlpool and Goat Island come into immediate contact with the subjacent Silurian limestone, no drift intervening, shows that the original valley of the Niagara was shaped out of limestone as well as drift. Hence he concludes that the rocks in the rapids above the present Falls had suffered great denudation while yet the Falls were at or below the Whirlpool.

Mr. Lyell thinks that the form of the ledge of rocks at the Devil's Hole, and of the precipice which there projects and faces down the river, proves the Falls to have been once at that point. An ancient gorge, filled with stratified drift, which breaks the continuity of the limestone on the left bank of the Niagara at the Whirlpool, was examined in detail by the author, and found to be connected with the valley of St. David's, about three miles to the northwest. This ancient valley appears to have been about two miles broad at one extremity, where it reaches the great escarpment at St. David's, and between two and three hundred yards wide at the other end, or at the Whirlpool. Its steep sides did not consist of single precipices, as in the ravine of Niagara, but of successive cliffs and ledges. After its denudation the valley appears to have been submerged and filled up with sand, gravel, and boulder clay, three hundred feet thick.

A description is next given of certain modern deposits, containing fresh-water shells, on the western borders of the Niagara, above the Falls, and in Grand Island, in order to show that the future recession of the Falls may expose patches of fluviatile sediment similar to those in and below Goat Island.

The author then passes to the general consideration of the boulder formation on the borders of Lakes Erie and Ontario, and in the valley of the St. Lawrence, as far down as Quebec. Marine shells were observed in this drift at Beauport, below Quebec, as first pointed out by Captain Bayfield, and also near the mouth of the Jacques Cartier river, and at Port Neuf and other places; also at Montreal, where they reach a height probably exceeding five hundred feet above the sea, the summit of Montreal mountain being seven hundred and sixty feet high, according to Bayfield's trigonometrical measurement, and the shells being supposed to be two hundred and forty feet below the summit. These shells, therefore, being more than three hundred feet above Lake Ontario, we may presume that the sea in which the drift was formed extended far over the territory bordering on that lake. The most southern point at which the author saw fossil shells belonging to the same group as those of Quebec was on the western and eastern shores of Lake Champlain, viz. at Port Kent and Burlington, in about lat. 44° 30′. Here, and wherever elsewhere the contact of the drift is seen with hard subjacent rocks, these rocks are smoothed, and furrowed on the surface, in the same manner as beneath the drift in northern Europe. The species of shells occurring in the drift, to which Mr. Lyell has made some additions, are not numerous, and are all, save one, known to exist, but are inhabitants, for the most part, of seas in higher latitudes. Many of them are the same as those occurring fossil at Uddevalla and other places in Scandinavia, and they imply the former prevalence of a colder climate when the drift originated. At Beauport there are large and far-transported boulders, both in beds which overlie and underlie these marine shells.

The author next describes the ridges of sand and gravel surrounding the great lakes, which are regarded by many as upraised beaches. He examined, in company with Mr. Hall, the "Lake ridge," as it is called, on the southern shore of Lake Ontario, and other similar ridges north of Toronto, which were formerly explored by Mr. Roy, (see Proceed. Geol. Soc. Vol. II, p. 537,) and which preserve a general parallelism to each other and to the neighboring coast. Some of these have been traced for more than one hundred miles continuously. They vary in height from ten to seventy feet, are often very narrow at their summit, and from fifty to two hundred yards broad at their base. Cross stratification is very commonly visible in the sand; they usually rest on clay of the boulder formation, and blocks of granite and other rocks from the north are occasionally lodged upon them. They are steeper on the side towards the lakes, and they usually have swamps and ponds on their inland side; they are higher for the most part and of larger dimensions than modern beaches. Several ridges, east and west of Cleveland in Ohio, on the southern shore of Lake Erie, were ascertained to have precisely the same characters. Mr. Lyell compares them all to the osars in Sweden, and conceives that, like them, they are not simply beaches which have been entirely thrown up by the waves above water, but that many of them have had their foundation in banks or bars of sand, such as those observed by Capt. Grey running parallel to the west coast of Australia, lat. 24° S., and by Mr. Darwin off Bahia Blanca and Pernambuco in Brazil, and by Mr. Whittlesey near Cleveland in Lake Erie. They are supposed to have been formed and upraised in succession, and to have become beaches as they emerged, and sometimes cliffs undermined by the waves. The transverse and oblique ramifications of some ridges are referred to the

meeting of different currents and do not resemble simple beaches.

The base-lines of the ridges east and west of Cleveland, are not strictly horizontal according to Mr. Whittlesey, but incline five feet and sometimes more in a mile. Those near Toronto are said by Mr. Roy to preserve the same exact level for great distances, but Mr. Lyell does not conceive that our data are as yet sufficiently precise to enable us to determine the levels within a few feet at points distant several hundred miles from each other. No fossil shells have been obtained from these ridges, and the author concludes that most of them were formed beneath the sea or on the margin of marine sounds. Some of the less elevated ridges, however, may be of lacustrine origin, and due to oscillations in the level of the land since the great lakes existed, for unequal movements, analogous to those observed in Scandinavia, may have uplifted fresh-water strata above the barriers which divide Lake Michigan from the basin of the Mississippi, or Lake Erie from Ontario, or the waters of Ontario from the ocean. Considerable differences of level may have been produced in the ancient beds of these vast inland bodies of fresh water, while the modern deposit and the subjacent Silurian strata may to the eye appear perfectly horizontal.

The author then endeavors to trace the series of changes which have taken place in the region of Lakes Erie and Ontario, referring first to a period of emergence when lines of escarpment like that of Queenston, and when valleys like that of St. David's were excavated; secondly, to a period of submergence when those valleys and when the cavities of the present lake-basins were wholly or partially filled up with the marine boulder formation; and lastly, to the re-emergence of the land, during which rise the ridges before alluded to were produced, and the boulder formation partially denuded. He also endeavors to show, how during this last upheaval the different lakes may have been formed in succession, and that a channel of the sea must first have occupied the original valley of the Niagara, which was gradually converted into an estuary and then a river. The great Falls, when they first displayed themselves near Queenston, must have been a moderate height, and receded rapidly, because the limestone overlying the Niagara shale was of slight thickness at its northern termination. On the further retreat of the sea a second fall would be established over lower beds of hard limestone and sandstone previously protected by the water; and finally a third fall would be caused over the ledge of hard quartzose sandstone which rests on the soft red marl, seen at the base of the river-cliff at Lewiston. These several falls would each recede further back than the other in proportion to the greater lapse of time during which the higher rocks were exposed before the successive emergence of the lower ones. Three falls of this kind are now seen descending, a continuation of the same rocks on the Genesee River at Rochester. Their union, in the case of the Niagara into a single fall, may have been brought about in the manner suggested by Mr. Hall, (Boston Journ. Nat. Hist., 1841,) by the increasing retardation of the highest cataract in proportion as the uppermost limestone thickened in its prolongation southwards, the lower falls meanwhile continuing to recede at an undiminished pace, having the same resistance to overcome as at first.

Mr. Lyell considers the time occupied by the recession of the Falls from the Whirlpool to be quite conjectural, but assigns a foot rather than a yard a year as a more probable estimate; thus he shows the Mastodon, found on the right bank near Goat Island, though associated with shells of recent species, to have claim to a very high antiquity, since it was buried in fluviatile sediment before the Falls had receded above the Whirlpool.

Lyell's overall vision of marine processes is clearly evidenced by his remarks about "marine boulder formation" and his view "that a channel of the sea must first have occupied the original valley of the Niagara." The marine theory pervades Lyell's (1845) work but his overall aim is to describe and explain the landforms. Although techniques and theories have changed, many recent publications have a similar aim. In a time period of about 130 years there have been many changes in the equipment available to scientists, in the communication between research workers, and in the means of travel. Lyell no doubt was conveyed by horse or horse and buggy for much of his North American travel. In contrast to this, the following paper by Dreimanis uses data accumulated by many scientists who probably used motorized transport, photocopied data, used extensive references, and employed automated analytical equipment in the preparation of their studies.

Dreimanis's paper has been selected to illustrate the change from the techniques of Lyell to the detailed information now available to students of landforms and sediments. In this paper evidence for the existence of various lakes is assembled from many studies, each of which represents the investment of considerable time and the development of modern techniques. Throughout the paper Dreimanis makes reference to the work of other scientists; this fact alone differentiates the work from that of Lyell who had few previous studies to draw upon.

The nature of the evidence which Dreimanis uses also merits a great deal of attention. Radio-carbon dates for the various deposits are widely used. References to the fauna and flora of cooler and warmer climates are also cited. These in many cases are the result of detailed analysis of fossil pollen grains extracted from sands, silts, and clays deposited thouands of years ago. Even the glacial clays have been analyzed and in Fig. 5 these are differentiated on the basis of their garnet content. Similarities between the carbonate content of varved clays and Dunwich Till are also quoted as evidence in reconstructing the glacial history of the area.

Both Dreimanis and Lyell are geologists. The common ground between physical geography and geology which exists in the study of landforms means that the work of either may be applicable to the other. We are pleased to present here a geological study which so closely integrates with physical geography, and students are referred to the bibliographies of works on Pleistocene geology and geomorphology where they should take note of the interdependence of geology and geography in these studies.

LATE-PLEISTOCENE LAKES IN THE ONTARIO AND THE ERIE BASINS[1] *

A. Dreimanis

Abstract. No undisputable information is available for the pre-Illinoian time, and the record from the Illinoian Glacial Stage is too fragmental for concluding on the proglacial lakes in the Ontario and the Erie basins.

A lake existed in the Ontario basin during the *Sangamon* Interglacial, and its level rose above the present lake towards the end of this interglacial, probably because of the isostatic uplift in the St. Lawrence outlet area. It is possible that a contemporaneous lake existed in the Erie basin, and the St. David's buried gorge was its outlet.

During the *Wisconsin* Glacial Stage, because of the recurring changes of the lake outlets by the advances and the retreats of the ice margin and the depression of the outlet areas by glacial load, several high- and low-level lakes alternated in the Ontario and the Erie basins.

In the Ontario basin high-level lakes (higher than at present) are known from the beginning of the Wisconsin Glaciation, for instance Lake Scarborough, then during the Port Talbot and Plum Point Interstadials and several towards the end of the last ice age. The lake level was below the present one during the St. Pierre Interstadial and for some time after the opening of the post-glacial St. Lawrence outlet.

In the Erie basin the number of lakes or their phases is greater than in the Ontario basin, because of more fluctuations of the ice margin and the development of outlets along two rather than one glacial lobe: the Erie-Ontario and Huron lobes. Lake levels lower than now are known from the Port Talbot and Plum Point Interstadials, the Port Huron-Cary Interstadial, and the Late-glacial. Lake levels were higher than at present whenever the Niagara area was blocked by the Ontario lobe.

Introduction

Most reports on the ancient Pleistocene lakes in the Ontario and Erie basins for instance those of Leverett (1902), Fairchild (1909), Leverett and Taylor (1915), Coleman (1937), Hough (1958, 1963, 1966), Karrow et al. (1961), Lewis et al. (1966, 1969), Dreimanis (1966), and others, deal with the period when the Wisconsin ice sheet retreated from this area approximately 10 to 14 thousand years ago. These lake phases range in the Erie basin from Lake Maumee to Early Lake Erie, and in the Ontario basin from various ice-marginal lakes to Early Lake Ontario. The outlines and the outlets of these lakes, their sequence, their ages have been deciphered fairly completely, though the following problems still require attention: (1) the position of the glacial margin during some of the lake phases; (2) their duration in absolute years; (3) the detailed history and the boundaries of those lake phases which preceded a rise of lake level, for instance the possible low water phases prior to Lake Whittlesey; (4) the history and extent of some of the minor marginal lakes.

Prior to the above lake phases, many others have existed. Though several of them occupied the Erie and the Ontario basins for periods of time much longer than all the post-glacial lakes together, they are not well known, and nearly all of them, except for one, Lake Scarborough, are nameless.

Scattered information may be found in the

[1] The research for the summary report has been supported by the National Research Council grant A-4215.

* Reprinted from *The Proceedings of the Twelfth Conference on Great Lakes Research*, 1969, pp. 170-180, by permission of the author and publishers (International Association for Great Lakes Research).

geologic literature on the deposits of some of the lakes which existed during the last ice age, and the Sangamon interglacial, for instance in the papers of Claypole (1887), Coleman (1894, 1933, 1941), Chalmers (1902), Leverett (1902), Antevs (1928), Dreimanis (1958, 1967), Terasmae (1960), Dreimanis et al. (1966), Karrow (1967, 1969), Duthie and Mannada Rani (1967). No deposits of undisputable pre-Sangamon lakes have been described so far.

The information on the older lakes is admittedly very fragmental. Their beaches, outlets, and bottom deposits are either buried underneath one or several till layers, or all these evidences have been destroyed by glacial erosion. However, scattered occurrences of littoral and lacustrine deposits have been found in natural and artificial exposures and concluded from drillings particularly in the Toronto area along the northern shore of Lake Ontario, and also along the north-central part of the Lake Erie shore. Their stratigraphic positions may be deciphered in relationship to the overlying and the underlying till layers. In some cases the till layers themselves, for instance the Sunnybrook Till at Toronto (Karrow, 1967) and the Port Stanley Till along the north shore of Lake Erie (Dreimanis, 1958) contain incorporated lacustrine sediments and thus indicate that proglacial lakes existed prior to their deposition; or the tills are interbedded with deposits of contemporaneous lakes. Radiocarbon ages have been obtained for some of the lacustrine and the associated terrestial deposits. Their number is slowly but steadily increasing, and presently more finite dates (at least 29) are known for the period of 23,000 to 67,000 radiocarbon years BP than for the post-Wisconsin lake phases in the Erie and the Ontario basins.

In spite of this gradually accumulating evidence from the stratigraphic investigations of the Pleistocene deposits around Lake Erie and Lake Ontario, it is still impossible to reconstruct the precise outlines of the old buried lakes and the contemporaneous positions of the glacial margins. Therefore the outlines of the selected lake phases and the Ontario glacial lobe are entirely hypothetical, and they indicate merely whether the levels of the lakes

shown were higher or lower than those of the present lakes. In several cases it has been possible to postulate how high the lake levels were, by comparing the elevations of the buried beaches, deltas, or lacustrine deposits with those of various postglacial lakes in the same area, and assuming that the uplift of land after each glacial retreat occurred at rates similar to those of the postglacial time. The ancient lake levels thus concluded are merely approximations with a variable precisity. In Fig. 1 they are given as relative elevations:

a) in the Lake Erie basin in relationship to the highest level of Lake Warren (with a plus sign—above it, with a minus—below it); Lake Warren was 120 ft above the present lake level in the area of horizontality (Hough, 1958);

b) in the Lake Ontario basin—in a similar relationship to the Lake Iroquois level; its zero isobase was about 85 ft above the present lake level (Hough, 1958).

The reference levels of Lake Warren and Lake Iroquois were chosen, because they permit further conclusions on the possible outlet directions. If the ancient lake levels in the Erie basin were higher than those of Lake Warren, it is probable that these lakes drained westward towards Mississippi. However, if the levels were lower than those of Lake Warren, eastward outlets via Ontario basin may be postulated. If the ancient lake levels in the Ontario basin were similar to that of Lake Iroquois or higher (but below the Warren level), the eastward drainage was probably towards the Hudson River. However, a level lower than that of Lake Iroquois suggests the St. Lawrence River as the outlet for a lake in the Ontario basin.

Figure 1 summarizes the presently available information on the major lake phases and glacial advances and retreats in the Erie and the Ontario basins since the end of the Sangamon Interglacial, up to the Two Creeks Interstadial (including). Though it is self-explanatory, some comments and references with detailed information will be given in the following paragraphs.

Sangamon Interglacial Stage

Erie basin. Several authors, for instance

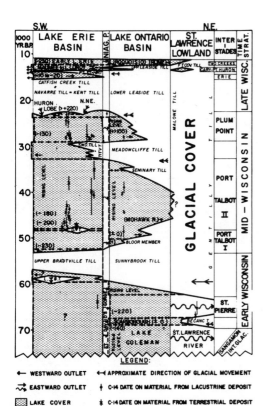

Fig. 1. A summary of the lake phases, glacial advances and retreats, and selected glacial deposits along a southwest-northeast line through Lake Erie, Niagara Peninsula (Niag. P.), Lake Ontario, and the upper St. Lawrence Lowland. Abbreviations in first column: YR. B. P. — years before present; second column: PT. ST. = HIR. T. — Port Stanley = Hiram Till, SWD. TILL — Southwold Till, TITV. T. — Titusville Till, D. T. — Dunwich Till, L. B. T. — Lower Bradtville Till; fifth column: FT. COV. T. — Fort Covington Till, BÉCANC. T. — Bécancour Till. Explanation for the lake levels (in parentheses) is in text.

Chalmers (1902) and Coleman (1941) postulate an interglacial lake in the Erie basin, but the lacustrine deposits listed by them are more probably post-Sangamon.

Saint David's buried gorge at Niagara, more than 10 mi long (Fig. 2), is considered by Hobson and Terasmae (1969) to be cut by the ancient Erie waters, draining towards the On-tario basin, during the Sangamon Interglacial or before it. In earlier works the age of this gorge has been assigned from the preglacial (Lyell, 1845; Spencer, 1910) to the Two Creeks Interstadial (Flint, 1957).

Ontario basin. Main evidence for a Sangamon interglacial lake is the Don Formation at Toronto (Coleman, 1933, 1941; Baker, 1931; Terasmae, 1960; Karrow, 1967, 1969; Duthie and Mannada Rani, 1967), rich in floral and faunal remains of warm temperate climate. This formation consists of deltaic, estuarine, lacustrine and alluvial deposits, formed along the northern shore of a gradually rising lake during the last half of the interglacial. The rise of the lake level from at least 20 ft below the present one to at least 60 ft above it (or 140 ft below the Lake Iroquois level) could have resulted from the uplift of the outlet area in the St. Lawrence Lowland at the northwestern end of the lake. It is analogous with the postglacial and recent uplift in the same area (Prince, 1954; Karrow et al, 1961).

In honour of the late Arthur Philemon Coleman, who named and studied the Don Formation, I am proposing to call the lake (Fig. 2-1) where this formation was deposited *Lake Coleman.*

Early Wisconsin

Ontario basin. The earliest glacial advance of the last ice age which deposited the Bécancour till in central St. Lawrence Lowland (Gadd, 1960), blocked the St. Lawrence outlet (Dreimanis, 1960; Karrow, 1967). A lake, about the same level as Lake Iroquois, and named *Lake Scarborough* by Coleman (1941) was formed (Fig. 2-II). It drained probably towards the Mohawk River valley (Karrow, 1967). At Toronto, a large delta known as the Scarborough Formation, was deposited in this lake. It contains abundant remains of cool climate flora and fauna (Coleman, 1941; Terasmae, 1960; Karrow, 1967, 1969; and further references mentioned therein).

When the glacier withdrew from the St. Lawrence Lowland during the *St. Pierre Interstadial* more than 67,000 radiocarbon years BP, St. Lawrence became again the outlet for the Ontario basin. As it had been downwarped by the load of the preceding glacial cover, the

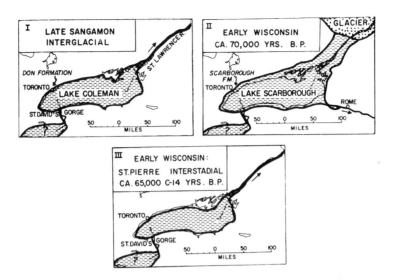

Fig. 2. Hypothetical outlines of selected Late Sangamon and Early Wisconsin lake phases.

level of the lake in the Ontario basin (Fig. 2-III) dropped below the present one, as suggested by several erosional valleys, cut into the Scarborough Formation (Spencer, 1910; Coleman, 1933, 1941; Karrow, 1967, 1969): one of the valleys extends about 20 ft below the present lake level at Toronto, or 220 ft below the Lake Iroquois level. The St. Pierre Interstadial beds in the St. Lawrence Lowland have been dated as 67,000 ±1000 years (GRO 1711) and 65,300 ± 1400 years (GRN 1799) old (Dreimanis, 1960; Muller, 1964), and a similar date (GRN 3212: 63,900 ± 17,000) has been obtained from the Otto interstadial beds in New York State, south of the eastern end of Lake Erie (Muller, 1964).

The St. Pierre Interstadial became terminated by the Early Wisconsin major glacial advance: the Ontario lobe blocked the St. Lawrence Lowland again, raising the lake level in the Ontario basin, and subsequently overrode the entire Ontario basin (Dreimanis, 1960; Karrow, 1967, 1969). Though no lake deposits have been reported from this time, the very clayey and silty texture of the Sunnybrook till which was deposited by this glacial advance at Toronto suggests incorporation of the overridden lacustrine deposits (Karrow, 1967).

Erie basin. At the beginning of the last ice age, while Lake Scarborough and the following low-level lake were formed in the Ontario

basin, the lake level probably did not change in the Erie basin. However, when the major Early Wisconsin glacial advances approached the Erie basin, high-level proglacial lakes must have developed here due to blocking of the outlet in the Niagara Peninsula. As suggested by the lithology of the Early Wisconsin Bradtville Tills and the interbedded glaciolacustrine deposits at Port Talbot, Ontario, an early glacial advance from the north which deposited the Lower Bradtville Till was followed by a later advance of the Ontario-Erie lobe, depositing the Upper Bradtville Till, also rich in incorporated lacustrine clays (Dreimanis et al, 1966). This glacial advance eventually extended over the entire Erie basin.

Mid-Wisconsin

Erie basin. Following the retreat of the Early Wisconsin ice sheet, high proglacial lakes, similar to the late-glacial Lakes Maumee to Warren, may have existed at the beginning of the Port Talbot Interstadial. No evidence of them has been found yet. The earliest nonglacial record from the north-central Erie basin is the *Port Talbot I Interstadial* Green Clay, a colluvial or low-level lake deposit (Dreimanis et al, 1966). The lake level was probably similar to that of the post-glacial Early Lake Erie, at least 230 ft below the Lake Warren level, and the outlet was across the glacially depressed

Niagara Peninsula (along the St. David's gorge?) towards the Ontario basin. Because of the isostatic uplift of the Niagara Peninsula, the lake level gradually rose, and it was already 30-50 ft higher (Fig. 3-IV) during the deposition of the *Port Talbot II Interstadial* shallow water and alluvial beds. Ten finite radiocarbon dates, ranging from 38 to 48 thousand years BP, have been obtained from these interstadial deposits in its type area at Port Talbot, Ontario.

Proglacial varved clays, consisting of about one hundred varves (the *Glaciolacustrine Unit I* of Dreimanis et al, 1966) separate the Port Talbot I from the Port Talbot II deposits. Similarities in the carbonate content of these varved clays and the Dunwich Till suggest that a short-time glacial advance from the north reached the northern shore of Lake Erie (Fig. 1). It is possible that this glacial advance did not block the outlet in the Niagara peninsula and did not raise the lake level in the Erie basin as assumed by Dreimanis et al. (1966).

Some time later, less than 35,000 years BP

(the youngest radiocarbon date on the Titusville Peat, OWU—315, Ogden and Hay, 1969 and more than 28,000 years BP (the oldest dates of the Plum Point Interstadial), another glacial readvance occurred in the eastern and central Erie basin, depositing the Titusville Till in Pennsylvania (White, 1969) and the Southwold Drift (Gravelly lower till No. 2 in Dreimanis, 1957) in the Port Talbot area. As the outlet across the Niagara area became blocked by the Ontario lobe (a spot-sample of the Titusville Till from Pennsylvania contained heavy minerals typical for the Ontario lobe), the lake level rose in the Erie basin and the outlet was probably westward. The *Glaciolacustrine Unit II* of Dreimanis et al. (1966) was deposited in this lake.

Recent discoveries of beach deposits at about 130 ft below the Lake Warren level between the Southwold Drift and the Catfish Creek Till in the Port Talbot-Plum Point area, Ontario (Dreimanis, 1967b and unpublished data) suggest a lowering of the lake level in the Erie

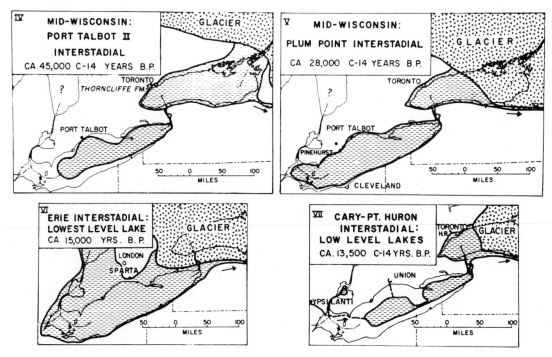

Fig. 3. Hypothetical outlines of selected Mid- and Late Wisconsin lake phases. In map VII: H. R. — Humber River, the location of Fig. 5.

basin during the *Plum Point Interstadial* and reopening of an eastward outlet (Fig. 3-V).

The lake level rose again at the transition from Mid- to Late Wisconsin. The extensive deltaic gravels in the Pinehurst area, Ontario (Fig. 3-V), deposited about 80 ft below the Lake Warren level and overlain by two Huron lobe tills, may belong to this transitional time during the Plum Point Interstadial.

Ontario basin. The uppermost or Bloor Member of the Sunnybrook Drift in the Toronto area contains varved clays (Lajtai, 1967; Karrow, 1967, 1969): over 1200 varves have been counted and correlated from section to section by Antevs (1928). They were deposited in a proglacial lake possibly during the *Port Talbot I Interstadial* while the Ontario lobe still oscillated in the Toronto area and the lake level was higher than that of Lake Iroquois.

During the following *Port Talbot II Interstadial* a lake, similar in elevation to the postglacial Lake Iroquois, existed in the Ontario basin (Karrow, 1967, 1969). The outlet area towards the Mohawk River was probably depressed by the Early Wisconsin ice sheet. Due to the isostatic uplift of the outlet during the Port Talbot Interstadial, the lake level gradually rose in the Ontario basin. That portion of the Thorncliffe Formation, which underlies the Seminary and the Meadowcliffe Tills (Karrow, 1967, 1969) was deposited in this lake. Five finite radiocarbon dates, ranging from about 34 to 49 thousand years BP have been determined from these deposits in the Toronto area (Karrow, 1969, and personal communication).

According to Karrow (1967, 1969), the varved clays, associated with the Seminary and the Meadowcliffe Tills, suggest presence of proglacial lakes at elevations above Lake Iroquois, while the glacial margin oscillated in the Toronto area during the deposition of the above tills. As already mentioned in the discussion of the Erie basin during the Mid-Wisconsin, the Titusville Till has an Ontario lobe heavy mineral composition, and therefore it is proposed here to correlate the Meadowcliffe Till, being the thickest of the above two Toronto tills, with the Titusville Till in the Erie basin (Fig. 1).

As suggested by a new radiocarbon date (GSC-1082: 28,300 ± 600 years BP, A. A.

Berti, personal communication) from the uppermost member of the Thorncliffe Formation, above the Meadowcliffe Till, a lake, higher than 100 ft above the Lake Iroquois, existed in the Ontario basin during the *Plum Point Interstadial*. Interbedding of silts and sands containing organic remains, with three units of varved clays (with 45, 300, and 90 varves at the Hi Section, Scarborough Heights, according to Karrow, 1967) suggest several readvances of the Ontario lobe close to the Toronto area, during this interstadial. The final advance, depositing the Leaside till over the 90 varves marked the beginning of Late Wisconsin.

Late Wisconsin

Erie Basin. Three lobes of the Late Wisconsin ice sheet entered the Erie basin (Fig. 1). Judging from radiocarbon dates, the Huron lobe advanced first, blocking the westward outlet 24 to 23 thousand years BP (Goldthwait et al, 1965, Fig. 3). The lake level rose to more than 220 ft above the Lake Warren level, which is more than 110 ft above the highest Lake Maumee level, about 23,000 years BP, judging from the radiocarbon dates at Garfield Heights (Cleveland, Ohio), determined from wood covered by varved clay and silt (White, 1953, 1968). The Ontario lobe was the last to advance: about the same time when the lake level rose in the Erie basin, the Niagara Peninsula was still free of ice, and wood, 22,800 ± 450 years old (GSC-816, Hobson and Terasmae, 1969) was deposited in sands overlain by 40 ft of (lacustrine?) silt and clay containing cool-climate pollen, in the buried St. David's gorge.

While the Late-Wisconsin ice-sheet advanced southward into Ohio and Pennsylvania, no lakes existed in the Erie basin for several thousand years (Fig. 1). They began to form again during the oscillating retreat of the Erie lobe. First major retreat occurred during the Erie Interstadial. Because of the high clay and silt content in the Port Stanley and Hiram Tills, and the presence of lacustrine deposits underneath (Claypole, 1887; Leverett, 1902), Dreimanis (1958) postulated a glacial retreat into the Erie basin after the deposition of the Catfish Creek = Kent Tills and prior to the deposition

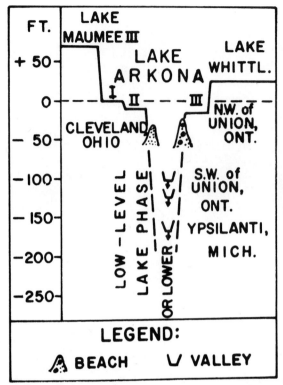

FT.
+ 50-
0-
- 50-
-100-
-150-
-200-
-250-

LAKE
MAUMEE III

LAKE
ARKONA

LAKE
WHITTL.

I

II   III

N.W. of
UNION,
ONT.

CLEVELAND
OHIO

S.W. of
UNION,
ONT.

LOW - LEVEL

LAKE PHASE

OR LOWER

YPSILANTI,
MICH.

LEGEND:

BEACH   VALLEY

Fig. 4. Evidence for a low level lake phase some time between the beginning and the end of Lake Arkona, in the Erie basin. References: Hough (1958): beach at Cleveland, Ohio; Kunkle (1963): buried valley at Ypsilanti, Mich.; Dreimanis and Karrow (1965): beach N. W. of Union, Ont.; Dreimanis (1967a): buried valleys S. W. of Union, Ont. All elevations are given in relationship to the level of the highest Lake Arkona (710 ft. above sea level in the area of horizontality).

of the Port Stanley = Hiram Tills, and existence of a high-level proglacial lake. Since 1958 beach and deltaic deposits, belonging to this interstadial, have been found in the London area (at elevations similar to those of Lake Maumee II and III) and at Sparta—about 15 ft below the Lake Warren level.

From these evidences several lake phases may be concluded in the Erie basin during the Erie Interstadial—with outlets mainly towards the west, and for some time probably also eastward, when the lake level was below that of Lake Warren (Fig. 3-VI). On Fig. 3-VI the

glacial margin between London and the eastern end of Lake Erie is based upon the most eastward occurrences of the Port Stanley Till over the Catfish Creek Till, as reported by Karrow (1963, 1968).

The following *Cary* readvance of the Erie lobe over the entire basin interrupted further development of lakes in the Erie basin for some time (Fig. 1). During the subsequent glacial retreat, first minor marginal lakes (Leverett, 1902; Forsyth, 1969), then the familiar sequence from Lake Maumee to the Early Lake Erie developed (Leverett, 1902; Leverett and Taylor, 1915; Hough, 1958, 1963 and 1966; Kunkle, 1963; Dreimanis, 1966, and 1967; Lewis et al, 1966; Wall, 1968; Lewis, 1969).

Ontario basin: Cary-Port Huron Interstadial. According to Karrow (1967, 1969), the first retreat of the Late Wisconsin ice occurred between the deposition of the upper and the lower units of the Leaside Till, as indicated by a sand and gravel layer between these two tills. Some of the sands in the Humber valley sections (Fig. 4) northwest of Toronto appear to be of beach origin (Dreimanis, 1967a), suggesting lowering of the lake level to about 120 ft above the Lake Iroquois level (or 400 ft below the north-eastward projected Lake Warren level). Such a lake level would permit development of a low-level lake phase in the Erie basin as suggested by Hough (1958, 1963, 1966), Kunkle (1963), Dreimanis (1967a) and Wall (1968). This lake must have drained eastward, towards the Mohawk River valley, but its outlet was at a higher level than during the Lake Iroquois time.

During the *Port Huron* substage the Ontario lobe readvanced beyond the southwestern end of Lake Ontario. When the glacial lobe retreated again, Lake Iroquois, the most prominent of the postglacial lake phases of the Ontario basin, came into existence (Spencer, 1890; Coleman, 1936, Karrow et al, 1961).

Summary

Though the post-glacial phases of Lake Erie and Lake Ontario are more familiar, many more existed during the last ice age, and the last interglacial. No record on the ancient lake phases is available for the pre-Sangamon time.

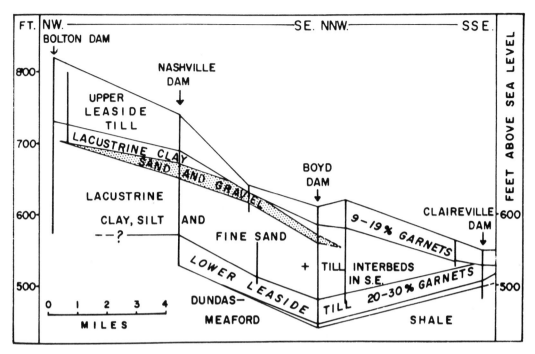

Fig. 5. Generalized geologic profile in the Humber River area. Vertical lines: wells, test drillings, sections. The sand and gravel layer in the middle is of alluvian and beach origin.

A condensed history of the late Sangamon and the Wisconsin lake phases, and their relationship to the fluctuating ice margin is given in Fig. 1.

ACKNOWLEDGEMENTS

Thanks are due to the radiocarbon laboratories of the Geological Suvery of Canada and Groningen, Netherlands, for the permission to use several unpublished radiocarbon dates, to James F. MacLaren Ltd. for releasing the test-drilling information from the Humber River area, to Dr. P. F. Karrow for comments on the Wisconsin stratigraphy in the Ontario basin, and to Mr. A. A. Berti for supplying new information on the Hi Section at Scarborough Heights, Toronto.

Even the extensive use of scientific literature exemplified by Dreimanis's paper is only one portion of a chain of inter-related studies. The studies referenced by Dreimanis each provide a great deal of local detail. This is synthesized into a regional study but the continental scale must also be considered. To view the work in this context, reference to the recent survey of the geology and economic minerals of Canada (Douglas 1970) will provide a wealth of information. Within this survey a section devoted to Quaternary Geology of Canada (Prest 1970) provides an interpretation of the same regions which Lyell studied. Prest's work is compiled at a national scale and integrates many studies which are essentially similar to the Dreimanis paper reprinted above. Also working at a regional and continental scale Bird (1972) provides an additional descriptive analysis of the landforms which Lyell discussed.

Lyell and Dreimanis are concerned with explaining the history of the landforms of the Lake Erie and Lake Ontario area. Dreimanis provides sketch maps and correlation diagrams to display his conclusions. Lyell draws on his own experience and provides little detail for his many assertions. Examination of the references used in Dreimanis's paper reveals the abundance of references to work published in the 1960s and the relatively few references to work of the previous thirty years. This reflects an overall expansion of knowledge in the mid twentieth century and the availability of precision apparatus which permitted more rapid analysis of sediments. Prior to this, analysis had been a lengthy procedure and the small number of results had not been of particular value.

A consideration of the differences between the work of Lyell and Dreimanis also indicates the nature of much recent research in geomorphology. Initially the study of landforms was regional in nature. It had to be so because the landforms being studied were located in a specific area. As the number of area studies increased, it became possible to generalize about the processes acting to create specific types of landform. As scientific and laboratory techniques developed, the detail of the processes could be more easily understood. At this point geomorphic studies began to develop a new form. Rather than describe the landforms of an area and speculate on their origin, studies could now be totally devoted to the evaluation of techniques or the development of methodology. Increasing numbers of studies included tables of data, and from these, relationships between measured variables could be inferred. Laboratory studies to further evaluate these relationships became necessary, and geomorphic studies included increasing proportions of laboratory data and experimental work. This was in great contrast to the physiography and landscape evaluation by descriptive interpretation which had been practised in the decades up to 1960.

In this description of the development of geomorphology, the generalizations do not deal adequately with the individual pieces of work containing many elements which are ahead of their time. There are studies published in the first half of this century which are scientific in every stage and which include laboratory data. For our purposes, however, the example of a data-based investigation where the measurements are recorded and evaluated without any detailed knowledge of processes is taken from a 1964 paper. The tone of many of the comments is particularly interesting, as there are several which are couched in terms such as "It is difficult at present to account for. . . ." or "The significance, if any, of the. . . ." In the context of the growth of geomorphology such papers are of great value. They serve to reinforce the fact that relatively little is known of the significance of many measurements which are presently obtained in geomorphic studies. The authoritative interpretations of Lyell can again be contrasted with the approach in the following paper which compiles data, states the results, and proceeds to interpret them with a realistic amount of caution.

Unwarranted assumptions have no real place in the increasingly rigorous scientific investigations now forming part of geomorphic research. In the paper reprinted here the first comment in the conclusion raises the question about the size and type of sample required for a representative sample. This is itself an enlightened approach because in the preceding era many studies were produced in which samples were not even collected and the representative nature of the results was never considered. Thus the paper by Andrews and Sim has been selected because it illustrates the nature of changes which took place in the 1960s in the whole field of geomorphology.

EXAMINATION OF THE CARBONATE CONTENT OF DRIFT IN THE AREA OF FOXE BASIN, N.W.T.*

J. T. Andrews and V. W. Sim

ABSTRACT: The Chittick apparatus has been used to examine the carbonate content of till from Melville Peninsula and the northern half of Baffin Island. Fifty-four samples have been examined. On Melville Island, east of the limestones, the fine-till fraction contains between 44 and 71 per cent carbonates; low values occur along the northern coast and south of Sarcpa Lake. On Baffin Island values are much lower, but it seems that they are slightly higher on the west coast than on the east coast. The high carbonate content across Melville Peninsula lends support to the hypothesis of a westward movement from Foxe Basin, but the Baffin Island samples, with the exception of one from Piling Bay (26-per-cent carbonate content), possibly indicate that the eastward movement across the island was followed by a later westward flow toward Foxe Basin.

Introduction

Dreimanis (1962) has described a method he has been using since 1935 to determine the carbonate content in the matrix of Pleistocene glacial deposits. He points out that such quantitative information can be used "in deciphering regional glacial movements, in distinguishing products of different glacial lobes, in stratigraphic and provenance studies, and in evaluation of depth of leaching in weathering profiles." The first of these uses seemed particularly applicable in the Foxe Basin area of the eastern Canadian Arctic, where regional ice movements that occurred during the Pleistocene are still imperfectly understood.

It has been suggested (Ives and Andrews, 1963, page 38) that during the "Wisconsin" maximum, ice moved radially outward from Foxe Basin across Baffin Island and Melville Peninsula and possibly southward across Foxe Peninsula. Later, the centre of dispersal shifted to lie west of the Barnes Ice Cap. This led to a dominant southwesterly movement across west

Baffin into Foxe Basin (Figure 1). Since the northern part of the Basin is surrounded and underlain by Palaeozoic limestones and dolomites, carbonate-content analysis of till samples might distinguish these movements, particularly where Precambrian igneous rocks were crossed as on Melville Peninsula and west Baffin Island.

The Method

Dreimanis gave a complete account of the application of the method (Dreimanis, 1962). It involves the use of the Chittick gasometric apparatus to measure the volume of carbon dioxide driven off when the finer fractions (<.074mm) of a till sample of precise weight are allowed to react with a 20-per-cent solution of hydrochloric acid. The weight percentage of the carbonates can then be determined by the use of simple graphic relation. It is possible to determine not only the total carbonate content but also, by a refinement in the analysis procedure, the weight percentage of calcite and dolomite, the two most important carbonate minerals in Pleistocene deposits. Other carbonate minerals, such as magnesite and siderite, are usually present in such small quantities that they can be ignored. The fine-size fraction of the sample is chosen for analysis because most till-matrix carbonates occur in the very fine sand, silt or clay fractions and because the speed of reaction between the acid and the carbonates can be compared if the grain size of all samples is similar (Dreimanis, 1962, page 521). With reasonable care the percentage of carbonates determined by the use of the Chittick apparatus should be accurate to ±0.3 per cent.

Results from the Foxe Basin Area

Chittick gasometric analysis was carried out on 54 samples of till from Melville Peninsula and Central Baffin Island in the geomorpho-

---

* From *Geographical Bulletin* 21 (1964), pp. 44-53. Reproduced by permission of Information Canada.

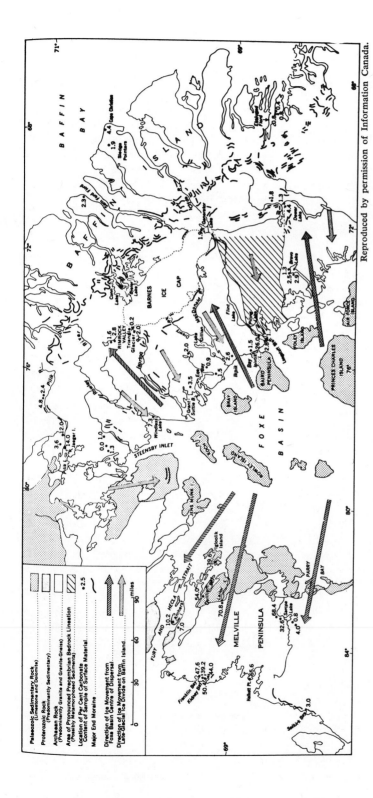

Fig. 1. The proposed radial directions of ice movement from the centre of Foxe Basin across Melville Peninsula and Baffin Island. On Baffin Island this was later followed by a westerly movement. The areas of limestone and dolomite have been drawn from existing geological maps

(Geol. Surv. Can.). In the Pilik River area old contacts have been revised from ground mapping (Falconer, Geog. Br.). Samples collected in 1963 just south of Windless Lake have up to 35 per cent carbonates in fine fraction.

logical laboratory at the Geographical Branch by means of a production-model Chittick apparatus manufactured by Fisher Scientific Company. The percentage, by weight, of calcite, dolomite and the carbonate total was determined for each sample (see table), and the last-mentioned was plotted on a bedrock geology map of the Foxe Basin area. What follows is a series of observations on the results obtained with the method.

**Carbonate content of samples of surface material from northern
Baffin Island and Melville Peninsula, N.W.T.**

Sample no.	Location	Weight percentage		
		Calcite	Dolomite	Carbonates (total)
Melville Peninsula				
VWS–58–8U	Baker Bay	14.6	29.4	44.0
VWS–58–9U	Kidney Bay (below marine limit)	20.8	18.4	39.2
VWS–58–10U	Kidney Bay	16.4	34.0	50.4
VWS–58–33U	Selkirk Bay (below marine limit)	2.5	0.5	3.0
VWS–58–38U	Mainland south of Amherst Island	2.8	4.2	7.0
VWS–58–41U	Franklin Bay	13.2	34.4	47.6
VWS–58–48U	Quillian Bay (below marine limit)	6.0	48.0	54.0
VWS–58–27U	North of Sarcpa Lake	37.2	31.2	68.4
VWS–58–28U	North of Sarcpa Lake	4.2	28.4	32.6
VWS–58–29U	South of Sarcpa Lake	0.0	0.8	0.8
VWS–58–31U	Southwest of Sarcpa Lake	0.9	3.1	4.0
VWS–58–32U	Amherst Island (below marine limit)	1.6	8.6	10.2
VWS–58–35U	Halkett Point (below marine limit)	26.0	39.6	65.6
VWS–58–43U	Igloolik Island (below marine limit)	0.8	38.8	39.6
VWS–58–45U	Lailor Lake	37.6	33.2	70.8
Baffin Island				
JTA–62–9U	Triangle Glacier	0.2	0.0	0.2
JTA–62–10U	Upper Isortoq River	0.9	1.1	2.0
JTA–62–20U	Grant-Suttie Bay	0.9	2.6	3.5
JTA–62–21U	Eqe Bay	0.5	1.0	1.5
JTA–62–22U	Eqe Bay	0.9	0.0	0.9
JTA–62–23U	69° 56′ N. 75° 32′ W.	0.7	1.3	2.0
JTA–62–24U	Windless Lake	1.4	5.9	7.3
JTA–62–25U	Longstaff Bluff (below marine limit)	1.0	1.8	2.8
JTA–62–26U	Longstaff Bluff (above marine limit)	1.1	0.6	1.7
JTA–62–27U	North of Steensby Inlet	0.8	0.5	1.3
JTA–62–28U	North of Steensby Inlet	0.0	0.0	0.0
JTA–62–29U	North of Steensby Inlet	0.3	0.7	1.0
JTA–61–20U	Rimrock Valley	1.0	0.6	1.6
JTA–61–27U	Rimrock Valley	1.1	1.7	2.8
VWS–61–3U	Generator Lake	1.3	0.6	1.9

**Carbonate content of samples of surface material from northern
Baffin Island and Melville Peninsula, N.W.T.** (concluded)

Sample no.	Location	Weight percentage		
		Calcite	Dolomite	Carbonates (total)
VWS–61–5U	Bravo Lake	1.4	1.1	2.5
VWS–61–6U	Ekalugad Fiord (below marine limit)	0.8	0.0	0.8
VWS–61–8U	Ekalugad Fiord (moraine in front of ice tongue)	0.8	0.0	0.8
VWS–61–9U	Ikpik Bay (below marine limit)	1.0	1.6	2.6
VWS–61–10U	Piling Lake (below marine limit)	16.4	9.6	26.0
VWS–61–11U	Bravo Lake (below marine limit)	1.0	1.9	2.9
VWS–61–13U	Dewar Lake	1.3	3.1	4.4
VWS–61–15U	Dewar Lake	1.5	0.3	1.8
VWS–61–17U	Bravo Lake	0.8	0.5	1.3
VWS–61–20U	Dewar Lake	0.8	0.5	1.3
VWS–61–25U	East of Ipiutik Lake	1.0	0.5	1.5
VWS–61–27U	Ekalugad Fiord	0.4	0.0	0.4
GF–61–1U	Pond Inlet	1.2	5.3	6.5
GF–61–2U	Pond Inlet	1.2	1.2	2.4
GF–61–4U	Pond Inlet	0.1	0.5	0.6
GF–61–8U	Arctic Bay	0.2	36.0	36.2
GF–61–10U	Sam Ford Fiord	1.3	1.0	2.3
GF–61–12U	Sledge Pointer	1.4	0.5	1.9
GF–61–15U	Cape Christian	2.0	2.4	4.4
GF–62–1U	Pilik River	2.0	10.0	12.0
GF–62–2U	Pilik River	0.7	1.7	2.4
GF–62–4U	Pilik River	2.0	2.8	4.8
GF–62–6U	Jaeger Island	2.6	11.4	14.0
GF–62–11U	Patlok Lake	1.2	2.4	3.6

First and most noticeable is the high carbonate content of the drift from Melville Peninsula. Of 15 samples tested only five contain less than 11 per cent carbonate. The one from the vicinity of Selkirk Bay, on the west coast of Melville Peninsula (3.0 per cent), and two from the area south of Sarcpa Lake (4.0 and 0.8 per cent) are the most southerly of the samples; two others, from Amherst Island and the mainland south of it (10.2 and 7.0 per cent carbonate respectively), are the most northerly samples tested. Of the remaining 10 samples, none contained less than 32.6 per cent carbonate, the greatest proportion being 70.8 per

cent in a sample collected at the west end of Lailor Lake, in the north-central part of the peninsula. In contrast, of the 39 samples from Baffin Island only four contained more than 10 per cent carbonate. Two of these, from Piling Lake and Arctic Bay (26.0 and 36.2 per cent respectively), are explained farther on. The other two, from the Pilik River and Jaeger Island (12.0 and 14.0 per cent respectively), are comparable to the minimum values for Melville Peninsula.

Precambrian volcanic and instrusive rocks usually contain less than 10 per cent calcium and magnesium oxides (Reiche, 1950, page

45). A large part of these oxides may be converted to carbonates during the weathering process (Lyon, Buckman and Brady, 1952, page 302), though this is perhaps doubtful under Arctic climate conditions. Any carbonate in glacial till overlying Precambrian bedrock in excess of this small proportion probably originated from nearby sedimentary areas, and the presence of the carbonates can be adduced as corrobative evidence of glacial movement determined by field techniques. Precambrian sedimentary rocks, of course, may contain considerable quantities of carbonate minerals, and known areas of these rocks must be considered in evaluating the source of carbonate tills (see map). Crystalline limestones have, for instance, been discovered in Isortoq Fiord, Baffin Island. Another possibility must be kept in mind: it is that high-carbonate tills may also result from the final glacial stripping of a younger limestone cover from a crystalline basement. Though no outliers of limestone are known to exist in the interior at present, Palaeozoic limestones once covered Melville Peninsula, and high-carbonate tills may have been derived from the final stripping of this cover from the Archaean basement during the glaciation. This would, however, involve perhaps greater glacial erosion than expected over this area of low relief. The same may be true of north-central Baffin Island. Palaeozoic outliers do, in fact, occur in the northern part of the region north of the Pilik River.

The carbonate content of the Melville Peninsula samples appears to be no higher than samples of limestone-derived till from southern Ontario examined by Dreimanis (1957, page 404).

It is believed that the carbonate content of Melville Peninsula samples supports the theory of relatively uncomplicated glacial transport from east to west across the peninsula (Sim, 1960). Swaths of carbonate-charged drift have been noted extending westward across the northern part of the peninsula from the geological contact between the Palaeozoic limestone and the Archaean granite-gneiss basement. It seems probable that the considerable amount of carbonate in the drift west of this contact and above the determined limit of post-glacial marine submergence originated on the northeastern sedimentary lowland of the peninsula. No significant east-west difference in carbonate content was apparent in the small number of samples tested. Indeed, the sample from Igloolik Island contained a smaller percentage of carbonate (39.6) than many of the samples from the interior and west coast. The high carbonate content of the west-coast samples is probably a reflection of a westerly movement from the limestone in Foxe Basin to the sample locations. Air-photo interpretation and field evidence of glacial features indicate it is unlikely that any of the other possible areas—for example, Simpson Peninsula, the east coast of Boothia Peninsula or Brodeur Peninsula—could be the source of the carbonates on Melville Peninsula.

It is difficult at present to account for the particularly low carbonate content of the samples south of Sarcpa Lake and on the west coast at Selkirk Bay. Drumlinoids and other streamlined till forms, as well as the visible swaths of limestone-charged ground moraine indicating ice movement from east to west, all occur north of the area of very low carbonate. This may indicate less intense ice movement and less effective glacial transport south of Sarcpa Lake. Even so, it still seems unusual that the carbonate content should be so low less than 20 miles from areas of limestone bedrock to the east. Dreimanis (1963: personal communication) has suggested that low values may result from the local incorporation of a friable Precambrian rock.

Similarly, the low carbonate content of the samples from Amherst Island and the mainland immediately to the south (10.2 and 7.0 per cent respectively) are difficult to explain if the glacial movement was from the southeast. It is suspected that the most recent ice advance in this area was from the north across Fury and Hecla Strait (Blackadar, 1958; Sim, 1960). The area of Baffin Island extending for a considerable distance inland from the north side of the strait, as well as the extreme northern tip of Melville Peninsula, is underlain by Proterozoic rocks in which local areas of limestone and dolomite occur and are separated from the north coast of Melville Peninsula by 60 miles of Precambrian bedrock. The carbonate in the two samples mentioned may derive from these Baffin Island sources.

As has been pointed out, the carbonate content of the 39 samples from Baffin Island is very much lower than the carbonate content of the Melville Peninsula samples. Thirty-five have a carbonate content of less than 10 per cent, and only two a carbonate content of more than 20 per cent. The sample from Arctic Bay (36.2 per cent) comes from an area where Ordovician limestone outcrops on the Precambrian surface of Borden Peninsula whereas all of Brodeur Peninsula, lying to the west across Admiralty Inlet, is underlain by both Ordovician and Silurian carbonate rocks. The sample from Piling Lake (26.0 per cent) was collected from marine-reworked deposits only a few feet above the present sea level. The relatively high carbonate content in this sample can perhaps be explained as carbonate material transported to the area from other parts of Foxe Basin by ice-rafting and longshore processes during submergence, but the carbonate rocks of Baird Peninsula are also a possibility, especially as there is no significant difference in carbonate content between samples collected from below and above the marine limit at Longstaff Bluff and Ekalugad Fiord. At the latter location, indeed, the carbonate content above and below is exactly the same. Only two samples have a carbonate content between 10 and 20 per cent, and both, one on the upper Pilik River (12.0 per cent) and the other on Jaeger Island, in Pilik Lake (14.0 per cent), were collected a few miles east of an extensive area of Palaeozoic sediments that have a considerable carbonate content. Glacial movement from the southwest or northwest could also account for the carbonate content.

Among the samples from Baffin Island, those with a carbonate content of less than 10 per cent predominate, and most have less than 5 per cent. It is noticeable, however, that the west-coast samples generally have slightly higher values than those on the east coast. It is possible that such a low content is not significant and cannot be used in the determination of the rocks that are the source of the drift. It was mentioned earlier that Precambrian vol-

canic and intrusive rocks, which underlie most of central Baffin Island, usually contain less than 10 per cent calcium and magnesium oxides. The fact that most of the Baffin Island samples contain less than this amount of carbonates means that the tills from which they were collected may have been derived from Precambrian crystalline country rock in the "waist" of the island and not from the carbonate-rich sedimentary rocks in Foxe Basin. This agrees with the preliminary interpretation of glacial events that has been proposed (Ives and Andrews, 1963). In particular, Andrews has stressed the interpretation that limited basal-ice movement in the central areas may be the cause of the highly angular and apparently locally derived characteristics of the bouldery till (Andrews, 1963).

The most recent glacial movement west of the Barnes Ice Cap was toward the southwest from an ice divide located between the present Foxe basin coast and the ice cap. Whether earlier ice advanced northeastward across the island from a Foxe Basin centre and deposited carbonate-charged drift on the waist of the peninsula is not yet known with certainty. If this deposition occurred, it may have been subsequently removed during the southwesterly movement and replaced or overlain by the present drift. The large number of erratic limestone fragments discovered throughout central Baffin Island (Ives and Andrews, 1963) may have been left by the earlier movement.

The significance, if any, of the relative amounts of calcite and dolomite in the samples remains to be determined. Dolomite exceeds calcite in 11 of the 15 samples from Melville Peninsula. Whether this is simply a reflection of the predominantly dolomitic character of the bedrock in Foxe Basin is not known. The samples from Baffin Island cannot be so clearly distinguished. Twenty samples are predominantly of dolomite, and 17 are predominantly of calcite, while two samples contained equal amounts of both.

Conclusion

The relatively small number of samples and the large area from which they have been collected limit the reliability of the results of the present study. It seems likely, however, that careful collection and analysis of till samples from other Arctic areas will yield valuable information on source areas and the direction of ice movement where this information cannot be derived from more conventional field techniques. A number of areas come to mind. Scanty data from the Foxe Peninsula of southern Baffin Island suggest that the peninsula was glaciated by ice from a dispersal zone over Foxe Basin. Determination of the carbonate content of till samples from the peninsula might provide additional support for this hypothesis. This method may also be used to determine glacial movement in southern Baffin Island eastward from the Ordovician bedrock area on the east side of Foxe Basin via Nettilling Lake to Cumberland Sound and via Amadjuak Lake to Frobisher Bay. The applicability of this method to Baffin Island will increase, however, as geological reconnaissance mapping progresses (Blackadar, 1956 and 1958). The depth of leaching in carbonate-rich glacial deposits may be useful in determining the relative age of tills in the Arctic in the manner suggested by Dreimanis (1957, pages 403-404). The use of this factor has so far not been attempted in the Arctic, where the presence of permafrost, low precipitation and a short frost-free season restrict the leaching process. Another complicating factor is the upward movement of carbonates in lime-rich Arctic areas that leads to a precipitate on stones. This might enrich the near-surface drift more than the parent drift.

Finally, layers of drift of different ages and origins may be distinguished by their varying carbonate content (Dreimanis, 1960, page 1853).

ACKNOWLEDGMENTS

Thanks are due to R. H. Kihl, who carried out the laboratory analysis of the samples, and to G. Falconer, who contributed a number of samples for analysis. Professor A. Dreimanis, of the University of Western Ontario, and Dr. J. Fyles and Dr. E. Miryneck, of the Geological Survey of Canada, have kindly read and commented on the paper.

Such investigations as that by Andrews and Sim produced sufficient data to create a need for more detailed analysis. This analysis became computerized as the speed and precision of data gathering improved and more individuals entered the field of "post-secondary" education.

Papers of the same type as the Andrews and Sim study provide measurements which facilitate comparison between studies from different regions or by different investigators. Many of these measurements can be used to evaluate the conclusions of the interpretive studies and the assertions about the apparent action of geomorphic processes. Thus inferred directions of ice movement can be re-examined using measures of lithology, carbonate content, trace element studies, fabric (at macroscopic or microscopic scale) and other properties of tills. The action of streams can be studied by accumulating measurements of the shear flow characteristics, the nature of the channel, and the size and type of the sediments moved by it. Similar remarks apply to the study of each geomorphic process and for all such studies there is a regional descriptive component which adds a further dimension to the study.

All of these comments are important because they illustrate the whole area of data analysis which developed from the measurement studies of the 1960s. Reviewing some of the current trends in geomorphology, Dury (1972) identified the major change during the 1960s as the use of quantitative methods, especially those methods which utilize data processing by computer. Dury (1972) is primarily concerned with trends in fluvial geomorphology, although in his abstract he does note that similar comments apply in the other major divisions of geomorphology. The Canadian example of this trend, reprinted here, is a publication by Carson

(1969) in which several statistical procedures are applied to a body of field measurements taken in a study of slopes. Carson's paper includes a great deal of useful discussion which places the use of statistical analysis in the full context of slope studies. Recent work by soil engineers is discussed, and the paper puts the various approaches to slope study into an overall framework which clearly illustrates the complementary nature of different types of study. By viewing the various approaches to slope studies in this way, Carson is providing a reason for integration of the different types of study.

The examples in Carson's paper are not specifically Canadian, but at the present time the publications stemming from Carson's work (Carson 1971, Carson and Kirkby 1972) constitute a significant addition to the literature and reflect the type of study currently included in the McGill curriculum and research activity in geomorphology. Slope studies are a basic component of all geomorphological studies and it can be argued that geomorphology is solely a study of slopes and slope processes. Canadian geomorphology has not been particularly concerned with such studies in the past, and Carson's paper provides a view of the complexity of slope studies which is pertinent to the present situation. It is clear that process-response models built on the results of engineering studies can provide the necessary functional link between the data collected in many geomorphological investigations and the attributes of landforms themselves. The topic of slope studies is often not included in introductory physical geography courses and thus the complexity of such study is not apparent to students in the early stages of their studies in physical geography. The following paper should be read with suitable comprehension of the importance of slopes in landform studies. An appreciation of the evolution of scientific thought and of the value of process-response models will also add to the significance of this work. Under these circumstances, the paper provides a valuable summary of the elements of current geomorphological investigation and it points the way to further research in an area which is fundamental to our understanding of landforms.

MODELS OF HILLSLOPE DEVELOPMENT UNDER MASS FAILURE*

M. A. Carson

An understanding of any scientific system demands knowledge at two different levels. One is the nature of the processes operative at any instant, the external variables which control these processes and the pattern of the resultant response of the system to the processes. This may be thought of as the construction of a process-response model [57] for the system. The second level relates to the way the external variables which control the system change through time, and makes it possible to set the present nature of the system into some sort of historical perspective.

A system may achieve equilibrium between form and process (assuming that the external variables which control the processes do not change) almost immediately in some cases; in other instances, the system may proceed so slowly towards equilibrium that an evolutionary approach is necessary to understand the nature of the system at any one point in time. In the situations where a system rapidly achieves equilibrium between form and process, an evolutionary model is unnecessary and a complete understanding of the nature of the system is furnished by a knowledge of the way in which

* Reprinted from *Geographical Analysis*, Vol. I, (Jan., 1969), pp. 76-100.

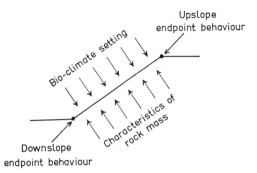

Fig. 1. External Variables Controlling Slope Form.

the equilibrium pattern depends upon the external variables. An exception occurs when the outside variables themselves change through time in a systematic manner: although it is still possible to understand the nature of the system at any one point in time by reference to the current state of the controlling variables, a more complete explanation is afforded by setting the system in a historical framework. These points have been emphasized previously by Chorley [14] in his discussion of general systems and geomorphology.

One of the unfortunate features of slope geomorphology is that early workers, with the exception of Gilbert [22, 23], have been preoccupied with the way in which the external elements controlling the slope system change through geologic time, and they have made no genuine attempt to construct sound models linking process and form. The outside elements controlling slope form (Fig. 1) are the nature of the underlying mass of rock, the climatic setting and the behavior of the endpoints of the slope. The climatic environment is important, because it is a major influence on the nature and intensity of the denudational processes acting on the slope. The behavior of the endpoints relates to both the upper and lower endpoints. The existence of a cap rock [20, 49] and whether the hilltop is a sharp divide or a plateau surface [55] are both important issues here. The most important feature however relates to the slope base. For example, material may accumulate at the slope base and lengthen the slope, debris may be removed by a stream and hence the same endpoint of the slope will

be maintained, or stream undercutting may be responsible for a moving endpoint.

The pattern of slope development outlined by W. M. Davis [18] is distinguished by its complete lack of reference to the denudational processes operating on the slope, except collectively in the term "agencies of removal." There is also little mention of the detailed nature of slope geometry in his writings. The Davisian treatment cannot be considered as a process-response model. Davis was more concerned with arguing that there must be a sequence in slope-forms throughout geologic time, because of a progressive change in the character of one of the external variables controlling the slope system, namely the behavior of the stream at the base of the slope. At an early stage in geologic time, side slopes are undercut by streams, while at a later date they are free to develop independently of the streams at their base. As is well known, Davis deduced this from his initial postulate that the base-level to which streams become graded is fixed.

Walther Penck appears to have been more interested than Davis in the formulation of process-response models for slopes, but the details of his models were never stated lucidly enough for them to make much impact. Moreover, most of Penck's efforts were not in the construction of slope models but, as with Davis, in discussing the ways the external variables change through geologic time. His major contribution was in challenging the two-phase sequence of undercut and stream-free slopes suggested by Davis. Penck [34] argued that a fixed base-level was the exception rather than the rule, so that any understanding of slope forms must involve the rate of stream-downcutting (via the medium of uplift) in a far more elaborate way than indicated by Davis. The fundamental point is, however, that neither worker offered a process-response model for understanding slope form, although both are thought of as pioneers in this field.

It cannot be overemphasized that the first task in understanding slope forms, as in understanding any system, is the derivation of a process-response model linking form (response) to a set of external variables which determine the processes at work. The nature of a physical system at any instant in time may be fully appreciated with reference to the model. An understanding of the ways in which the outside variables change through time, and therefore the way in which the system changes through time, is a luxury which must succeed rather than precede the development of the model. The preoccupation of many workers with the influence of base level changes in the landscape has hindered the development of process-response models in slope geomorphology.

Process-Response Models

The distinction between the inductive and the deductive approaches has long been evident in geomorphology, especially in the field of slope studies. This was emphasized by Sparks [48] in discussing the development of slope profiles. The outstanding feature of the last two decades is not that the balance has shifted from one approach to the other, but that both approaches have become increasingly quantitative and more precise.

In the inductive approach, the relations between slope-form and variables which may control slope form, such as the nature of the waste mantle and the energy of the stream at the base of slope, have been analyzed by multiple correlation analysis [33] and samples of hillslopes have been chosen according to the principles of statistical theory [32] rather than personal judgment. On the deductive side, the pioneer effort in mathematical expression afforded by Horton [25], and the contrast between the ideas of Wood [58] and King [26] and the work of Bakker and Le Heux [2, 3, 4, 5] show the progression in mathematical sophistication involved here.

The emphasis on statistical models (the inductive approach in its most sophisticated form) in geography is perhaps such that insufficient thought now is being given to the functional relations between different aspects of the landscape. It is recognized that both approaches are necessary for the development of the subject. Amorocho and Hart [1] note this in connection with hydrology and make the distinction between "systems investigations" (the statistical

approach) and "physical-sciences research" (the deductive approach). It is questionable, however, whether the systems approach is a valid end in itself. Strahler [51] points out that "Aside from predicting the magnitude of one form element when the other is known, this mathematical statement is of limited value for it does not improve understanding of the genesis of the landform." In many instances, the statistical model will *suffice*, but it is preferable, if possible, to go further and develop a functional model, that is, a process-response model. Such models, developed by engineers, often are commonplace in the study of stream forms [30], but are relatively few in the case of slope studies.

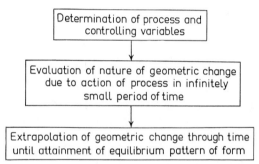

Fig. 2. Stages in the Construction of a Process-Response Model.

The stages in the formulation of a process-response model are depicted in Figure 2. The initial task is the identification of a particular process or set of processes and a determination, theoretically or empirically, of the variables which control the rate at which the processes operate. Some of these elements are completely external to the system and may not change during the operation of the process. Others may depend upon the form of the system itself and thus change during the time in which the process operates. This phenomenon is termed 'feedback.' A change which accelerates the process is labelled positive feedback and one which acts to retard the process is termed negative feedback. Negative feedback is very important in slope systems. Most processes of slope debris transport depend partly on slope angle, and in the case where the result of debris transport is to lower the angle of slope there exists a mechanism which retards the operation of the process. Such ideas were employed by Melton [32] in the derivation of a model for slopes where stream undercutting and soil wash on the slopes were the most important processes.

The second stage is the discovery of the exact way in which the process operates to change the form of the system in an infinitely small period of time. The nature of this change depends upon the ratio of inflow and outflow of material at different points in the system. In slope development three main types of geometric change are well known: slope retreat, slope decline and slope rounding. In the case of slope retreat—or parallel rectilinear recession, as suggested by Bakker and Le Heux [2]— the outflow is greater than the inflow at all points on the initial slope, and the ratio of the two is constant. In slope decline (central rectilinear recession), outflow equals inflow at the slope base, but there is a gradual linear increase in the outflow-inflow ratio up the initial slope. In the case of slope rounding, outflow is equal to inflow at all points on the initial slope except at the uppermost point where there is no inflow at all. These three types do not exhaust the possible geometric changes.

The third stage involves the extrapolation of the geometric change in an infinitely small period of time, allowing possible feedback, until the system reaches equilibrium. The nature of this problem is such that it has lent itself very readily to the use of the integral calculus. The new equilibrium form may be static or dynamic in type. A simple process-response model ending in static equilibrium is illustrated by the retreat of a cliff face under the process of rockfall with a platform at the base of the slope. This is depicted in Figure 3. The process operates until a new static equilibrium slope, a talus veneered straight slope, has replaced the initial cliff slope. This simple model assumes, of course, that the intensity of weathering is uniform over the exposed cliff face and that the talus itself does not weather into a finer mantle and thus re-precipitate mass failure and retreat of the new slope. The possibility of a sequential

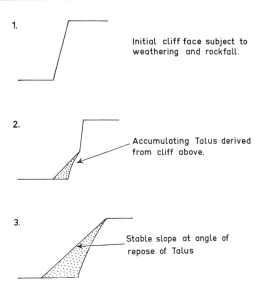

Fig. 3. Stages in the Formation of a Talus Slope through the Parallel Retreat of a Cliff Face.

pattern of retreat and emergence of static equilibrium slopes is discussed later. It has been emphasized by Chorley [14] that, at a larger scale, the Davisian sequence ending in a peneplain is essentially a movement towards a static equilibrium form. In contrast, if it is assumed that retreat of slopes and stream undercutting at the base proceed at a constant rate, then it is possible to derive an unchanging form in dynamic equilibrium as postulated by Penck [34] and more recently by Hack [24].

A limitation of slope geomorphology is that few models have satisfactorily considered all three stages in the construction of a process-response model. Early model builders rapidly went from the first to the last stage without carefully considering the geometric changes associated with particular processes. In recent decades a great deal of attention has been focused on the last stage in the development of models. On the assumption that the change in geometry in a small period of time conforms to a particular type, the path to an equilibrium form for the slope has been formulated mathematically. An early attempt was that by Lehmann [29] on the retreat of a cliff face leaving a talus cone. This has been developed by Bakker and Le Heux [5], who varied the initial assumptions to accommodate slope decline as well as parallel retreat. Mathematical models based on the changes in slope geometry which are thought to be valid for soil creep, have been considered by Culling [15, 16, 17] and a simpler treatment has been given recently by Young [60].

Although the development of this stage of a process-response model is a laudable one, it is essential that it be seen in perspective. It represents only one phase in the development of an appropriate model.

It is important to be able to extrapolate small-scale changes in geometry until equilibrium is attained, but it is at least as important to identify a particular change in form with a particular physical process. Scheidegger [39] argues, "In every case, a comparison between theoretical and observed slope profiles will ascertain the physical conditions that produced the latter." (Page 37.)

This is qualified on two grounds. First, it is not yet possible to identify a specific type of change in slope geometry with a particular denudational process. Moreover, it is often not known which particular processes are operating on a given hillslope. As Sparks [48] points out, most of the mathematical approach is "really equivalent to assuming what one wishes to find out." A great deal of work is currently being undertaken to establish what controls the rate of operation of certain processes, so that it may be possible to infer which processes dominate in a particular area and to select the most appropriate process-response model. There still remains, however, the issue of determining what type of change in geometry is associated with specific processes, and this is still unanswered in current research.

Statistical Models

The dilemma involved in the inductive versus the deductive approach led to the suggestion of Chamberlain [13] that the most sensible approach to scientific enquiry in the early stages of a particular problem is through the "multiple-working hypothesis." Data must be gathered which are pertinent to existing ideas but not geared solely to the testing of one specific model. Instead, it is more fruitful to formulate as many initial working models as possible and select data so that there is some general indication of the relative importance of each model in our study, even though the data are inadequate to give a detailed assessment of any one model. This approach in its current more sophisticated form is the method of the statistical model. It is not a goal in itself, but it is tremendously useful in the 'sorting out' of the initial working hypotheses. It allows the subsequent detailed testing of any individual model, with the confidence that the model is to some degree relevant to the problem. The exact relevance of the particular model is the object of the subsequent phase of research.

In the case of hillslope development, the investigator is at once perplexed by the multitude of existing hypotheses and the absence of relevant data which might indicate which type of model is most appropriate in any given set of circumstances. In humid temperate areas the three processes of soil wash, soil creep, and mass failure have all been suggested by different schools as the dominant processes. Others have mentioned that present processes may have no relevance at all to current slope forms, since these have been inherited from former periglacial conditions. The problem is clearly one to which the "multiple-working hypothesis" approach is ideally suited.

A particularly interesting problem in slope geomorphology is the determination of the elements which control the angle of straight hillside slopes. This problem [11] is considered here to illustrate the potential of the statistical approach. A sample of less than fifty straight hillslopes was taken from different rock types in the areas of northern Exmoor and the southern Pennines, England, and at each site measurements were made on variables which were thought to be relevant to various hypotheses. As an example, the silt-clay fraction of the surface mantle may be an indicator of the relevance of soil wash on slopes: the shear resistance of soil to overland flow is basically a function of cohesion and this depends strongly on the silt-clay content of the soil. Similarly, the angle of shearing (frictional) resistance of loose material is closely related to the over-all particle size of the debris, and if mass failure is important then it should be revealed in the correlation between slope angle and the rock fraction in the waste mantle. The slope and bankfull discharge of the stream at the base of the hillside should cast some light on the importance of the balance between stream energy and slope processes. Finally, the over-all degree of interdependence among the variables might act as an indicator of how closely the existing slope-forms are related to present attributes of the landscape.

The variables measured at each site were as follows:

x_0 = angle of straight slope
x_1 = extent of straight slope vertically
x_2 = mean rock fragment fraction of the mantle: the proportion of the mantle just under the vegetation mat containing material coarser than a $\frac{3}{8}''$ sieve by weight
x_3 = mean rock fragment size: intermediate axis
x_4 = mean thickness of the waste mantle
x_5 = mean slit-clay content of the soil component: the fraction of the mantle less than $\frac{3}{8}''$ which passed through the No. 200 B.S. sieve

x_6 = stream gradient at base of slope
x_7 = estimate of bankfull discharge of stream at slope base
x_8 = average boulder size on stream bed
x_9 = width of floodplain at slope base.

The methods involved in the measurement of these variables are discussed in detail elsewhere [10]. Special attention was paid to the accuracy of the site sample means relative to the population means, and this has been discussed in part in Carson [11].

The data were subjected to a number of standard procedures of statistical analysis. One was the determination of the simple correlation matrix for the ten variables and a principal axes factor analysis with Varimax rotation. A second was a sequential multiple-linear regression of slope angle on the nine other variables in a manner similar to the one described by Krumbein, Benson, and Hempkins [28]. A third was a principal components analysis of the nine variables excluding slope angle, and multiple regression of slope angle on the nine resulting components. The results are given in Tables 1 to 5.

TABLE 1

Sample Correlation Matrix for Slope Angle and Nine Other Variables Based on a Sample of Forty-Six Sites

X_1	X_2	X_3	X_4	X_5	X_6	X_7	X_8	X_9	
.55	.67	.58	−.87	−.36	.28	.43	.69	−.34	X_0
	.36	.40	−.43	−.42	.24	.67	.60	−.51	X_1
		.35	−.71	−.18	.17	.45	.61	−.44	X_2
			−.51	−.25	.26	.25	.45	−.31	X_3
				.24	−.06	−.35	−.65	.27	X_4
					−.41	−.02	−.40	.25	X_5
						.19	.52	−.58	X_6
							.49	−.27	X_7
								−.51	X_8

TABLE 2

Principal Axes Factor Analysis of Slope Angle and Nine Other Variables Based on a Sample of Forty-Six Sites

Variables	Components		
	1	2	3
X_0	−.867	−.271	−.232
X_1	−.748	.091	.353
X_2	−.743	−.328	.007
X_3	−.635	−.079	−.293
X_4	.784	.475	.264
X_5	.474	−.463	.444
X_6	−.489	.733	−.021
X_7	−.607	−.173	.674
X_8	−.871	.085	.006
X_9	.627	−.456	−.209
Eigenvalue	4.866	1.418	1.029
Percentage	48.66	14.18	10.29
Cumulative	48.66	62.84	73.12

TABLE 3

Varimax Solution of Slope Angle and Nine Other Variables Based on a Sample of Forty-Six Sites

Variables	Components		
	y_1	y_2	y_3
X_0	−.881	.208	.243
X_1	−.349	.338	.675
X_2	−.711	.050	.390
X_3	−.640	.283	.067
X_4	.937	−.001	−.182
X_5	.318	−.712	.166
X_6	−.004	.859	.196
X_7	−.233	−.024	.893
X_8	−.598	.466	.437
X_9	.141	−.632	−.476

TABLE 4

The Per Cent Reduction of the Variance of Slope Angle Attributable to the Nine Other Variables

Per Cent Reduction	Variables
76.2	4
81.7	4, 6
83.6	4, 6, 1
84.6	4, 6, 1, 9
85.1	4, 6, 1, 9, 3
85.5	4, 6, 1, 9, 3, 2
85.8	4, 6, 1, 9, 3, 2, 8
85.9	4, 6, 1, 9, 3, 2, 8, 7
85.9	4, 6, 1, 9, 3, 2, 8, 7, 5
85.9	All Variables

TABLE 5

The Per Cent Reduction of the Variance of Slope Angle Attributable to the Principal Components of the Nine Other Variables

Component	Per Cent Reduction
1	64.0
2	6.2
3	7.8
4	0.0
5	0.0
6	0.7
7	1.1
8	5.7
9	0.5
All Components	86.0

It is not proposed here to discuss in detail the statistical or geomorphic significance of each phase of the analysis. It is hoped rather that the general role of this analysis as a 'sorting out' technique may be illustrated. The first outstanding feature is the fairly high interdependence among the variables and especially with slope angle. The regression of slope angle on the first principal component of the remaining variables (Table 5) alone indicates that 64 per cent of the sample variance of slope angle is predictable in terms of one component underlying the other variables. This general pattern might well be taken as contradicting the view that the slope angles in the present landscape are inexplicable in terms of other attributes of the present landscape. A second feature is the relative independence of stream gradient and slope angle in this picture; this contrasts strongly with the ideas of Strahler [50], although the scale of investigation is rather different. Thirdly, the negative correlation between slope angle and silt-clay content of the soil fraction is especially interesting and, along with other evidence, suggests that the strength of the soil mantle to withstand soil wash is relatively unimportant in determining the angle of straight hillsides. A feature of equal importance is the positive association between slope angle and the rock fragment fraction of the waste mantle: steeper slopes possess coarser mantles. Classically, this would be interpreted as indicating that soil erosion is greater on steeper slopes, so there would inevitably be a smaller soil fraction on these slopes. Alternatively, it might be taken to indicate that there is a genetic connection between the angle of straight slopes and the angle of shearing resistance of the material on the slope. The general pattern of the statistical results suggested that a more detailed study of the relation between straight slope angle and angle of shearing resistance of the material on the slope would be a worthwhile project. The results of these subsequent investigations are to be described by Carson and Petley [12]. Some of the material is included in the subsequent section.

The important point is the efficiency of the statistical approach in detecting the relevance of alternative models, and in the particular case discussed above it was an invaluable prelude to a more detailed analysis of the role of physical process-response models based on mass failure in the determination of straight slope angles.

Slope Forms and Mass Failure

The mechanism of mass failure expressed in its simplest form is as follows: on an initially stable slope (Fig. 4), the inherent shear stresses (which result from the downslope component of the weight of the material on the slope) along any potential failure surface are not greater than the shear resistance of the slope material provided by cohesion and friction. Although the shear stresses depend on the moisture content of the material (since this affects the weight of the material), the changes in the shear stresses through time are usually far less important than the change in the shear strength of the material to withstand failure. The shear strength of any material depends upon the intrinsic cohesion of the material, the coefficient

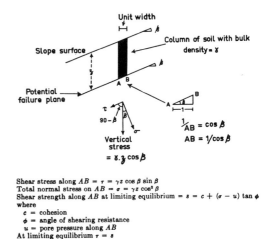

Shear stress along $AB = \tau = \gamma z \cos \beta \sin \beta$
Total normal stress on $AB = \sigma = \gamma z \cos^2 \beta$
Shear strength along AB at limiting equilibrium $= s = c + (\sigma - u) \tan \phi$
where
   c = cohesion
   ϕ = angle of shearing resistance
   u = pore pressure along AB
At limiting equilibrium $\tau = s$

$$\gamma z \cos \beta \sin \beta = c + (\gamma z \cos^2 \beta - u) \tan \phi$$

Fig. 4. Shear Stresses and Shear Strength on a Plane Failure Surface parallel to the Surface Slope.

or angle of shearing resistance, and the stress acting normal to the potential surface of failure which is effective in mobilizing friction.

All three components are subject to marked change over time. Weathering will alter the cohesion and angle of shearing resistance of any material during a geologic time scale. In the case of *cohesion* the most obvious instance is when joint surfaces are enlarged through frost and chemical action and detached boulders fall under the influence of gravity. A steep slope of cemented rock may be stable but the loose material derived from the weathering of such a rock will not be stable except at a much lower angle of slope. Similarly, a newly-created bluff of clay may stand initially at a vertical angle, but slowly the cohesion in the mass of clay is dissipated near the surface and the material will eventually fail under the stress of gravity. The type of failure (commonly a slide) may differ from the previous case (loose rock fall) but this is simply due to the nature of the surface of failure, and the mechanics involved are substantially the same.

The *angle of shearing resistance* (analogous to the coefficient of friction in arctangent form) of any material will also change during geologic time as the material is subjected to continual weathering. A scree mantle must ultimately weather into a mantle of soil particles. In this process of change it is to be expected that the angle of shearing resistance will differ in the intermediate stages of weathering from the initial scree and the eventual soil mantle. Indeed, even when a true soil mantle has been produced, mineralogical changes may alter the angle of shearing resistance even though there may be little change in the grain size pattern. In Hawaii, it has been noted by Wentworth [56] that the initial debris from rock fall may be stable at about 45 degrees, but as it weathers into "taluvium" (a mixture of talus and colluvium), the material becomes unstable at this angle and mass failure occurs. The mechanics of failure of loose debris by weathering into a mixed mantle of talus and soil particles here are the same as the instability of the initial rock slope when it breaks down into loose debris along joint planes.

Although the changes in both cohesion and angle of shearing resistance during geologic time are very significant, the *effective stress* active in mobilizing the frictional strength changes even more markedly over time and on a much smaller time scale. The effective stress normal to a failure surface is the difference between the total component of the weight of the material above that point acting perpendicularly to the failure surface, and the water pressure in the pores of the mass at that point. The former variable (dependent on the bulk density of the soil and the angle of the failure surface) changes little over time, but the water pressure in the pores is highly variable. In a partially-saturated soil mass, the pore pressure relative to atmospheric pressure is negative and acts to supplement the total normal stress. In a dry soil the pores are occupied entirely by air at atmospheric pressure and the effective normal stress is now equal to the total normal stress. In a mass of soil which is saturated with freely-draining water, the pore pressures are positive and depend upon the height of the water table above the point and the direction of the ground-water flow. Thus, whether or not a slope is stable depends on the moisture content of the material. A mass of soil may be stable in a partly-saturated state, but it may fail if it dries out completely, or if it becomes completely saturated with freely-draining water.

The effect of mass failure is to produce a change from a steep slope to a slope at a lower angle which is stable against this type of failure. The mathematical model suggested by Lehmann [29] relates to this in the change from a cliff slope to a slope at the angle of repose of the loose debris. This may be thought of as a one-phase instability model in the sense that one change in slope angle is involved. It is possible, however, that the initial debris from rockfall may weather into material which is no longer stable at this angle and that there is further failure and conversion to another limiting slope.

In examining the role of mass failure on slope geometry and hillslope development there are three features in the model which are especially important.

1. The "initial" slope may change to a new stable slope through hinge decline (Fig. 5a) or parallel retreat of the initial slope

Fig. 6. Slope Angle Decline through a Rotational Slide.

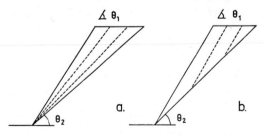

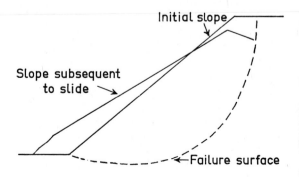

Fig. 5. The Change from One Limiting Slope to a New Limiting Slope through a. Hinge-Decline and b. Parallel Retreat of the Initial Slope.

leaving a stable basal slope (Fig. 5b) or a combination of these two. What determines the nature of these changes?

2. Most models of mass failure have considered only one phase of instability: the change from a rocky cliff slope to a talus slope. Other phases may, however, occur. What controls the number of phases of instability which occur in the history of any slope? The concept of a "phase of instability" as used here is not the same as the concept of an "unstable phase" as used by Butler [8, 9] in connection with soil periodicity on slopes. A large number of "K-cycles" or "unstable phases" might occur within a single "phase of instability" as used in this article.

3. The whole concept of mass failure implies certain limiting angles [59] above which rapid movement occurs and below which the slope is stable. What determines the angles of these temporarily or permanently limiting slopes?

The type of change in slope profile geometry.
—There is little doubt that in many instances the change from one stable slope to another is achieved through parallel retreat of the initial slope rather than hinge decline. This is especially true with the weathering of a strong rock slope and the creation of a loose rock debris slope. In other instances the mere presence of a cap rock may achieve the same effect. It is pertinent here to discuss the forces which determine the nature of the geometric changes in general terms. One item is clearly the nature of the surface of mass failure. In the case of tall slopes in clay, the failure surface is commonly curved and the sliding is rotational (Figure 6), so that inevitably the effect is to flatten the slope rather than induce retreat. In contrast, there are many examples of mass failures along planes parallel to the surface of a slope. This is especially the case where the hillmass is characterized by a mantle of soil or rock waste over solid rock [54, 31], but it has also been observed in a relatively homogeneous mass such as the London Clay [46]. In the example of a rocky cliff where loose rock is produced by weathering, the 'average' surface of failure may be considered also as parallel to

the slope. The nature of the sliding or detaching surface is only one feature. The existence of parallel retreat demands not only that there is stripping parallel to the slope when mass failure occurs, but also that the transport of the material downslope is rapid. The fundamental point is that mass failure usually owes its immediate origin to a built-up of pore water pressure in a mass. Once sliding occurs, the opportunity then exists for the water to seep out on the surface at the top of the slide so that the excess pore pressures may rapidly become dissipated. When this happens, the moving mantle will slow down and eventually stop. The probability that this will happen on the lower part of the hillslope depends upon the steepness and length of the hillside. In the case of a rock face, boulders released by joint enlargement may fall rapidly to the slope base. In contrast, a detritus slide on a gentle slope may transport the mantle only a short distance downslope before it becomes temporarily stable again. In this way we would expect hinge decline rather than parallel retreat. An example of slope flattening through shallow planar landslides has been described by Skempton [44] in an area of boulder clay in Northern England.

Carson [10] notes that the issue of retreat and decline in the context of mass failure may well depend upon the vegetation cover and the thickness of the waste mantle. On strong rocks on Exmoor and in the southern Pennines it has been observed that angles of 33–35 degrees and 25–27 degrees are especially common and represent limiting angles of stability for scree and mixed soil-talus mantles respectively. In two cases in these areas the mantle on the hillsides is so thick that the extra strength against shearing provided by the vegetation mantle is negligible and landslides take place, effecting a change from the scree slope to the soil-talus slope by parallel retreat. In two other areas the mantle is much thinner and the binding effect of the vegetation is relatively much greater. This seems to prevent rapid landslides, and terracettes are the usual expression of instability. Although the surface of failure may still be parallel to the slope, the rate of movement downslope is so slow that the amount of net loss on the slope may be expected to increase upslope and produce hinge decline rather than retreat. The evidence for this rests on the discontinuity between the two limiting angles in the frequency distribution curve for the areas of thicker mantle (Fig. 7a) and the two-segment slope profiles for the area of thicker mantle (Fig. 7b), in contrast to the single-segment slope in the other area. It is undeniable that both hinge decline and parallel retreat may take place in a phase of instability and this issue is an interesting sphere for future research.

The number of phases of instability.—The number of phases of instability in a particular landscape depends largely on the history of disintegration of the material involved. The simplest case would be an unjointed sandstone in a desert area. Weathering on a cliff face results in the detachment of soil grains which

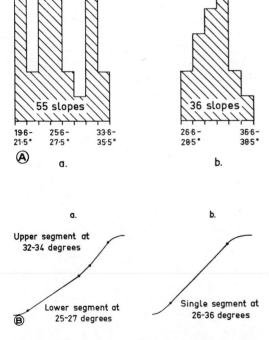

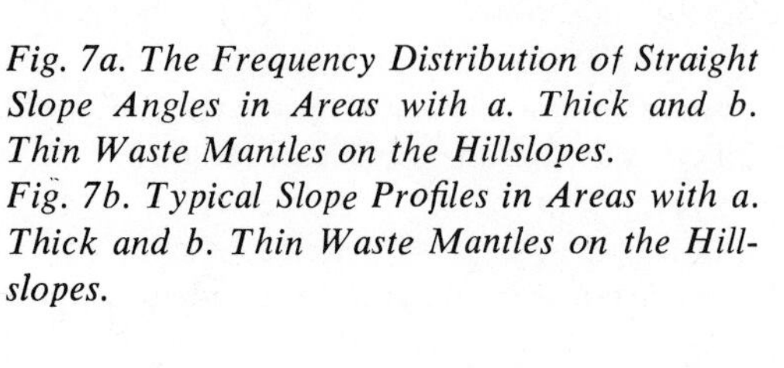

Fig. 7a. *The Frequency Distribution of Straight Slope Angles in Areas with a. Thick and b. Thin Waste Mantles on the Hillslopes.*
Fig. 7b. *Typical Slope Profiles in Areas with a. Thick and b. Thin Waste Mantles on the Hillslopes.*

then accumulate at the base at the angle of repose. The fact that in practice this often does not occur in semi-arid areas is due to the operation of other processes, such as soil wash, which may remove material from the base of the slope before there is a chance for it to accumulate into a talus heap. A more complicated case is a history of a rock cliff in a jointed mass. Weathering of the cliff face will produce a talus slope at the angle of repose of the scree material. This loose debris represents only one phase in the weathering of the initial material. Some fragments will soon weather into soil particles, and a mixture of talus and soil will result. This is unlikely to be stable at the angle of initial talus. Although the greater interlocking may increase the angle of shearing resistance of the material, the much smaller pore spaces will more easily become saturated at times of prolonged rainstorms and the slope will then be unstable. A new stable slope for the mantle of taluvium will thus arise. This mixture of coarse and fine material will eventually break down entirely into soil particles and a new phase of instability will prevail until a slope which is stable for the soil mantle has emerged. Such a sequence has been described in detail by Carson and Petley [*12*] in the case of two areas in England. The complexity of this sequence may depend also on the climatic setting. It has been suggested by Lester King that in the semi-arid area the mantle on a 'talus slope' will not change into a mixture of talus and soil grains through weathering, since the fines are washed off the slope almost as soon as they are produced. This is a consequence of the intensity of rainstorms and the sparse vegetation cover in such areas.

In a clay mass the sequence should be simpler. A steep bluff in clay might be expected to flatten rapidly to an angle which is stable but which depends upon both cohesion and frictional resistance. Skempton [*45*] notes, however, that over geologic time cohesion tends to zero and the only permanent source of strength is friction. It might be expected, therefore, that the initial stable slope will eventually flatten to a permanently stable slope at a lower angle. A possible instance of this in northern England has been described by Skempton [*44*].

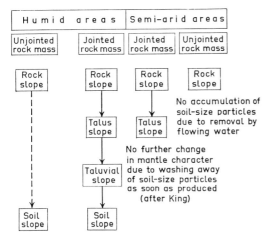

Humid areas		Semi-arid areas	
Unjointed rock mass	Jointed rock mass	Jointed rock mass	Unjointed rock mass

Fig. 8. The Number of Phases of Instability in Relation to the History of Disintegration of the Rock Mass and the Climatic Environment.

The number of unstable phases therefore depends fundamentally on the pattern of disintegration of the rock mass (Fig. 8), but it also depends upon the climate.

The steepness of limiting angles of slope.—The most interesting and most complicated of the three issues noted earlier is probably the determination of the limiting angles of slope in different situations. The relationship between shear stress and shear strength on a plane failure surface is summarized in Figure 4. The general equation used to determine the limiting slope, β is

$$\gamma z \cos \beta \sin \beta = c + (\gamma z \cos^2 \beta - \mu) \tan \phi$$

where γ is the bulk density of the soil
z is the depth to the failure plane
β is the angle of slope and angle of the failure plane
c is the soil cohesion
μ is the pore pressure on the failure plane
ϕ is the angle of shearing resistance.

A derivation of this equation may be found in Skempton and De Lory [46]. In the case

where there is no material cohesion in the soil ($c = 0$) and the soil is dry ($\mu = 0$), the equation reduces to

$$\tan \beta = \tan \phi$$

which is the basis for the concept of the repose slope. Many scree slopes stand about 35 degrees, which approximates the angle of shearing resistance of rough talus. This is possible since the voids on a talus slope are very large and are unlikely to become completely saturated so that positive water pressures never develop. On the other hand, the pores are likely to drain completely so that the worst condition of stability is likely to occur when the pore pressure is atmospheric.

Material which is of a finer calibre is a different case. In times of prolonged and heavy rain the pores will become saturated and the water table will rise or perched water systems will form and pore pressures will exceed zero. The actual pore pressure depends on the height of the water table above a point and the direction of groundwater flow. A wide variety of groundwater flow patterns is possible (Deere [19]), but especially important is the case of flow parallel to the surface and the water table at the surface. In this instance, the value of μ is given by $\gamma_w z \cos^2 \beta$ (where γ_w is the density of water) and the maximum stable slope is given approximately by

$$\tan \beta = \tfrac{1}{2}\tan \phi$$

A mixture of talus and soil with an angle of shearing resistance at about 43 degrees will thus be stable only at an angle of about 25 degrees. The frequency of this slope angle in nature suggests that the situation is very important. Indeed, it would be expected to be common since many rocks are jointed and such rocks must weather into a mantle of this description.

It is possible, in contrast, to have situations where no free-draining water system ever occurs and the mantle never completely dries out; in this case the pore pressures will be negative and the limiting slope will exceed the angle of shearing resistance of the material. Such high cohesion slopes [50], where there is a permanent source of capillary cohesion due to the nature of the debris and the climatic and geo-

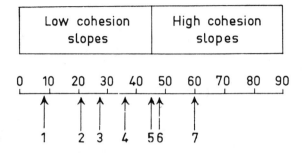

Fig. 9. Typical Limiting Slope Angles under Different Environmental Conditions.

1. Stable slopes in London Clay (see text) with $c = 0, \phi = 16$ degrees, and $u = \gamma_w z \cos^2\beta$
2. Stable slopes of sandy material (see text) with $c = 0, \phi = 35$ degrees, and $u = \gamma_w z \cos^2\beta$
3. Stable slopes of taluvium (see text) with $c = 0, \phi = 43$ degrees, and $u = \gamma_w z \cos^2\beta$
4. Stable talus slope (see text) with $c = 0, \phi = 35$ degrees, and $u = 0$
5. Stable Brule clay slopes (see Schumm [40]) with u assumed negative
6. Stable Perth Amboy 'fill' slopes (see Schumm [41]) with u assumed negative
7. Stable fine-textured, soil-mantled slopes with strong root development observed by Bailey and noted by Strahler [50]

logic setting, have been described by Smith [47] and Schumm [40] in the Brule formation of South Dakota and by Schumm [41] in the Perth Amboy badlands.

A simple relation between the angle of shearing resistance and the limiting slope demands that there be no material cohesion in the waste material. In clays there is definitely cohesion, and, depending upon the amount of cohesion and the height of the slope [43], a clay slope may stand at any angle. In the case of the London Clay it has been suggested that over a long time scale the cohesion may dissipate and a simple relation between angle of internal friction and limiting slope may arise.

It is interesting to consider the question of whether particular angles of limiting slope are especially common in nature. As shown in Figure 9, it is possible to have a wide range from a 90-degree cliff in strong rock to extremely gentle slopes in clays where there is no longer any material cohesion and the material is subject to artesian porewater pressures. Nevertheless, it has been suggested that certain angles are particularly common. Straight hillsides with a veneer of talus commonly exist at angles of 33–37 degrees in a wide variety of climatic environments, as indicated by the reports of Bryan [7], Simpson [42], Terzaghi [52], Rapp [35], and Tinkler [53]. Angles of 43–47 degrees may occur with 'taluvial' mantles under dry conditions [7, 56], but under conditions where such mantles become saturated with

freely-draining water flowing slowly parallel to the hillslope surface, the limiting angle of slope seems to approximate 25–27 degrees. Straight hillsides at angles in the mid-twenties have been noted by Savigear [38], Young [59], Ruxton [37], Koons [27], and Robinson [36], in addition to those discussed by Carson [10]. A sandy mantle with an angle of shearing resistance at about 35 degrees and saturated with freely-draining water parallel to the hillslope should be just stable at about 19–20 degrees and this seems to be 'characteristic' of many sandstone areas in humid-temperate climates. In the case of clays, the picture is more complicated: material cohesion may distort the relationship between angle of limiting slope and angle of shearing resistance, abnormal porewater pressures are possibly more common and mineralogy may be more important than grain size pattern here. The work of soils engineers, especially Skempton and DeLory [46] and Skempton [45], nevertheless, suggests that angles of shearing resistance of 15–20 degrees are common for clays, and, for the residual state where there is no cohesion in the clay, this may explain the frequency of straight clay hillsides at 9–10 degrees, assuming 'normal' porewater pressures.

These three aspects of slope stability—*the type of change in slope profile geometry, the factors determining the number of phases of instability, and the steepness of "limiting" angles of slope*—form the central issues in the development of process-response models based on mass failure. Slowly, especially through the work of soils engineers, models of hillslope development due to the process of mass-wasting are beginning to emerge, and physical explanations may now be offered for hitherto perplexing observations.

Summary

The purpose of this paper has been, firstly, to summarize some of the more interesting contributions made by soils engineers in the area of slope studies, and, secondly, to place this physical approach to the subject into some type of framework along with alternative modes of model-building. The latter objective is at least as important as the former one; the pendulum of fashion exists in geomorphology as in other pursuits of knowledge. It has swung suc-

cessively from the vague, qualitative theories of early workers in the field, over to highly mathematical treatments as typified, for example, by the work of Bakker and Strahler [6], and back to studies of a more inductive nature, stressing the need for field observation. The fruits of over a decade of field study are, however, almost as indiscernible as the contributions of the purely mathematical approach. The construction of a genuine model demands not only field observation and mathematical expression, but also, as Strahler [51] notes, an appreciation of the basic mechanics of the situation in order that a genetic link may be placed between form and process. In recent years this gap has been bridged by soils engineers, and, for the first time, there now exist genuine process-response models of slope development under mass failure.

LITERATURE CITED

1. AMOROCHO, J. and W. E. HART. "A critique of methods in hydrologic systems research." *Transactions of American Geophysical Union*, 45 (1964), 307–21.
2. BAKKER, J. P. and J. W. N. LE HEUX. "Projective-geometric treatment of O. Lehmann's theory of the transformation of steep mountain slopes." *Koninklijke Nederlandsche Akademie van Wetenschappen*, Series B, 49 (1946), 533–47.
3. _____. "Theory on central rectilinear recession of slopes." *Koninklijke Nederlandsche Akademie van Wetenschappen*, Series B, 50 (1947), 959–66 and 1154–62.
4. _____. "Theory on central rectilinear recession of slopes." *Koninklijke Nederlandsche Akademie van Wetenschappen*, Series B, 53 (1950), 1073–1084 and 1364–74.
5. _____. "A remarkable new geomorphological law." *Koninklijke Nederlandsche Akademie van Wetenschappen*, Series B, 55 (1952), 399–410 and 554–71.
6. BAKKER, J. P. and A. N. STRAHLER. "Report on quantitative treatment of slope recession problems." *International Geographical Union, 1st Report on the study of slopes.* 1956.
7. BRYAN, K. "Erosion and sedimentation in the Papago country, Arizona." *United States Geological Survey, Bulletin 730.* 1922.
8. BUTLER, B. E. "Periodic phenomena in land-

scapes as a basis for soil studies." *CSIRO Australian Soil Publication*, No. 14. 1959.
9. _____. "Soil periodicity in relation to landform development in south-eastern Australia." In *Landform Studies From Australia to New Guinea* (Eds. J. N. Jennings and J. A. Mabbut), (Canberra, 1967).
10. CARSON, M. A. "The Evolution of Straight Debris-Mantled Hillslopes." Unpublished Ph.D. dissertation, Cambridge University, 1967.
11. _____. "The magnitude of variability in samples of certain geomorphic characteristics drawn from valley-side slopes." *Journal of Geology*, 75 (1967), 93–100.
12. CARSON, M. A. and D. J. PETLEY. "The existence of threshold hillslopes in the denudation of the landscape." In preparation.
13. CHAMBERLIN, T. C. "The method of multiple working hypotheses." *Journal of Geology*, 5 (1897), 837–48.
14. CHORLEY, R. J. "Geomorphology and general systems theory." *United States Geological Survey Professional Paper*, 500-B. 1962.
15. CULLING, W. E. H. "Analytical theory of erosion." *Journal of Geology*, 68 (1960), 336–44.
16. _____. "Soil creep and the development of hillside slopes." *Journal of Geology*, 71 (1963), 127–61.
17. _____. "Theory of erosion on soil-covered slopes." *Journal of Geology*, 73 (1965), 230–54.

18. DAVIS, W. M. "The geographical cycle." *Geographical Journal*, 14 (1899), 481–504.
19. DEERE, D. U. "Effect of pore pressures on the stability of slopes." Paper presented to meeting of Geological Society of America at New Orleans, November 20–22, 1967.
20. EVERARD, C. E. "Contrasts in the form and evolution of hill-side slopes in central Cyprus." *Transactions Institute of British Geographers*, 32 (1963), 31–47.
21. FISHER, O. "On the disintegration of a chalk cliff." *Geological Magazine*, 3 (1866), 354–56.
22. GILBERT, G. K. "Report on the Geology of the Henry Mountains." *United States Geological Survey Report*. 1877.
23. _____. "The transportation of debris by running water." *United States Geological Survey Professional Paper*, 86 (1914).
24. HACK, J. T. "Interpretation of erosional topography in humid temperate regions." *American Journal of Science*, 258A (1960), 80–97.
25. HORTON, R. E. "Erosional development of streams and their drainage basins: hydrophysical approach to quantitative morphology." *Bulletin of the Geological Survey of America*, 56 (1945), 275–370.
26. KING, L. C. "Canons of landscape evolution." *Bulletin of Geological Society of America*, 64 (1953), 721–52.
27. KOONS, D. "Cliff retreat in southwest United States." *American Journal of Science*, 253 (1955), 44–52.
28. KRUMBEIN, W. C., B. BENSON, and W. B. HEMPKINS. "Whirlpool, a computer program for 'sorting out' independent variables by sequential multiple linear regression." *Office of Naval Research, Technical Report 14, ONR Task No. 389-178*. Northwestern University. 1964.
29. LEHMANN, O. "Morphologische Theorie der Verwitterung von Steinschlag wänden." *Vierteljahrsschrift der Naturforschende Gesellschaft in Zurich*, 87 (1933), 83–126.
30. LEOPOLD, L. B., M. G. WOLMAN, and J. P. MILLER. *Fluvial Processes in Geomorphology*. San Francisco. 1964.
31. LUMB, P. "The residual soils of Hong Kong." *Géotechnique*, 15 (1965), 180–94.
32. MELTON, M. A. "An analysis of the relations among elements of climate, surface properties and geomorphology." *Office of Naval Research, Report No. 11*, Columbia University. 1957.
33. _____. "List of sample parameters of quantitative properties of landforms: their use in determining the size of geomorphic experiments." *Office of Naval Research, Report No. 16, ONR Project 389-042*. Columbia University. 1958.
34. PENCK, W. *Morphological Analysis of Landforms*. (1924). Translated into English by H. Czeck and K. C. Boswell. London. 1953.
35. RAPP, A. "Recent developments of mountain slopes in Kärkevagge and surroundings, northern Scandinavia." *Geografiska Annaler*, 42 (1960), numbers 2 and 3.
36. ROBINSON, G. "Some residual hillslopes in the Great Fish River Basin, S. Africa." *Geographical Journal*, 132 (1966), 386–90.
37. RUXTON, B. P. "Weathering and subsurface erosion in granite at the Piedmont angle, Balos, Sudan." *Geological Magazine*, 95 (1958), 353–77.
38. SAVIGEAR, R. A. G. "Technique and terminology in the investigation of slope forms." *International Geographical Union, 1st Report on the Study of Slopes*, (1956), 66–75.
39. SCHEIDEGGER, A. E. "Mathematical models of slope development." *Bulletin of Geological Society of America*, 72 (1961), 37–50.
40. SCHUMM, S. A. "The role of creep and rainwash on the retreat of badland slopes." *American Journal of Science*, 254 (1956), 693–706.
41. _____. "Evolution of drainage systems and slopes in Badlands at Perth Amboy, New Jersey." *Bulletin of Geological Society of America*, 67 (1956), 597–646.
42. SIMPSON, S. "The development of the Lyn drainage system and its relation to the origin of the coast." *Proceedings of the Geologists Association*, 64 (1953), 14–28.
43. SKEMPTON, A. W. "Earth pressure and the stability of slopes." in *The Principles and Application of Soil Mechanics*. Institution of Civil Engineers, London. 1945.
44. _____. "Soil mechanics in relation to geology." *Proceedings of Yorkshire Geological Society*, 29 (1953), 33–62.
45. _____. "The long-term stability of clay slopes." (The Rankine Lecture.) *Geotechnique*, 14 (1964), 75–102.
46. SKEMPTON, A. W. and F. A. DELORY. "Stability of natural slopes in London Clay." *Proceedings of 4th International Conference on Soil Mechanics and Foundation Engineering*, (London). 2 (1957), 378-81.
47. SMITH, K. G. 1958. "Erosional processes and landforms in Badlands National Monument, South Dakota." *Bulletin of Geological Society of America*, 69 (1958), 975–1008.

48. Sparks, B. W. *Geomorphology*. London (1960), p. 60.
49. Sparrow, G. W. A. "Some environmental factors in the formation of slopes. *Geographical Journal*, 132 (1966), 390–95.
50. Strahler, A. N. "Equilibrium theory of erosional slopes approached by frequency distribution analysis." *American Journal of Science*, 248 (1950), 673–696 and 800–814.
51. _____. "Dynamic basis of geomorphology." *Bulletin of Geological Society of America*, 63 (1952), 923–38.
52. Terzaghi, K. "Landforms and subsurface drainage in the Gaika Region in Jugoslavia." *Zeitschrift für Geomorphologie*, NF Band 2 (1958), 76–100.
53. Tinkler, H. J. "Slope profiles and scree in the Eglwyseg Valley, North Wales." *Geographical Journal*, 132 (1966), 379–85.
54. Vargas, M. and E. Pichler. "Residual soil and rock slides in Santos, Brazil." *Proceedings of 4th International Conference on Soil Mechanics and Foundation Engineering, (London)*, 2 (1957), 394–98.
55. Van Dijk, W. and J. W. N. Le Heux. 1952. "Theory of parallel rectilinear slope recession." *Koninklijke Nederlandsche Akademie van Wetenschappen*, Series B, 55 (1952), 115–129.
56. Wentworth, C. K. "Soil avalanches on Oahu, Hawaii." *Bulletin of the Geological Society of America*, 54 (1943), 53–64.
57. Whitten, E. H. T. "Process-response models in geology." *Bulletin of Geological Society of America*, 75 (1964), 455–64.
58. Wood, A. "The development of hillside slopes." *Proceedings of the Geologists Association*, 53 (1942), 128–40.
59. Young, A. "Characteristic and limiting slope angles." *Zeitschrift für Geomorphologie*, Band 5 (1961), 126–31.
60. _____. "Deductive models of slope evolution." *Nachrichten der Akademie der Wissenschaften in Göttingen II. Mathematisch-Physikalische Klasse* (1963), 45–66.

References to fashion in types of studies also provide important material. Careful reading of Carson's paper illustrates the historical perspective in slope studies, and he cites the work of Davis, Penck, Scheidegger, Bakker, and Strahler as the important influences to 1960. The application of mathematical techniques by Bakker, Scheidegger, and Strahler was in marked contrast to the qualitative work of the earlier research workers. Availability of the computer, as noted by Dury (1972), made more extensive application of mathematical techniques possible. Processing large data matrices through a series of statistical procedures also became fashionable and this itself is reflected in Carson's papers where Tables 1 to 5 are the product of precisely this type of work. By further extending the work on slopes through application of soils-engineering results, Carson reaches the conclusion that process-response models of slope development under mass failure do exist. The important implication of this is that many geomorphic studies of slope processes represent data collection for some intermediate stage of descriptive work on which inductive reasoning is based. Unless these studies are related to basic mechanisms so that form may be related to process (Strahler 1952), they are unlikely to bring about major advances.

Studies of this type are predicated on a need to understand the processes acting in the creation of a landform. If these processes are fully understood the nature of the materials on which they act must also be fully understood. This point is central to the following paper by Yatsu (1971) who, in a series of publications, has stressed the need for a much greater understanding of the materials of which landforms are made (Yatsu 1966, 1967, 1969, 1971). His paper is reprinted here to provide a further dimension in our consideration of the nature of geomorphology in Canada.

LAND FORM MATERIAL SCIENCE: ROCK CONTROL IN GEOMORPHOLOGY*

Eiju Yatsu

Introduction

This paper is concerned with the methodology of studying Rock Control in Geomorphology. A new term, "Landform Material Science", is proposed in order to focus our keenest attention on the materials of which landforms are composed.

Of course some aspects of landform materials have been studied for many years. For example the mechanical, chemical and mineralogical composition of sediments and soils have long aided the identification of the origin and nature of landforms. Sediments and soils result from the action of processes, therefore they are very helpful keys to the understanding of processes which have taken place.

Another important approach is the hydraulic and mechanical interpretation of geomorphic processes. These two approaches are now being co-ordinated to overcome the geomorphological fudge, full of endless speculation, initiated by the Davisian school and some equivalents. Through them momentous changes have been accomplished in Geomorphology since the last World War.

The Davisian impact is still felt however, especially in the area of rock control studies. Instead of studying the resistance of rocks to the forces of various processes, the Davisian school replaced the word "rock" by "geological structure", and indulged in endless invention of jargon—such as homoclinal ridge, anticlinal valley, fault valley, fault-line valley and so on. By so doing, they made a substantial contribution to obscuring and blurring over the basic concept of our subject as a whole.

Strahler (1952) classified the landform materials, their properties, and types of stress and strain, etc. The second part of his paper, which is concerned with dynamic open systems and mathematical models in Geomorphology, has stimulated many discussions along these lines. However, it was very unfortunate that the first part, dealing with landform materials, escaped the public attention it deserved. Since then, studies on landform materials have not been widely developed, either by Strahler or other people. Unless the fundamental properties are understood, the various geomorphic processes are very difficult to grasp. Geomorphologists should focus their careful attention on the nature of landform materials. This is why the concept of "Landform Material Science" is being proposed strongly here.

Landform Material Science

What is Landform material science? Before entering into this discussion, it may be instructive to consider Engineering Material Science. Every engineer—mechanical, civil, electrical, or other is vitally concerned with the materials available to him. Whether his product is an automobile, a space vehicle, a computer or a bridge, he must have a detailed knowledge of the properties and behavioural characteristics of the materials he proposes to use. The system of this knowledge as a whole is called Engineering Material Science.

Figure 1 compares the engineering processes and those of landforms.

The landform material is composed of solid rock and/or unconsolidated rock, including the liquid and gaseous phases, and work is done on it by various processes to produce landforms. Even though the materials, processes and products of Engineering are quite different from those of Geomorphology, the principle of the production system is identical. Therefore, why should not the system of the knowledge of landform materials be called "Landform Material Science"?

Landform materials must be studied in terms of processes and the conditions where work is done on them, because the properties and behaviour of materials greatly depend upon processes and environmental conditions. Examples of research work along this line are very few (Yatsu, 1957, 1966; Chorley, 1959, 1964). This field seems to be one that clearly reveals a great weakness of our science.

---

* Reprinted from *Research Methods in Geomorphology* (Geo Abstracts, Norwich 1971), pp. 49-56, by permission of the author.

FIGURE 1

Comparison of Processes

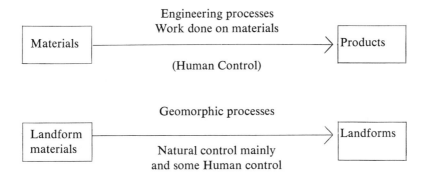

FIGURE 2

Classification of Research Methods

Scale	Field work	Indoor work
Macrosopic Mountain ranges	air reconnais-sance	Laboratory Office
	Field observation	
	ground inspection	
	car	Map work
	boots & hammer	Air-photo intepretation
	Field measurement	Measurement & Testing of samples
	Seismic wave Young's modulus measurement by water chamber etc.	Analysis
		mechanical petrological mineralogical microscope electron-microscopes
Microscopic Atoms and molecules	Chemical analysis	X-ray
		chemical

Research Methods of Landform Material Science

There are many ways to carry out research work in the field of landform material science.

The present writer tentatively classifies these methods in Figure 2.

For collecting the data concerning the properties and behaviour of landform materials

Fig. 3. A cliff of Kanto Loam in Kawasaki city, suburb of Tokyo (photograph by Dr. T. Suzuki).

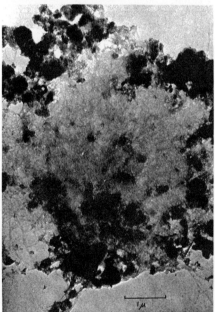

Fig. 4. Allophane in Kanto Loam.

Fig. 5. Filmy Allophane in Kanto Loam.

there are too many ways to list, because they range from ultramicroscopic to macroscopic. Figure 2 shows just an idea.

Hack (1960) in his paper concerning erosional topography, denies the cyclic concept of landform evolution and emphasizes the dynamic equilibrium theory, in which forces and resistance of rocks are stressed. However, it

seems that his arguments are not derived from precise measurements of the resistant strength of materials, but from simple field observations. Karst provides a good example of rock controlled landforms, and many karst geomorphologists are using methods ranging from macroscopic to microscopic.

The methods of measuring and testing of

solid and unconsolidated rocks, and the geo-morphological significance of rock properties have been discussed elsewhere by the present writer (Yatsu, 1966).

In a presidential address 20 years ago to the A.A.G., Professor R. J. Russell pointed out that geomorphologists in their concern with structure, process and stage had failed to cover their field adequately. As a result, their land-form research was not providing factual speci-fic information but instead was producing hazy theoretical suggestions and conclusions which were often erroneous (Russell, 1949). Very recently more geomorphologists have concerned themselves with Russell's questions of "what, where, and how much". And moreover, some of them want to know "why." As we inquire more deeply, seeking the answer to this ques-tion why, our research work usually becomes more microscopic in its nature. Let us look at an example. Figure 3 shows a cliff of Kanto Loam near Tokyo. Cliffs of this material are often almost vertical and typically seven to ten metres in height. During or after the heavy rainfall caused by typhoons, slope rupture (abrupt failure) sometimes takes place but never mass movement of earth-flow type. This is very strange. Why should it be so? The soil is described as weathered volcanic ash, high void ratio 3-5, high water content of from 100 to 180%, cohesion 0.3-0.5 kg/cm^2, internal

friction angle 25-30°, content of clay-sized particles more than 40%. (Yatsu, 1964).

In order to clarify the behaviour of this material, closer examination is necessary. Simple optical techniques are unable to supply a satisfactory explanation of the mystery of this soil. By X-ray diffraction however, it can be determined that the dominant clay mineral is allophane with an auxiliary component of hydrated halloysite. With the electron micro-scope it can be further seen that fibrous clays (imogolites) and filmy membranes of allo-phane entangle with each other.

Perhaps it is this fact which explains the strange characteristics of this soil and which is reflected in the unusual stability of the vertical cliffs of Kanto Loam.

Perhaps the speaker has conveyed the im-pression that microscopic work is the most im-portant type of research. This is not true. Geomorphologists still need "boots and ham-mer". Our research work must be organised so that we can see both the wood *and* the trees.

It is strongly hoped that many geomorpholo-gists will be keenly interested in the highly analytical approach of Landform Material Sci-ence. Many roads lead to Rome however, and many different approaches are necessary for the harmonious development of Geomorph-ology.

Yatsu's views are strongly endorsed by Carson (1971, 1972) in his volumes which are concerned with the mechanics of erosion and the processes of hillslope development. There is a fundamental importance to the understanding of process and an associated understanding of the materials on which the process acts. This, at the present time, constitutes an important item in the international impact of Canadian geomorphology because both these geomorphologists have published their work from within Canadian departments of geography.

Canadian geomorphology is noticeably introspective. Two papers in the "Geographica" section of *The Canadian Geographer* exemplify this point. The first of these is a summary of the recommendations made to the Solid Earth Science Study Group by a committee of Canadian geomorphologists. This summary (Jop-ling 1970) is of particular interest to potential geomorphologists since it documents the status of geomorphology at the end of the 1960s. It is clear from this document that in Canada too little attention has been paid to geomorphology as a science which can contribute to the planning of the country. Understanding landform-soil relationships and the geotechnical properties of soil are important areas which geomorphologists in Canada have not developed to the same degree as geo-morphologists in other countries. The brief considers that up to 1969 Canadian studies had concerned themselves too much with descriptive study and had too little concern with methodology and scientific method.

This situation is now changing and the papers by Carson and Yatsu must be viewed in the context of the comments by the committee of geomorphologists. A similar message was contained in a paper by Foster (1969) which was entitled *"Geomorphology: Academic exercise or social necessity?"* This paper also provides excellent material for evaluating the status of Canadian geomorphology at the end of the 1960s. Foster concludes that Canadian geomorphology should concern itself with the more densely populated zones of the south, where there are more appreciable applications of the results, rather than continue with an orientation towards the more sparsely peopled areas of the north.

Discussions of this type continue, and they are documented in part by the Guelph symposium of geomorphology which includes a discussion section in its published proceedings (Yatsu et al 1971, Yatsu and Falconer 1972). A more important ingredient in these comments must be some reference to the nature of the growth of knowledge. At no point in time is our knowledge complete. Each of the papers reprinted here exemplifies the results of an investigation which reflects the fashions and the limitations of its time. Carson (1969) comments on this in his paper reprinted above.

This evolutionary component cannot be too greatly stressed as many introductory physical geography texts present geomorphology as a completed study. They do this by delimiting processes and regions, defining terms, and stating in categorical terms that certain processes act in a precisely formulated manner. In reality, the delicate balance of forces maintaining many landforms is *not* so fully understood. The intricacies of the study required to further this understanding deserve much more attention. The challenge to increase our knowledge is one which should not be lightly dismissed.

A recently published study by Goodchild and Ford (1971) provides a good example of the challenge of geomorphology. A small-scale feature of the walls of limestone caves and limestone boulders is examined in a detailed manner to provide an illuminating result.

ANALYSIS OF SCALLOP PATTERNS BY SIMULATION UNDER CONTROLLED CONDITIONS*

M. F. Goodchild and D. C. Ford

Abstract

Working with plaster of paris in an experimental flume, the authors have simulated the formation of scallop patterns, an intriguing feature of eroded limestone, under controlled conditions of velocity and viscosity. Analysis of the resulting scallop length as a frequency distribution has shown that certain of the statistical parameters are well correlated with the hydrodynamic conditions. Length is inversely related to velocity and directly to viscosity. These results are similar to those found by Curl (1966) in a theoretical dimensional analysis of the simpler flute problem. Work in limestone caverns has confirmed that these results apply to scallops generated on limestone. Certain lithologic effects have been noted, however, and these are believed to be correlated with the physical structure of the material.

Introduction: The Form of Cave Scallops

The flow of a viscous fluid over a modifiable bed can produce a variety of small-scale relief

* Reprinted from *Journal of Geology* 79 (1971), pp. 52-62. © 1971 The University of Chicago. Reproduced by permission of The University of Chicago Press.

patterns. Sediment ripples and dunes, which form on unconsolidated, non-cohesive beds, are perhaps the best known. Allen (1969) has studied "flute marks" in weakly cohesive clays. Features in many respects similar to these can result from the solution of limestone by flowing water. In the literature of limestone geomorphology, the terms "scallop" and "flute" have been used to refer to these small-scale, ripple-like features (fig. 1A) commonly developed on cavern walls. The features resemble a mosaic of inlaid scallop shells and are therefore treated as individuals defined by the ridges rather than the depressions of the surface.

As Bretz (1942, p. 731) noted many years ago, scallops (which he called "flutes") are generally steeper on one side, again in analogy to scallop shells. Further, the orientation of this steeper side is usually the same for all scallops on a cave wall. Scallops are in this respect similar to some sedimentary structures, such as current ripples in sand.

Observations of scallop patterns on limestone surfaces in the beds of streams have led many authors to the conclusion that they are the result of erosion by flowing water. The orientation of the scallop pattern is found to be such that scallops are always steeper on the upstream side. The pattern in figure 1A would therefore be interpreted as having been formed by a stream flowing in the direction indicated. Two definitive dimensions of a scallop are used: the width, the greatest extent perpendicular to the stream flow in the plane of the surface; and the length, the greatest extent between crests parallel to the flow.

Scallops are found in open channels, on the walls, ceilings, and floors of cave conduits, and on boulders or cobbles in stream courses. Although on any one surface there is little variation in scallop dimensions, these range widely between scattered locations. To quote extremes measured by the authors, scallop lengths of 2 mm are typical of steeply sloping conduits developed in Cambrian marbloid rocks in the Nakimu Caves, high in the Selkirk Mountains of British Columbia. They are commonly 2 m long in the great horizontal trunk aquifers of Mammoth Cave in the Central Kentucky Karst, developed in Mississippian limestones.

Theories of Scallops

Theories of speleogenesis recognize two processes of subterranean limestone removal, both dependent on flowing water. Limestone may be removed by the action of dissolved carbon dioxide on calcium carbonate; alternatively, the erosion may be by abrasion by bedload or suspended sediment in the cave stream. Much has been written concerning the process of limestone removal responsible for the creation of a scalloped surface. Maxson (1940) observed scallop-like features on boulders in the Colorado River and believed their formation to be due to abrasion. Although most of the examples were on limestone boulders, he identified some on noncarbonate rocks. It is evident from the illustrations, however, that the features classified by Maxson cover a much wider range of form than those considered here.

Maxson attempted an analysis by calculating a Reynolds Number, Re, based on an arbitrary dimension B of the scalloped boulder: $Re = (Bv\rho/\eta)$, where ρ is the fluid density, v the velocity, and η the viscosity. A quantitative relationship was found between Re and the size of the features, smaller features being formed at lower Reynolds Number and thus at lower velocities, *ceteris paribus*. Maxson considered that the qualitative nature of the relationship between scallop size and Reynolds Number was due to difficulty in the definition of the velocity parameter v. He further noted similarity between these features on river boulders and the faceting of desert pebbles by blown sand and proposed the same analysis, using appropriate values of density and viscosity.

Several speleologists have offered hypotheses to account for the wide variation in mean scallop length. Glennie (1963), following a suggestion of Ashwell (1962), proposed that a small mean length is the result of a fast, turbulent flow. A large mean length results from an extended period of steady flow. If allowed to remain undisturbed, such a steady flow would eventually produce a set of infinitely long scallops. Davies (1963) reported a conversation with Hsuan Yeh in which the latter suggested that the size variation is a matter of age, the length of a scallop increas-

Fig. 1a. A scalloped boulder in the Nakimu Caves, British Columbia. Water flow direction indicated by arrow.

ing with time, apparently indefinitely. However, if the scallops grow continuously, there must, on the same surface, be other features which continuously shrink. The nature of this competition between neighboring scallops for the available space must be considered.

From general observation, Ford (1964) proposed that large features are produced by turbulent, sediment-laden water corroding the limestone, whilst small features are produced by solution in slowly moving water. Eyre (1964) suggested the opposite relationship between stream turbulence and scallop length, noting that scallops which had formed on a bulge of the cave wall in Gaping Ghyll Caverns, Yorkshire, England, were smaller on the upstream side of the bulge than on the downstream side. This inverse velocity-size relationship was also the theme of a first suggestion by Curl (1959). Moore and Nicholas (1964, p. 11) published a graph of the relationship between length and velocity, showing an inverse power law of the form: $l = 56\, v^{-0.75}$, where l is a measure of scallop length in centimeters, v is the stream velocity in centimeters per second. Moore and Nicholas give no source for their data. The first published experimental work is that of Rudnicki (1960), who produced scallops by immersing blocks of plaster of paris in natural streams and found that the number of scallops per unit area diminished in faster waters.

A hydrodynamic solution of the scallop problem is not feasible, whether on the basis of solution or abrasion. Curl (1966) reduced the problem to a two-dimensional one by eliminating all variation of form transverse to the flow. The features thus become similar to the swales of sand ripples. Curl called these features "flutes," a term used throughout this study. Approximate flute forms have been reported from several cave localities but are very much less common than scallops. Curl provided several illustrations of the form in nature and then proceeded to a dimensional analysis based on parameters likely to affect the flute pattern on a surface produced by solution alone.

There is much empirical evidence in support of a wholly solutional origin of scallop patterns. Many eroded, scalloped cave walls have insoluble material standing out in sharp relief from the surface. The material is frequently fragile, responding to the slightest touch. In the Bonnechère Caves, Renfrew County, Ontario, shale beds protrude from the limestone walls of the passage and yet are readily crumbled. Oxidation may have severely reduced the mechanical stability of these materials after the passage was drained, but the presence of scallops on the roofs of other passages produced under very slow water flow adds weight to the solution hypothesis, as does the lack of contrast between scallops on roofs and those on the floors at the same sites. Finally, the authors have produced excellent patterns by eroding plaster of paris under controlled, sediment-free conditions in the laboratory. None of this evidence, however, completely rules out the possible influence of abrasion in the formation of such patterns.

Curl considered the following parameters to be those controlling the periodic length λ of a flute pattern: $\lambda = f(v, \rho, \eta, L, D)$, where L is a channel dimension, D is the combined diffusivity of the molecular species involved in the reaction. The reaction is controlled, in a simplified view, by the rate of diffusion of fresh CO_2 to the surface through the boundary layer and the rate of diffusion of bicarbonate back into the main stream.

Three dimensionless numbers may be constructed from these parameters, yielding $f[\lambda/L, \eta/D\rho, (v\lambda\rho)/\eta] = $ Constant. Curl suggested that the term λ/L could be ignored for flutes small compared with the channel dimensions and that the Schmidt Number, $\eta/D\rho$, could be ignored for extremely slow solution such as that of limestone. The equation becomes $(\lambda v\rho)/\eta = N_f$, a dimensionless "Flute Reynolds Number." From two brief field observations Curl derived an approximate value of 22,500 for N_f.

It is essential to this theory of flute formation that the flute pattern be stable, maintained on the limestone wall without change of form as the rock is eroded. There must therefore be a relationship at every point on the surface between the rate of rock removal, controlled by the diffusion equation and the Navier Stokes equations, and the slope of the surface at that point. The specific flute cross section results from this requirement.

The equations governing the diffusion of the molecular species involved in limestone solution differ only in their constants from those governing diffusion of any other molecular species, and are also identical in form with those governing the diffusion of heat. The theory of flute formation therefore predicts that similar patterns will occur on all surfaces eroded by solution in fluids obeying the hydrodynamic equations, and on surfaces eroded by heat transfer also. Similar patterns are well known on ablating snow surfaces. Snow scallops are generated when the dominant mode of removal is by warm moving air, rather than by direct solar heating. They are common on high alpine

snowfields in late summer (where we have measured mean lengths ranging from 10 to 30 cm), and often occur during mild winter spells in temperate latitudes. Jahn and Klapa (1968) provide good illustrations of this phenomenon.

Curl suggested in his paper that the breakdown of the ideal flute pattern into the more common, transversely asymmetrical scallop pattern is due to natural fluctuation of the controlling parameters, notably the stream velocity. His flute analysis has had great success, both in predicting periodic lengths under known hydrodynamic conditions and in predicting the development of similar forms on snow and ice. The great majority of erosion patterns observed by speleologists have no such cross-flow symmetry, however. But information on ancient flow rates in dry, abandoned cave passages can be of great value in understanding both the underground and the related surface land forms. It is for this reason that the authors investigated the feasibility of extracting hydrodynamic data from the record preserved by scallop patterns.

Laboratory Experiments: Design

A direct analysis by solution of the relevant hydrodynamic and molecular diffusion equations is clearly infeasible. The presence of boundary layer separation in the flow pattern requires the inclusion of viscous terms in the equations, and there is no indication that the flow is steady.

Scallop patterns were therefore investigated by laboratory simulation under controlled conditions. Plaster of paris was selected as base material rather than limestone because the material could be more readily standardized, and because the erosion process would be speeded up by several orders of magnitude. The dimensions of the experimental flume cross section were constrained by the pumping capacity and yet had to be much greater than the size of scallops produced.

The flume is shown in figure 1B. Overall length is 6 m, width 2 m, and circuit length 12 m. The Plexiglas working section is 4 m in length. The cross section is a constant 32 cm

Fig. 1b. The experimental flume.

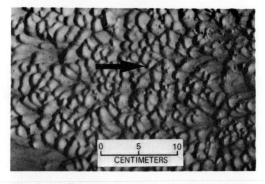

Fig. 2. Scallops generated on plaster of paris in the experimental flume, at 35°C and 114 cm/second on the low gradient block. The arrow indicates flow direction.

square. Bends in the flume are semicircular and contain no baffles. Water is driven around the duct by a propellor directly in the flow. Since the processes relevant to scallop and flute formation occur very close to the surface, conditions in the main stream flow can be largely

ignored. Therefore, smooth flow over the experimental area was not of major importance in the flume design.

The solubility of calcium sulphate in water is low; therefore it was necessary to continuously replace the flume water with a fresh supply, at a rate of 10 liters per minute. Temperature control, and hence viscosity control, was achieved by switching the incoming water between hot and cold by the use of solenoid-operated faucets. In this manner temperature was maintained to ±0.05° C over 100-hour periods.

Scallops were produced on upstream-facing, wedge-shaped blocks of plaster which were cast in the flume to ensure a tight seal. Well-scalloped surfaces (fig. 2) developed in 12–60 hours, depending on flow velocity and temperature, after the removal of up to 1 cm of material from the surface. In all cases the initial surface was plane.

The flume was carefully kept clear of sediment during the experiments. Initial scallops were observed to originate about small-size imperfections in the base material, either insoluble fragments or voids. Such imperfections, on the surface of the eroding block, cause a boundary layer separation immediately downstream. In this area the erosion rate is increased allowing a shallow pit to develop. If sediment is present as a bed load, such incipient scallops are observed to fill with material, the erosion rate is reduced, and the surface remains flat. Absence of sediment also ensured that erosion was by solution alone. Early experiments, using commercial grade plaster of paris, failed because of a high silt content. A change to reagent grade led to immediate success.

Characteristic velocities used were those 3 cm above the surface of the wedge in midstream, measured by a Pitot probe. The flume was operated at three temperatures in the range 10°–30°C, and four velocities between 30 and 150 cm/second. Because there was an acceleration of fluid over the experimental surface, the influence of this variable was investigated by repeating the experiments with wedges of different slopes. The wedges were 65 cm long, rising to 6 cm or 10 cm at the downstream end.

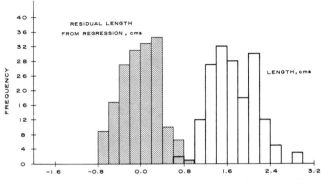

Fig. 3. Distribution of scallop lengths from a typical experimental run, shown before and after correction for position on the plaster block.

At the highest velocity (150 cm/second) and highest temperature (30° C) no scallops developed on the steeply sloping block. With all other combinations of velocity, temperature, and slope, scallops were produced, amounting to a total of 23 successful experiments.

Because conditions varied over the wedge surface, due to the acceleration of the fluid, it was necessary to adjust measured scallop lengths according to their position on the wedge, before the frequency distribution of scallop lengths could be determined. Accordingly the distance of each scallop from the downstream end of the wedge was recorded in addition to its length. The effect of location was removed by fitting a least-squares regression line to the lengths and distances and removing that variation explained by the regression line. In this way all scallop lengths were adjusted to a value appropriate to a location at the downstream end of the wedge. In this way any systematic, linear variation in scallop lengths was eliminated. The velocity used in the analysis was the extrapolated velocity 3 cm above the downstream end of the wedge. Figure 3 shows an experimental length distribution, before and after adjustment.

This procedure assumes only that the variation in scallop lengths over the wedge is a linear function of the distance from the downstream end. It does not imply the assumption of a linear relationship between size and velocity, nor does it preclude an absence of systematic variation over the wedge.

Analysis of Experimental Results

Eleven length parameters were derived from each experiment: the mean length, and 10 percentiles of the length distribution, the tenth, twentieth, thirtieth, etc., through one hundredth. It was hypothesized that scallop patterns are the result of a breakdown of an ideal flute pattern due to fluctuation in the controlling parameters, and that the distribution of scallop lengths is related to the flute periodic length through one or more of its statistical parameters. The mean lengths are tabulated in table 1 for each set of operating conditions.

The particular model investigated was
$$l = a_0 \, v^{a1} \, \eta^{a2} \, a^{a3},$$
where l is a length parameter, a is the acceleration term and the a's denote constants. The acceleration term a was given a value of 1 for the steeply sloping wedge configuration, and 2 for the gentler slope.

This model can be linearized to the form $\log l = \acute{a_0} + a_1 \log v + a_2 \log \eta + a_3 \log a$, where $\acute{a_0} = \log a_0$. It is equivalent to the Curl flute model if $a_1 = -1.0$, $a_2 = +1.0$, and $a_3 = 0.0$.

The model was evaluated by a least-squares multiple regression for each length parameter. Results are given in table 2. In all cases the acceleration term was found to be insignificant at the 95 percent level (if the conventional assumption of a multi-variate normal distribution may be made). This may be due to one of two factors. First, acceleration may play no part, independently, in the formation of scallops. Second, the actual value of the corresponding coefficient a_3 may be so low that differences in scallop measurements due to this factor are masked by sampling error and the probabilistic nature of the distributions themselves.

TABLE 1

Experimental Operating Conditions and Resulting Mean Scallop Lengths

TEMPERATURE (° C)	VISCOSITY (CGS UNITS)	VELOCITY (CM/SEC)		MEAN SCALLOP LENGTH (CM)	
		High Gradient	Low Gradient	High Gradient	Low Gradient
10.0	0.0131	129.3	112.1	1.305	1.633
10.0	0.0131	114.1	91.2	1.503	1.404
10.0	0.0131	99.2	80.2	1.786	1.755
10.0	0.0131	79.8	64.6	2.402	1.946
21.1	0.00988	129.3	112.1	0.905	0.984
21.1	0.00988	114.1	91.2	0.759	1.199
21.1	0.00988	99.2	80.2	1.169	1.374
21.1	0.00988	79.8	64.6	1.317	1.847
32.2	0.00761	....	112.1	....	0.767
32.2	0.00761	114.1	91.2	0.753	1.040
32.2	0.00761	99.2	80.2	0.711	1.008
32.2	0.00761	79.8	64.6	1.016	1.321

The values of a_1 and a_2 selected by the regression procedure are shown in table 2. A value of 1.0 (or −1.0) is expected in the case of linear dependence on the variable. The values determined differ from 1.0 (or −1.0) less in the case of velocity, which was directly measured, than viscosity, which was varied indirectly through the flume operating temperature. It should also be noted that four values of velocity were used, compared to three temperatures.

Although a full display of the five-dimen-

TABLE 2

Regression Analysis of Flume Data

Length Parameter	Pure Constant a_0	Velocity Coefficient a_1	Viscosity Coefficient a_2	Multiple Correlation Coefficient	Constant of Constrained Model $y = ax_2/x_1$
Mean	3.149	−0.9740	1.2997	.9353	11476
Percentile 10	3.464	−1.1981	1.4824	.9270	8187
Percentile 20	3.540	−1.1692	1.4967	.9343	9457
Percentile 30	3.550	−1.1999	1.4753	.9203	10404
Percentile 40	3.526	−1.1075	1.4713	.9275	11026
Percentile 50	3.313	−1.0319	1.3701	.9331	11508
Percentile 60	3.213	−0.9669	1.3233	.9199	12148
Percentile 70	3.005	−0.8949	1.2250	.9157	12631
Percentile 80	2.983	−0.8366	1.2008	.9154	13226
Percentile 90	2.836	−0.8346	1.1264	.9004	14192
Percentile 100	2.519	−0.4872	1.0074	.8673	16947

sional data is not possible, figure 4 has been included to illustrate the intersection of the mean length regression plane with planes of constant temperature for the high gradient block configuration.

A second model of the form $l = a\eta/v$, where a is a constant, was also investigated by a similar least-squares technique. This model is less critical, but allows estimates of the constant a to be compared with Curl's Flute Reynolds Number, N_f. These estimates are given in table 2, and should be compared with the value 22,500 determined by Curl. Although the estimates converge to this value, they are lower by as much as a factor of 2. These estimates are discussed later in a speleological context.

Field Investigations in Bonnechère Caves, Ontario

The relationship between flow parameters and the scallop pattern cannot be investigated in the field in dry cave passages because the flow rates are never known. Relative flow rates may be deduced from geometrical considerations (for example, if it is known that a particular passage was full of water when the scallops were produced), and in this way the form of the relationship can be found. The laboratory results from plaster of paris were confirmed by comparison with limestone cave scallops at diverse sites in British Columbia, Ontario, California, Kentucky, and New York State. In one respect and in one instance, there was serious disagreement.

One area of the Bonnechère Caves, Renfrew County, Ontario, has been formed by the simultaneous erosion of five distinct limestone beds of quite different characteristics. The geometrical configurations of the passages are such that they can only have been formed by a stream which filled them completely (Ford 1961). It is now abandoned and drained, except for standing pools of filtration water. Because the passages are rectangular and horizontal, and because the beds are almost horizontal, the five limestones must have been subjected to similar hydrodynamic conditions. Comparison of scallop length distributions on the separate beds (fig. 5) shows a wide, systematic variation, however. Beds were numbered 0 through 4 from the uppermost downward, and samples of 25 scallops were measured on each bed at eight different sites. The mean lengths are given in table 3 as a ratio relative to the mean length in Bed 2. Although the mean lengths in each bed vary widely between sites, the ratios deviate by less than 15 percent from the mean values shown in table 3.

Several possible explanations for this lithologic control of the scallop patterns in Bonnechère Caves were considered. The lithology may affect the solution kinetics of the system or its hydrodynamics. Under given conditions of temperature, pH and CO_2 concentration, limestone solution depends primarily on the mineralogical composition of the carbonate rock. For example, the solution of dolomite proceeds at a much slower rate than the solution of calcite. Alternatively, the physical structure of the rock, the presence of insoluble fragments and voids, may produce local varia-

TABLE 3

Mean Scallop Lengths and Chemical Analyses of Various Limestones in Bonnechère Caves, Ontario

Bed	Thickness (Feet)	Mean Ratio	CaCO₃ (%)	MgCO₃ (%)	Insoluble (%)
0	2.0	1.30	95.6	1.3	3.1
1	0.4	1.95	95.1	0.1	4.8
2	1.1	1.00	93.4	1.7	4.9
3	2.0	1.53	91.9	5.0	3.1
4	3.4	1.60	93.9	2.5	3.6

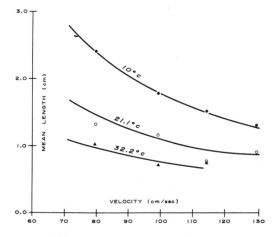

Fig. 5. Experimental data for the high gradient blocks. Mean length is plotted against velocity for the three experimental temperatures.

Fig. 4. Scallops developed on different beds under similar hydrodynamic conditions in the Bonnechère Caves, Renfew County, Ontario.

tion in solution rates, altering wall friction, the boundary layer thicknesses, and hence the mass transfer rates. Again, irregularities in the material may cause a larger number of initial vortices, leading to development of a larger number of scallops, with consequent increase in the competition between scallops for the available space.

Samples of the beds were analyzed by buffered EDTA titration to determine the $CaCO_3$, $MgCO_3$, and insoluble contents. These chemical parameters showed no clear correlation with scallop sizes. Apparently, variations in the mean lengths cannot be explained simply by a variation in the rate of solution.

Thin polished slices were cut from samples of each limestone, at random orientations to the geological structure. The slices showed widely varying physical characteristics. Some beds showed greater irregularity, which could have led to a greater likelihood of scallop initiation and hence greater competition between scallops, but such a concept would be difficult to quantify and test empirically.

The textures of the eroded surfaces of these limestones showed great variation. The exposed surface of Bed 1 had the smoothest appearance, while Bed 2 was roughest. Again, the texture of the eroded surface is responsible for scallop initiation, so that the greatest competition would occur on the roughest surface, leading to smaller scallops. But again any empirical test of this notion would be difficult.

Conclusions

On a standardized material such as a single, uniform bed of limestone or a block of plaster of paris, certain statistical parameters of scallop length distributions are well correlated with the fluid dynamic parameters in a manner similar to the model developed by Curl for the special flute form. Of the 11 parameters tested in this study, the mean length appears to be most highly correlated with the fluid parameters, with a multiple correlation coefficient of .9353.

The lower percentiles of the length distribution show a greater correlation with the fluid parameters than the higher percentiles, implying that other, unknown factors play a larger part in the generation of the latter.

The best-fit values of the constant analogous to Curl's Flute Reynolds Number differ considerably from 22,500. The value corresponding to mean length was found to be 11,476, while the tenth percentile gave a value of 8,147 from flume observations. Measurements made on limestone showed, however, that the value of this constant depends on the nature of the eroded material, in particular upon some indefinite quality of its physical structure.

ACKNOWLEDGEMENTS.—We wish to thank the National Research Council of Canada and the vice-president (science) of McMaster University for grants to help this research.

Goodchild and Ford provide evidence of the vitality of Canadian geo-morphology. Only two years before their publication, Jopling's report (1970) of the recommendations to the Solid Earth Science Study group had noted that Canadian geomorphology exhibited insufficient basic scientific background, insufficient methodological discussions, and too great a concern with the descriptive aspects. None of these deficiencies are visible in the Goodchild and Ford paper.

If the comments and recommendations of the Committee of the Canadian Association of Geographers are examined in detail, it seems that Canadian geo-morphology is both adaptable and self-critical. Within the years which have elapsed since the conclusions of the committee were published, there has been an increase in the amount of methodological discussion and basic scientific background visible in recently published work. Symposia on geomorphology are more frequently held and in total there seems to be a strengthening of geomorphology as a scientific endeavour.

Conclusion

No attempt has been made to select material which provides a regional balance in Canada and no attempt has been made to provide an evaluation of each subject area. The information on Canadian geomorphology presented here is not assembled to permit ready reference to publications about the glacial geomorphology of western Canada or the fluvial geomorphology of the Arctic. There is no discussion about fluvial or coastal or aeolian processes in Canada. This is a deficiency of the selection presented here but the intent is to provide a summary of the character-istics of the rather nebulous concept, "Canadian geomorphology." The nature of the studies, their context in the literature and their illustration of the variety of study in progress should provide both a stimulus and a rationale for proceeding beyond the introductory level in this division of physical geography.

BIBLIOGRAPHY

Allen, J. R. L.: 1969, "Erosional current marks of weakly cohesive mud beds," *Journal of Sedimentary Petrology (Jl. Sedim. Petrol.)* 39, p. 607.

Andrews, J. T.: 1963, "The cross-valley moraines of north-central Baffin Island: a quanti-tative analysis," *Geog. Bull.* 20, pp. 82-129.

Antevs, E.: 1928, "The last glaciation," *Research Series of the Am. Geog. Soc. 17.*

Ashwell, A. W.: 1962, "Flow markings," *Newsletter of the Cave Research Group of Great Britain (Newsl. Cave Res. Grp. Gt. Br.)* 84.

Association of Official Agricultural Chemists: 1950, "Official methods of analysis," Washing-ton 910 pp.

Baker, F. C.: 1931, "A restudy of the interglacial molluscan fauna of Toronto, Canada," *Trans. of the Illinois Academy of Science (Ill. Acad. Sci.)* 23, pp. 358-366.

Bird, J. B.: 1972, *The Natural Landscapes of Canada,* Wiley, Toronto, 191 pp.

Blackadar, R. G.: 1956, *Geological reconnaissance of Admiralty Inlet, Baffin Island, Arctic Archipelago, Northwest Territories,* Geol. Surv. Can. Paper 55-6, 25 pp.

Blackadar, R. G.: 1958, "Fury and Hecla Strait," Geol. Surv. Can. Preliminary Series Maps 3-1958, 4-1958.

Blackadar, R. G.: 1958, "Patterns resulting from glacier movements north of Foxe Basin, N.W.T." *Arctic,* 11, no. 3, 157-165.

Bretz, J. H.: 1942, "Vadose and phreatic features of limestone caverns," *Jl. Geol.* 50, pp. 657-811.

Carson, M. A.: 1971, *The Mechanics of Erosion,* Pion, London, 174 pp.

Carson, M. A. and M. J. Kirkby: 1972, *Hillslopes Form and Process*, Cambridge University Press, London, 475 pp.

Chalmers, R.: 1902, "On borings for natural gas, petroleum and water; also Notes on the surface geology of part of Ontario," Geol. Surv. Can. Summary Report 1901, pp. 158-169.

Chorley, R. J.: 1959, "The geomorphic significance of some Oxford soils," *American Journal of Science (Am. Jl. Sci.)* 257, pp. 503-515.

Chorley, R. J.: 1964, "Geomorphological evaluation of factors controlling shearing resistance of surface soils in sandstone," *Journal of Geophysical Research (Jl. Geophys. Res.)* 69, pp. 1507-1516.

Chorley, R. J., A. J. Dunn and R. P. Beckinsale: 1964, *The History of the Study of Landforms, Vol. 1*, Methuen, London, 678 pp.

Claypole, E. W.: 1887, "The lake age in Ohio," Transactions of the Edinburgh Geological Society (Quoted from Leverett, 1902, p. 604).

Coleman, A. P.: 1894, "Interglacial fossils from the Don Valley, Toronto," *American Geologist (Am. Geol.)* 13, pp. 85-95.

Coleman, A. P.: 1933, *The Pleistocene of the Toronto region*, (accompanied by Map 41 g.) Ontario Dept. Mines, 41: pt. 7.

Coleman, A. P.: 1937, *Lake Iroquois,* (accompanied by Map 45f.) Ontario Dept. Mines, 45: 1-36, pt. 7.

Coleman, A. P.: 1941, *The Last Million Years.* Univ. of Toronto Press, Toronto, 216 pp.

Curl, R. L.: 1959, "Flutes—geometry and propagation," *National Speleological Society Newsletter (Nat. Spel. Soc. Newsl.)* 17, p. 94.

Curl, R. L.: 1966, "Scallops and flutes," *Trans. Cave Res. Grp. Gt. Br.* 7, pp. 121-162.

Davies, R. N.: 1963, "Flow markings," *Newsl. Cave Res. Grp. Gt. Br.* 87, p. 8.

Douglas, R. J. W.: (Ed.): 1970, "Geology and economic minerals of Canada," *Geol. Surv. Can. Economic Geology Report* 1, 838 pp.

Dreimanis, A.: 1957a, "Stratigraphy of the Wisconsin glacial stage along the northwestern shore of Lake Erie," *Science*, 126, pp. 166-168.

Dreimanis, A.: 1957b, "Depths of leaching in glacial deposits," *Science*, 126, pp. 403-404.

Dreimanis, A.: 1958, "Wisconsin stratigraphy at Port Talbot on the north shore of Lake Erie, Ontario," *Ohio Journal of Science (O. Jl. Sci.)* 58, pp. 65-84.

Dreimanis, A.: 1960a, "Pre-classical Wisconsin in the eastern portion of the Great Lakes region, North America," *International Geological Congress*, 21st, Copenhagen, 1960, Rept. Session, Norden 4, pp. 108-119.

Dreimanis, A.: 1960b, "Significance of carbonate determinations in till matrix," *Geological Society of America Bulletin (Geol. Soc. Am. Bull.)* 71, p. 1853 (abstract only).

Dreimanis, A.: 1962, "Quantitative gasometric determination of calcite and dolomite by using Chittick apparatus," *Jl. Sedim. Petrol.* 32, pp. 520-529.

Dreimanis, A.: 1966, "Lake Arkona-Whittlesey and post-Warren radiocarbon dates from 'Ridgetown Island' in southwestern Ontario," *O. Jl. Sci.* 66, pp. 582-586.

Dreimanis, A.: 1967a, "Cary-Port Huron Interstade in eastern North America and its correlatives," *Program of the Geological Society of America, N.E. Section 2nd Ann. Meeting*, Boston, Mass., p. 23.

Dreimanis, A.: 1967b, "Pre-Maumee lake stages of Wisconsin Ice Age in Lake Erie basin." *Abstracts*, Tenth Conference on Great Lakes Research, p. 33.

Dreimanis, A. and P. E. Karrow: 1965, "Southern Ontario," in *Guidebook for field conference G*, Great Lakes-Ohio River Valley, Nebraska Academy of Science, pp. 90-110.

Dreimanis, A., J. Terasmae and G. D. McKenzie: 1966, "The Port Talbot Interstade of the Wisconsin Glaciation," *Canadian Journal of Earth Sciences (Can. Jl. Earth Sci.)* 3, pp. 305-325.

Dury, G.: 1972, "Some Current Trends in Geomorphology." *Earth-Science Review (Earth-Sci. Rev.)* 8, pp. 45-72.

Duthie, H. C. and R. G. Mannada Rani: 1967, "Diatom assemblages from Pleistocene interglacial beds at Toronto, Ontario." *Canadian Journal of Botany (Can. Jl. Bot.)* 45, pp. 2249-2261.

Eyre, J.: 1964, "Flow markings in Southeast Passage, Gaping Ghyll," *Newsl. Cave Res. Grp. Gt. Br.* 90/91.

Fairchild, H. L.: 1909, "Glacial waters in central New York." *Bulletin of the New York State Museum*, 127, pp. 5-66.

Flint, R. F.: 1957, *Glacial and Pleistocene Geology,* Wiley, New York, 553 p.

Ford, D. C.: 1961, "The Bonnechère Caves, Renfrew County, Ontario: a note," *Can. Geog.* 5, pp. 22-25.

Ford, T. D.: 1964, "Flow markings," *Newsl. Cave Res. Grp. Gt. Br.* 90/91.

Forsyth, J. L.: 1959, "The beach ridges of northern Ohio." State of Ohio, Department of Natural Resources, Division of Geological Survey, *Information Circular No. 25*, 10 pp.

Forsyth, J. L.: 1969, "Evidence for a pre-Maumee lake in northwestern Ohio—a progress report." *Abstracts*, Twelfth Conference on Great Lakes Research, Ann Arbor, Mich., p. 61.

Foster, H. D.: 1969, "Geomorphology: Academic Exercise or Social Necessity?" *Can. Geog.* 13, pp. 283-288.

Gadd, N. R.: 1960, *Surficial geology of the Beancour map-area, Quebec, 31 1/8,"* Geol. Surv. Can. Paper 59-8, 34 pp.

Glennie, E. A.: 1963, "Flow markings," *Newsl. Cave Res. Grp. Gt. Br.* 87, p. 9.

Goldthwait, R. P., A. Dreimanis, J. L. Forsythe, P. E. Karrow, and G. W. White: 1965, "Pleistocene deposits of the Erie lobe." in H. E. Wright, Jr. and D. G. Frey (ed.), *The Quaternary of the United States.* Princeton University Press, Princeton, pp. 85-97.

Hack, J. T.: 1960, "Interpretation of erosional topography in humid temperate regions," *Am. Jl. Sci.* 258, pp. 80-97.

Hobson, G. D. and J. Terasmae: 1969, *Pleistocene geology of the buried St. Davids gorge, Niagara Falls, Ontario: geophysical and palynological studies,"* Geol. Surv. Can. Paper 68-67.

Hough, J. L.: 1958, *Geology of the Great Lakes.* University of Illinois Press, Urbana, 313 pp.

Hough, J. L.: 1963, "The prehistoric Great Lakes of North America." *Am. Jl. Sci.* 51: 84-109.

Hough, J. L.: 1966, "Correlation of glacial lake stages in the Huron-Erie and Michigan basins." *Geology*, 74, pp. 62-77.

Ives, J. D. and J. T. Andrews: 1963, "Studies in the physical geography of north-central Baffin Island, N.W.T." *Geog. Bull.*, 19, pp. 5-48.

Jaiin, A. and M. Klapa: 1968, "On the origin of ablation hollows (polygons) on snow," *Jl. Glac.* 7, pp. 299-312.

Jopling, A. V.: 1970, "Recommendations on the future development of geomorphology in Canada," *Can. Geog.* 14, pp. 167-173.

Karrow, P. F.: 1963, "Pleistocene geology of the Hamilton-Galt area." Ontario Department of Mines, *Geological Report 16*, 68 pp.

Karrow, P. F.: 1967, "Pleistocene geology of the Scarborough area." Ontario Department of Mines, *Geological Report 46*, 10 pp.

Karrow, P. F.: 1968, "Pleistocene geology of the Guelph area." Ontario Department of Mines, *Geological Report 61*, 38 pp.

Karrow, P. F.: 1969, "Stratigraphic studies in the Toronto Pleistocene." *Proceedings of the Geological Association of Canada* 20, pp. 4-16.

Karrow, P. F., J. R. Clark and J. Terasmae: 1961, "The age of Lake Iroquois and Lake Ontario," *Jl. Geol.* 69, pp. 659-667.

Kunkle, G. R.: 1963, "Lake Ypsilanti: a probable Late Pleistocene low-lake stage in the Erie basin," *Jl. Geol.* 71, pp. 72-75.

Lajtai, E. Z.: 1967, "The origin of some varves in Toronto, Canada." *Can. Jl. Earth Sci.* 4, pp. 633-639.

Leverett, F.: 1902, "Glacial formations and drainage feature of the Erie and Ohio basins." *U.S. Geological Survey Monogr. 41*, 802 pp.

Leverett, F. and F. B. Taylor: 1915, "The Pleistocene of Indiana and Michigan and the history of the Great Lakes." *U.S. Geographical Survey Memoir 53*, 529 pp.

Lewis, C. F. M., T. W. Anderson and A. A. Berti: 1966, "Geological and palynological studies of Early Lake Erie deposits," *Ninth Conference Great Lakes Research Proceedings, Univ. Michigan*, pp. 176-191.

Lewis, C. F. M.: 1969, "Late Quaternary events in Lake Huron and Lake Erie basin," *Abstr. Twelfth Conference on Great Lakes Research*, pp. 31-32.

Lyell, C.: 1845, *Travels in North America in the years 1841-42; with geological observations on the United States, Canada and Nova Scotia*, Wiley and Putnam, New York.

Lyon, T. L., H. O. Buckman and N. C. Brady: 1952, *The nature and properties of soils.* Macmillan, New York, 591 pp.

Maxson, J. H.: 1940, "Fluting and facettings of rock fragments." *Jl. Geol.* 48, pp. 717-751.

Moore, G. W. and G. Nicholas: 1964, *Speleology.* Heath, Boston, 120 pp.

Muller, E. H.: 1964, "Quaternary section at Otto, New York," *Am. Jl. Sci.* 262, pp. 461-478.

Ogden, J. G. III and R. J. Hay: 1969, "Ohio Wesleyan University natural radiocarbon measurements IV." *Radiocarbon* 11, pp. 137-149.

Parry, J. T.: 1967, "Geomorphology in Canada," *Can. Geog.* 11, pp. 280-311.

Prest, V. K.: 1970, "Quaternary Geology," in R. J. W. Douglas (Ed.) "Geology and economic minerals of Canada," *Can. Geol. Surv., Economic Geology Report 1*, pp. 675-764.

Price, C. A.: 1954, *Crustal movement in the Lake Ontario-Upper St. Lawrence River basin*, Canadian Hydrographic Service.

Reiche, P.: 1950, "A survey of weathering processes and products," University of New Mexico *Publications in Geology* 3, 95 pp.

Rudnicki, J.: 1960, "Experimental work on flute development," *Speleologia*, 2, pp. 17-30.

Russell, R. J.: 1949, "Geographical Geomorphology," *Annals of The Association of American Geographers* 39, pp. 1-11.

Sim, V. W.: 1960, "A preliminary account of late 'Wisconsin' glaciation in Melville Peninsula, N.W.T." *Can. Geog.* 17, pp. 21-34.

Sissons, J. B.: 1960, "Subglacial, marginal and other glacial drainage in the Syracuse-Oneida area, New York," *Bull. Am. Geol. Soc.* 71, pp. 1575-1588.

Spencer, J. W. W.: 1890, "The deformation of Iroquois beach and birth of Lake Ontario," *Am. Jl. Sci.* 40, pp. 443-451.

Spencer, J. W. W.: 1910, "The fall of Niagara," *Geol. Surv. Can. Pub. 970.*

Spencer, J. W. W.: 1910, "Relationship of Niagara River to the glacial period," *Bull. Am. Geol. Soc.* 21, pp. 433-440.

Strahler, A. N.: 1952, "Dynamic basis of Geomorphology," *Bull. Am. Geol. Soc.* 63, pp. 923-937.

Terasmae, J.: 1960, "A palynological study of Pleistocene interglacial beds at Toronto, Ontario," *Geol. Surv. Can. Bull.* 56, pp. 24-40.

Wall, R. E.: 1968, "A sub-bottom reflection survey in the central basin of Lake Erie," *Bull. Am. Geol. Soc.* 79, pp. 91-106.

White, G. W.: 1953, "Sangamon soil and early Wisconsin loesses at Cleveland, Ohio," *Am. Jl. Sci.* 251, pp. 362-368.

White, G. W.: 1968, "Age and correlation of Pleistocene deposits at Garfield Heights (Cleveland), Ohio," *Bull. Am. Geol. Soc.* 79, pp. 749-752.

White, G. W.: 1969, "Pleistocene deposits of the northwest Alleghany Plateau, U.S.A." *Q. Jl. Geol. Soc. London* 124, pp. 131-151.

Yatsu, E.: 1957, "On the application of methods of soil mechanics to the investigation on the erosibility of the bare mountain area in Japan.—An approach to dynamic geomorphology," *Miscellaneous Reports of Research Institute of Natural Resources*, Tokyo.

Yatsu, E.: 1964, "Sur l'observation de l'allophane par le microscope electronique," *Applied Geography (Appl. Geog.)* 5, pp. 48-56.

Yatsu, E.: 1966, *Rock control in geomorphology,* Sozosha, Tokyo, 135 pp.
Yatsu, E.: 1967, "Some problems on mass movement," *Geografiska Annaler (Geog. Annal.)* 49, pp. 396-401.
Yatsu, E., F. A. Dahms, A. Falconer, A. J. Ward and J. S. Wolfe (Eds.): 1971, *Research methods in geomorphology.* Geo Abstracts, Norwich. 140 pp.
Yatsu, E. and A. Falconer, (Eds): 1972, *Research methods in Pleistocene Geomorphology.* Geo abstracts, Norwich, 205 pp.

Section 3
Weather and Climate

Chapter 7

The Status of Canadian Climatology

Climatological research in Canada is undertaken by geographers, soil scientists, agriculturalists, hydrologists, and numerous physical scientists who are all concerned with atmospheric processes and phenomena as the main controlling agents of a selected physical environment. Consequently, reference to climatological research findings involves consultation with an extensive range of scientific journals of national and international repute, together with reference to numerous government publications. For example, Canadian climatology is represented by government reports from the Department of the Environment (Atmospheric and Inland Waters Branches) and the National Research Council. Also, climatological research in Canada provides the basis of publications in a wide range of journals, eg., *Canadian Journal of Earth Sciences, Canadian Geographer, Canadian Journal of Plant Science, The Agronomy Journal, Scientific Agriculture,* as well as the well-known international meteorological and geographical journals published regularly around the world.

As far as the physical geographer is concerned, Canada has internationally known climatological projects organized by the geography faculty at a number of the main university centres. However, reference to *The Canadian Geographer* (which represents the major national outlet for geographical research findings) over the past decade reveals that only a small number of weather and climate articles have been published, despite the fact that climatological research is conducted by most geography departments. The most disturbing example of this serious omission occurred in 1967 when the centennial issue of *The Canadian Geographer* was issued, which referred extensively to the progress and developments in most of the geographical systematics (including geomorphology and biogeography) in Canada but which did not include a climatological contribution. It is apparent that the research findings of Canadian climatologists (including geographers) are published in the non-geographical journals and reports mentioned earlier. Indeed, these publications provide a great wealth of excellent material devoted to most atmospheric processes and climatic parameters, related to the macro, meso, and micro scales, and applied to every possible physical and cultural interface. To conclude this introductory statement on the status of climatology in Canada, reference must be made to the excellent rapport and cooperation between the climatologists of every relevant discipline. Ontario, for example, is represented by the Friends of Climatology and the Ontario Climatologist groups which meet regularly to exchange ideas and to discuss common interests (particularly associated with instrumentation).

The assessment of the progress and developments of Canadian climatology over the last few decades is an extremely difficult task since a comprehensive statement of the research in every process and parameter, in all the political and physical environments, would fill several volumes. The selection of climatological topics for discussion in this section has been governed by the main objective of illustrating the development of significant research in this field, in terms of processes and problems of national and local importance to the livelihood of the inhabitants of this country. For a more complete coverage of the development of every aspect of Canadian climate up to 1957, the reader should refer to the detailed *Bibliography of Canadian Climate 1763-1957* (Thomas 1961). The study commences with a general statement on macro-, and meso-, and microclimatological research in Canada, emphasizing examples of the outstanding progress over the last two decades. It continues with an assessment of four climatic themes related to phenomena and processes which will have a vital role to play in the utilization of the Canadian environment in future decades.

The themes are:

1) *Arctic climatology.*   The Canadian North has a very important future with the proposed exploration and exploitation of mineral resources and oil and gas reserves. Therefore it is essential for the geographer to understand the climatic controls and weather genesis of this vital area, particularly the climate-permafrost relationships and the modification of the aerial and soil environments associated with surface construction.

2) *Snow and ice climatology.*   Canada is dominated by seasonal and perennial snow, land ice and floating ice, which consequently represent a significant part of the natural environment. Numerous economic and cultural activities are controlled by the snow and ice cover (*viz*, water power, irrigation and drinking water, shipping, and mining to name only a handful of activities) and therefore it is essential for the geographer to understand the climatology of snow and ice.

3) *Agroclimatology.*   The development of Canadian primary industries will depend on knowledge of the relationship between crops and climate on every scale, so that researchers will be able to introduce new varieties to meet the demands of changing agricultural requirements over future decades. It is most important for data to become increasingly available on the energy and moisture fluxes from soil and plant surfaces, using heat budget and aerodynamic models.

4) *Urban climatology.*   Most Canadians are urban dwellers, and it is apparent that urbanization is expanding at a rapid and alarming rate. It is important for climatologists to examine the climatic implications of the Canadian urban sprawl, particularly the structure and advantages of the urban core heat island development and the characteristics of chronic air pollution, which is such a serious environmental problem in metropolitan and industrial areas.

The Character and Dynamics of the Canadian Climate

The Atmospheric Environment Service of the Department of the Environment (formerly the Meteorological Branch of the Department of Transport) has published a great number of climatological analyses over the past 100 years and the data include monthly summaries for all climate stations, together with regional analyses of the major cities and all the provinces and territories. The material is mainly in the form of descriptive reports of regional climatic controls and data "bookkeeping" in the form of tables and distribution maps of dominant macroclimatic parameters, for example, the Climatological Studies Series of the Atmospheric Environment Service and the classic *Climatological Survey of Southern Ontario* published by Putnam and Chapman in 1938. This, of course, is a valuable service in terms of understanding the climatic characteristics of Canada but interpretation of the data regarding weather genesis or dynamic climatology was fairly uncommon before 1950. For example, in his 1951 paper, "Some Climatological Problems of the Arctic and Sub Arctic," F. K. Hare was largely concerned with the ecological climatology of the area and made only brief reference to the role of the atmospheric circulation on the climate of the northlands in the section of Greenland. This dynamic approach was largely an assessment of Greenland climatology in terms of the Hobbs glacial anticyclone theory of 1926-45, which influenced meteorological thinking quite considerably. Consequently, climatologists built up a very static model of the balanced circulation over the Canadian Arctic and, indeed, this anticyclonic model was assumed to affect the entire country during the winter season. The model had two main facets with permanent anticyclonic divergence (thermal origin) at the surface layers and a circumpolar vortex convergence aloft. The upper and lower systems are linked by more or less permanent subsidence over the Arctic area which is reflected in the characteristic polar aridity. However, the existence of glaciers and ice caps in the north emphasized the role of precipitation and a periodic breakdown of the thermal high with the migration of depressions and fronts. Of course, Hobbs demonstrated that the ice was maintained by the hoar frost development from radiative cooling, despite the adiabatic warming and increasing dew point remoteness in subsiding air. The glacial anticyclone theory was seriously overemphasized and this was mainly on account of the few climatic stations located in the Arctic. For example, Hare (1955) stated that the role of migrating depressions and upper perturbations was once confined to the south of Canada (i.e., supporting the Hobbs' theory). The original storm trajectories correlated closely with the main line of the Canadian Pacific Railway across Canada, which represented for a long time the northernmost line of telegraphic stations reporting weather conditions.

Over recent decades telegraphic synoptic weather stations have been established in the high Arctic (eg., Eureka on Ellesmere Island, 80° north), and consequently the storm paths have shifted considerably northwards—an example of man-induced climatic change! Frequency maps published by Hare in 1955 were pioneer attempts to determine the genesis of the Arctic weather and indicated that the eastern archipelago was dominated by more than 50 percent frequency of cyclonic flow in each January between 1942 and 1947, which was the glacial anticyclone season. Also, in each July between 1942 and 1947, the Maritime Atlantic air had a 30 percent frequency over eastern Baffin Island. These maps made a great impact on climatological thinking and went a long way to minimize the preconception of the alleged anticyclonic dominance over the Arctic archipelago.

In the 1955 article, Hare challenged the developing interest in weather genesis as a geographical tool and queried its value to the geographer with his preoccupation with areal differentiation of the earth's surface and his prime concern for the needs of human society. However, in recent years Canadian geographers have placed increasing emphasis on the study of process, which has resulted from improved instrumentation and new methodology, together with the trend toward a more analytical approach in research projects. As a result of these changes, the geographer has recently specialized in systematic studies of processes (*viz*, dynamics) of the physical and cultural environments, e.g., the geomorphic agency, the settlement motivation, and the weather genesis of a climatic region. Today macro- and mesoclimatological research is characterized by a dynamic methodology which emphasizes the role of atmospheric circulation in terms of weather types, depression tracks, air masses, Rossby waves, and jet streams.

Examples of the post 1950 dynamic approach to Canadian climatological problems are outlined below:

Firstly, the study of F. E. Burbridge (1951) related to the modification of the continental polar air over Hudson Bay, when it was demonstrated that the actual moisture and thermodynamic changes of an air current were particularly dependent upon the specific trajectory of the air over open water or the completely frozen bay.

Secondly, J. R. Villimow (1956) studied the nature and origin of the Canadian dry belt of Alberta-Saskatchewan. The drought conditions were analyzed in the genetic terms of surface cylonic/anticyclonic frequency, moist air mass trajectories, jet stream location (with the inflow subsidence on the northern periphery) and anticyclonic ridge subsidence at the 700 mb and 300 mb contour levels.

The third example refers to the research of R. G. Barry (1960) who analyzed the synoptic climatology of Labrador-Ungava in terms of the precipitation and temperature distributions related to airflow types, with particular application to palaeoclimatology and the deglaciation of the area.

It should be emphasized that the dynamic or genetic approach is not the sole concern of current macro- and mesoclimatological studies in Canada. In fact, physical climatological research is vital in Canada and a most interesting trend here is the application of climatonomy (H. Lettau, 1969) to traditional parameter problems, which represents the response of the atmosphere–soil–vegetation system to the forcing function of solar radiation. For example, evapotranspiration climatonomy represents a new approach to numerical prediction of monthly evapotranspiration, run off, and soil moisture storage and has been adapted by Hare and Hay (1971) to the annual water balance of northern North America. Also, radiation and heat balance research on a regional scale is a useful study, particularly in the Arctic (Vowinckel and Orvig 1962-1967) where micro-energy balance investigations are limited to a very small number of sites.

Microclimatological research in Canada has been traditionally concerned with the assessment of energy and moisture fluxes over a variety of environmental interfaces, *viz*, crops, forest, grass cover and, (more recently) snow, ice, permafrost, and the urban fabric. The original research was hindered by inadequate instrumentation when the fluxes were assessed by the application of empirical formulae to easily measured parameters, such as traditional mean temperature and precipitation data, for example, the Thornthwaite estimation of evapotranspiration flux which was applied to Canadian climatic data by Sanderson in 1948. Changes over the last two decades have been mainly associated with the greater sophistication of microclimatological instrumentation. Also, the methodology has changed markedly from empirical formulae (based on mean data) to more accurate, specific measurements of the energy and moisture balances over a wide range of surfaces, including the aerodynamic transfer and eddy correlation studies of recent years. These changes have been most apparent in agroclimatological research and will be discussed at length in that section.

Chapter 8

The Climate of the Canadian Arctic

Climatological research in this inhospitable region, under arduous physical conditions, will be vital over future decades owing to the enormous economic potential of the Arctic region and the desire to exploit these valuable resources without too much disturbance of the fragile physical environment. Studies over past decades have been traditionally concerned with easily measured Arctic parameters and the assessment of major climatic controls and basic environmental characteristics. Consequently, the pre-1950 research publications were essentially listings of specific and relevant data for selected stations in the Arctic region, and this approach applies to one of the latest statements of the Atmospheric Environment Service entitled "The Climate of the Canadian Arctic" (1967). However, since the 1951 paper by F. K. Hare on climatological problems of the northlands, there have been numerous attempts to determine the actual dynamics of the Arctic climate in terms of weather genesis in the lower and upper troposphere. For example, Namais (1958), Hare and Orvig (1958), and Lee (1960) were concerned mainly with the general circulation of the Arctic in terms of surface and upper troposphere distributions and jet stream characteristics. Namais and Lee analyzed the 700 mb flow over the area and located the circumpolar vortex and wave features (troughs/ ridges) over northern Canada, especially in the winter season. They discovered that the Hudson Bay trough extended to the 25 mb level (24,000 m) and represented a forced perturbation associated with the Rocky Mountain barrier.

Dynamic climatological research in the 1960s included work by Reed and Kunkel (1960), Boville (1961), Tissot (1963), Bryson (1966), Barry (1967) and was associated with the main features of the atmospheric circulation in the Arctic, particularly fronts, air masses, and the Arctic jet stream. The findings are summarized in the "Atmospheric Distributions" section of the reprinted paper by F. K. Hare (1968) "The Arctic," which contrasts markedly with his earlier 1951 findings when tropospheric data were not available. "Since then the situation has been transformed. Almost two decades of regular synoptic observations have come in from land stations and long radiosonde series are available in many areas" (Hare, 1968).

It should be emphasized that a great deal of the above research was undertaken by the Arctic Meteorology Research Group at McGill University, Montreal. It undertook work in two main areas, dynamical and synoptic studies (outlined above) and the physical climatology of the Arctic. The latter research was mainly associated with the energy balance of the Arctic Basin and the principal investigators were Vowinckel (1962, 1964), Vowinckel and Orvig (1964, 1965a, 1965b, 1966), and Vowinckel and Taylor (1965). The McGill energy balance data represent the most comprehensive regional heat flux studies of the Arctic and the methodology and results have been concisely summarized by Orvig (1965). The program of work was to determine the heat flux between 300 mb and the earth's surface (at several areas in the Polar Ocean) and to examine each term in the energy balance equation *viz,* ocean current heat flux, solar radiation/albedo, long-wave terrestrial radiation, sensible and latent heat flux, sea ice export, greenhouse effect, and the heat conduction through the ice. The results indicate that:

The Arctic has a strongly negative radiation balance during most of the year. Part of the required energy import is fulfilled by ocean currents and the remainder by atmospheric advection (especially latent heat). . . . The surface energy budget shows that the radiative terms are far greater than all other influences and the long-wave components are the greatest in all areas and months. The sensible heat flux from atmosphere to ground is negligible. In winter all energy expenditure is radiative from the Polar Ocean. . . . Looking at the tropospheric energy budget over the Arctic, there is a sharp increase in importance of non-radiative terms on the income side, and an even more pronounced decrease on the expenditure side. . . . Calculations for the earth-atmosphere energy budget show that the result of no advection into the north polar regions would be a temperature drop of . . . about 50°C over the central Polar Ocean (Orvig, 1965).

The Vowinckel–Orvig estimates for the annual heat exchange over the Arctic Ocean have been summarized in section 4c of the Hare (1968) reprinted article. The combination of dynamic-synoptic and energy balance approaches in this reprint emphasize the value of the Hare statement, in terms of genetic and physical controls of present day climates in the area.

The macroclimatic and heat budget research of Fletcher (1965, 1966) represents a quantitative study concerning the input, transformation, and output of thermal energy, related to the general circulation of the atmosphere. "An assessment is made of each component of the heat budgets at the surface and of the earth–atmosphere system in the Central Arctic, both for an ice-covered and for an ice-free ocean. The annual patterns of atmospheric heat loss for both conditions are obtained as residuals; the relation of these patterns to general atmospheric circulation and glacier accumulation is discussed" (Fletcher, 1965). Perhaps the most useful aspects of Fletcher's studies are the heat balance and atmospheric circulation correlation and the realization that less extensive pack ice contributes to a more vigorous atmospheric circulation, despite the reduced albedo and a decreased heat deficit (i.e., a difference of about 15 Kcal/cm² or 20 percent from present conditions). The stronger circulation would be highly favourable for accumulating continental ice sheets at high latitudes (for example, on Ellesmere and Baffin Islands above 200 m). This accumulation would result from the additional moisture brought into the region by cyclonic activity as the thermal anticyclone over the Arctic Basin was removed.

THE ARCTIC*

F. Kenneth Hare

1. Introduction

Seventeen years ago I was commissioned by the American Meteorological Society to write a review of Arctic climatology for the new *Compendium of Meteorology* (Malone 1951). My review (Hare 1951) dealt with such things as the definition of the Arctic, the climatic control of the northern forest distributions, the sea ice, the Greenland Ice Cap and permafrost. It was accompanied by an article on Arctic circulation patterns by Dorsey (1951) that took account of Russian experience on the drifting pack-ice stations. The two articles to-

* This paper is the text of the presidential Address delivered before the Royal Meteorological Society, 24 April 1968. The paper was subsequently published in the *Quarterly Journal of the Royal Meteorological Society*, Volume 94, No. 402, October 1968, pp. 439-459, and it is reproduced here by permission of the Royal Meteorological Society.

gether showed how little we really knew about northern climates and their bearings on biotic distributions.

Since then the situation has been transformed. Almost two decades of regular synoptic observations have come in from land stations, and long radiosonde series are available in many areas. Both surface and radiosonde series are also available from drifting stations on the pack-ice of the Arctic Ocean and on the remarkable ice-islands—fragments of thick Ellesmere shelf-ice that drift with the Arctic pack. There has been extensive geophysical, glaciological, oceanographic and biogeographic exploration. The basal topography of the Arctic basin is well-known, and submarines have even examined the underside of the permanent pack-ice.

I have been directly involved in several aspects of this work as meteorologist and as geographer. In the early 1950's I directed an interdisciplinary survey of land-forms and vegetation in Labrador-Ungava (Hare 1959), and in association with Ilmari Hustich of Helsinki have since attempted hemisphere-wide syntheses of the northern forest distributions. In 1954 McGill University set up its Subarctic Research Laboratory at Schefferville, Quebec, to enable this work to be intensified in the field. In the same year the Arctic Meteorology Research Group, established three years previously at the University of California at Los Angeles, was transferred to McGill. It undertook work in two main areas—dynamical and synoptic studies, chiefly in the stratosphere, by myself, A. D. Belmont, B. W. Boville and others; and the physical climatology of the Arctic basin, by S. Orvig, E. Vowinckel and others. For much of the time we worked in close association with a group at the University of Washington under R. J. Reed, and with Wayne Hering and his associates at the Air Force Cambridge Research Center. Our work was greatly aided by W. L. Godson and the research section of the Canadian Meteorological Service.

Throughout this period I tried to see how the various northern distributions related to one another. We are now well ahead of the *Compendium* articles of 1951, so it seems useful to attempt another synthesis of what has been learned. In this address I shall deal mainly with low-level, surface and sub-surface phenomena, having recently reviewed our higher level results (Murgatroyd, Hare, Boville, Teweles and Kochanski 1965). I intend my remarks as a tribute to the various colleagues from whom I have learned so much, and to the organizations named at the end of my address who have contributed notably to the scientific exploration of the northern world. My address will go far beyond what my personal research can justify, and will involve much speculation. For this I hope that I, and not they, will be blamed.

2. Atmospheric Distributions

(a) *The Arctic core* We begin with the mean mass distribution fields, now known with some precision. Figs. 1 and 2 show mean 500-mb topographies for January and July, superimposed on certain other distributions to which we shall turn later. These charts are quite similar to the surface to 500-mb thickness topographies, and hence give a broad indication of the thermal field as well as the mean motion. Within the ring of westerlies there is a sluggish inner core of Arctic air that generally lacks organized circumpolar motion. This cold, nearly barotropic core is often deformed by up to three cyclones that may or may not be old occluded centres drifting up from the westerlies, and that may sometimes disconcertingly plunge southwards into inviting parts of these same westerlies. The uniformity of the Arctic core (at least above 800 mb) was first noted by Dzerdzeyevskii (1945) and Flohn (1952), and the cold lows critically examined by Scherhag (1958) in what is by now the old stone age of Arctic meteorology. The core is distinct from the westerly ring only up to 15 to 20 km. The stratospheric polar night westerlies are more tightly coiled round the pole—and in summer there is no westerly vortex above 20 km.

Flohn thought that the barotropic core owed its uniformity to horizontal mixing. It is certainly true that the middle and upper troposphere of the core have very similar tempera-

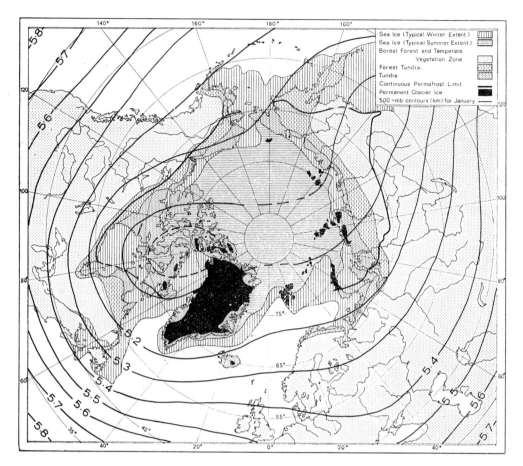

Fig. 1   Mean 500-mb topography for January (geopotential km), with associated sea-ice distribution for midwinter, the location of glaciers (solid black), the southern limit of continuous permafrost (after Brown 1967; Stearns 1966) and the extent of tundra and forest-tundra.

tures over the whole Arctic, in spite of the widely differing radiation temperatures of the surface below—from 272°K over the Norwegian and Barents Seas to below 230°K in the interior of Siberia, Alaska and parts of Canada in winter.

The mean sea-level pressure fields reflect the effects of thermal differences below 800 mb. In winter low pressure extends into the high Arctic over the warm Norwegian-Barents sea, and a belt of relatively high pressure across the Arctic Basin connects the continental high pressure systems over Siberia and the Yukon. In spring, when the inner Arctic remains very

cold relative to its surrounds, highest mean pressure lies over the permanent pack-ice and the Canadian Arctic Islands. In summer the entire Arctic has uniform sea-level pressures north of the shallow sub-Arctic lows. The complexity of these sea-level charts has for long obscured the surprising simplicity and uniformity of the Arctic core above the shallow layers responding directly to thermal contrasts at the air-sea, air-ice and air-land surfaces.

(b) The Arctic frontal zone   The pioneers of air mass analysis, notably Tor Bergeron, were aware of the existence of this cold core, calling it the Arctic air mass. They also postulated an

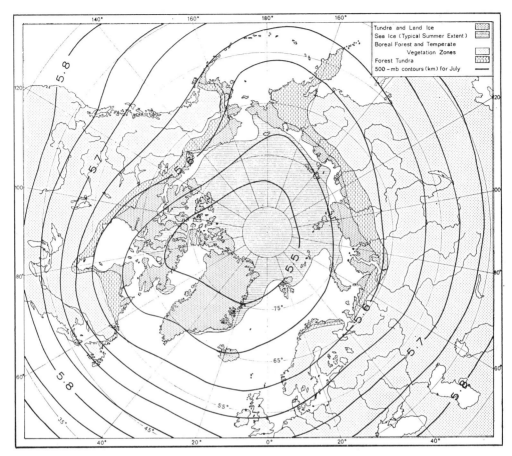

Fig. 2   Mean 500-mb topography for July (as in Fig. 1), with associated sea-ice distribution for midsummer, and the extent of tundra and forest-tundra.

Arctic front bounding this air mass discontinuously, with separate strands over the Eurasiatic and Siberian-Alaskan sectors. These views rested on qualitative arguments concerning surface source regions and their differing radiative régimes (Petterssen 1940).

Within the westerly ring of the Northern Hemisphere there are habitually two or three concentrations of troposphere-deep baroclinity, and in Siberia and North America one of these concentrations lies along the Arctic fringe of the westerlies. The Arctic front is a reality in these longitudes, though it is not simply related to surface controls, nor is it continuous round the pole. Daily high-latitude analysis shows that this front lies within a baroclinic westerly current in which wave disturbances may amplify to produce cyclonic effects at lower levels.

Canadian operational analysis schemes give considerable prominence to the triple structure of the westerlies (Godson 1950; McIntyre 1950, 1955, 1959; Penner 1955). Their three-front model identifies conventionally three fronts and four air masses, the northernmost of which are continental Arctic and maritime Arctic separated by the continental Arctic front. The air masses are assumed in the classical fashion to be differentiated over surface source regions: McIntyre (1959) writes, for example, "continental Arctic air . . . . . is rapidly modified into maritime Arctic air when it passes over northern ice free oceans." In

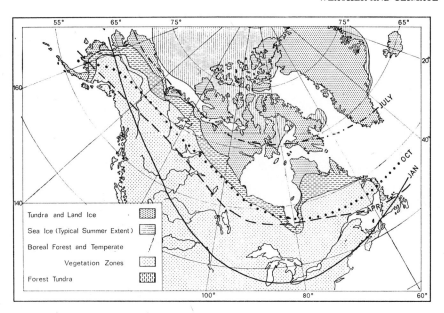

Fig. 3 Modal positions of the continental Arctic front at 850-mb in 1961-65, as defined by Canadian operational practice, in mid-season months (from data published by Barry 1967).

application, however, the model assigns specific values of temperature on isobaric surfaces, and neglects origins. This permits circumpolar location of the front, whose existence is often assumed even in the absence of strong baroclinity at that temperature. Barry (1967) presented a climatology of the continental Arctic front, as thus arbitrarily defined, for the 850 mb surface over North America. Fig. 3 gives the locus of the front in the mid-season months. If the front is real and deep, it must correspond to an Arctic jet stream. Tissot (1963) used the McGill University 80 W daily cross-sections for 1959-60 to test this hypothesis. At this longitude he found the Arctic jet to be distinct except in summer, and to lie on the average near 60 deg N. He prepared mean cross-sections using the jet-core as origin to demonstrate its distinctness from other jets further south (Fig. 4).

Using less conventionalized methods, Reed and Kunkel (1960) found that in summer there is a distinct belt of maximum frontal frequency along a line from north-central Siberia across northern Alaska and Mackenzie District to James Bay in Canada (Fig. 5). Their position is confirmed almost exactly by

streamline fields or surface winds over Canada (Bryson 1966), though it cannot be reconciled with Barry's result. They found the baroclinity to extend to Arctic jets near the tropopause, which suggests that surface conditions alone could not account for its presence and intensity. Finally they prepared composite cloud charts for the front and its cyclonic waves for a period in July 1956 (when the Ptarmigan flights crossed it daily). As expected, the major concentrations were near wave-crests or occlusions. Fig. 5 shows estimated mean net radiation at the surface for July (Vowinckel and Orvig 1965a and unpublished data) in relation to the locus of the Reed-Kunkel frontal maximum. A broad minimum of net radiation extends roughly parallel to the locus and 600 to 1,000 km north of it. Presumably this reflects the distribution of cloud associated with the frontal belt, which is in turn reinforced by the surface radiation field—a typical climatological non-linearity.

In winter the front shifts southward so as to lie well over the interior of North America, from Alaska down the Rockies, and then eastwards to the Great Lakes and the Canadian Maritime Provinces. At this season there is also

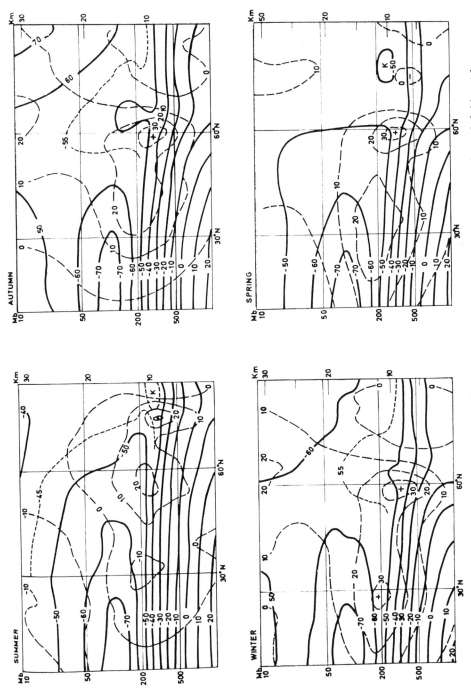

Fig. 4 Mean cross-sections for 1959-60 along 80 W meridian, showing temperature and zonal wind fields with respect to the Arctic jet (taken as origin of co-ordinate grid) (after Tissot 1963).

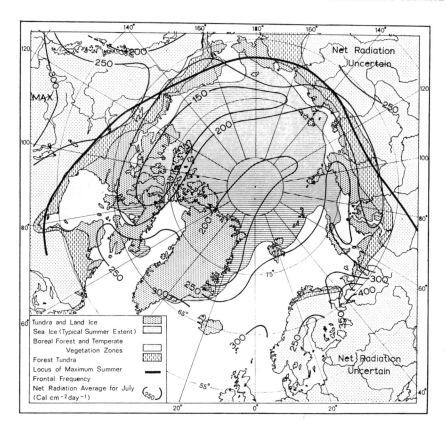

Fig. 5   Locus of Reed-Kunkel (1960) maximum of summer frontal activity in relation to estimated net radiation (1y day⁻¹) for July (after Vowinckel and Orvig 1965a, also unpublished data).

intense baroclinity over the Norwegian, Barents and Laptev Seas, though it is hard to draw a frontal locus through it. I confess I have had little analytical experience in this area, and this opinion may simply be due to ignorance.

(*c*) *Wave structure* Figs. 1-2 show that the mean circumpolar flow is eccentric, and also is deformed by standing long waves. The amplitude of these waves in the mean height field (and hence in the resultant geostrophic flow) has been calculated by Van Mieghem (1961). The eccentricity has peak amplitude outside the Arctic, affecting primarily the mid-latitude westerlies, although the effective centre of the Arctic circulation is well off the geographical pole. Wave number 2 (Fig. 6), sometimes called the bipolarity, has maximum amplitude in the latitude belt 55 to 70 deg and undergoes a substantial phase-shift between summer and winter. This implies that the bipolarity reflects longitudinal heating inequalities. Wave number 3 (Fig. 6) has a less prominent maximum in latitudes 50 to 55 deg and undergoes no seasonal phase shift. Van Mieghem reads this as implying orographic control.

What is clear from Figs. 1 and 2 is that the physical and biotic distributions—sea-ice distribution, the position of the permafrost limit and the forest-tundra—also show the strong bipolarity. They are closely in phase with the atmospheric motion at 500 mb. Since both treeline and mid-tropospheric flow are natural integrators of thermal effects near the ground, this resemblance is understandable.

Recent attempts by Lamb, Lewis and Woodroffe (1966) to reconstruct the 500 mb topography (actually thickness) at various dates since the final glaciation imply that the strong

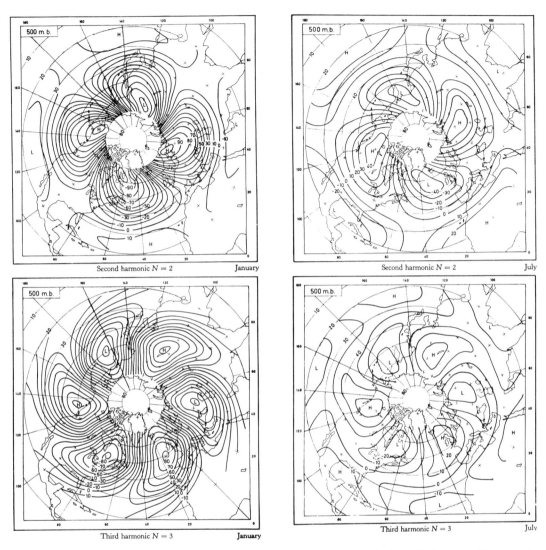

Second harmonic N = 2   January

Third harmonic N = 3   **January**

Second harmonic N = 2   July

Third harmonic N = 3   July

Fig. 6   Second and third harmonics (geopotential m), 500-mb surface, January and July (after Mieghem 1961). The second harmonic shows a seasonal shift of phase, and has maximum amplitude in the latitude of the Arctic jet.

seasonal phase shift of wave number 2 was somewhat reduced in late glacial times, when amplitudes were greater. But all available evidence—including the history of deglaciation over North America (Bryson 1968)—implies that the standing waves so characteristic of the circumarctic circulation and the physical-biotic distributions of the surface have a very long history, whatever phase-shifts may have occurred.

The most remarkable feature of these deforming waves is that they extend high into the winter stratosphere. The mean topography and temperature of the 30 mb surface (roughly 23 km) shows a strong warm ridge over Alaska (Fig. 7) in winter, with troughs over eastern North America and eastern Siberia—with substantial amplitudes at wave numbers two and three. These deformations certainly get stronger as one rises. Teweles (1965) gives

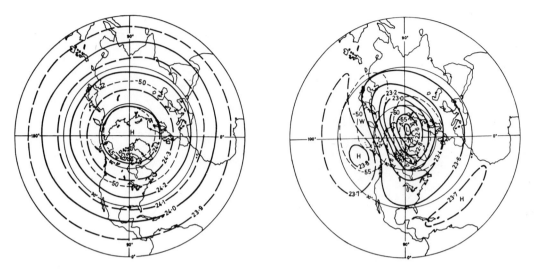

Fig. 7   Mean 30-mb topography for summer and winter season, 1958-61 (geopotential km), Northern Hemisphere.

instances when on individual days the Alaskan ridge was detectable at 55 km (though at this level cold), and Frith (1968) confirms from rocket evidence that the ridge is still present in the mean flow at 55 km. The great depth of the long-wave structure of the westerlies is maintained in spite of the fact that transient wave activity (including amplitude and phase variations at low wave numbers) in the polar-night westerlies of the stratosphere often looks separate from events at tropospheric levels, or may even appear to be in antiphase at certain wave numbers (Boville 1961; Hare and Boville 1965).

On the other hand the summer easterlies of the polar stratosphere (Fig. 7) appear quite undisturbed by standing waves. The resultant wind vectors at 20 km and above are almost due east at all available wind stations in high latitudes, and the best representation of the mean topography of the 30, 20 and 10 mb surfaces is by means of circles centred on the geographical pole. Much of the variance of observed winds lies within the probable instrumental error.

The eccentric baroclinic westerly ring round the Arctic core is thus systematically deformed by standing waves at circumpolar numbers two and three. Wave number two appears to under-

go seasonal phase changes associated with longitudinal heating variations, whereas three seems to be orographically linked. These features are familiar in the mid-latitude westerlies, and here I wish to stress three circumstances usually overlooked: (i) that the bipolarity is primarily a feature of the Arctic jet and frontal belt; (ii) that the long standing waves of winter (but not summer) extend, without drastic phase differences, to above 30 km, and probably to above 50 km; and (iii) that the bipolarity is also present in the physical and biotic distributions at the air-land and air-sea interfaces and below.

The Antarctic continent sits fairly and squarely in its pole, and the asymmetry and eccentricity of the Arctic distributions have no Southern Hemisphere equivalents.

(d) Synoptic scale motion   I have mentioned already the existence within the Arctic core of drifting cold lows, each with a strong tropopause depression and baroclinic ring round an extremely cold troposphere and warm lower stratosphere. There are also many cases where anticyclones and frontal cyclones cross the inner Arctic. The climatology of these systems has been treated for winter by Keegan (1958) and for summer by Reed and Kunkel (1960). Their studies confirm the impression

left by daily analysis that conditions within the mean Arctic core are by no means undisturbed, and that synoptic-scale motion systems other than cold lows commonly affect the core region, some of them not migrants from mid-latitudes. For the most part they are Arctic front disturbances during periods of extreme eccentricity of the westerly vortex.

In the late 1950's the Arctic Meteorology Research Group devoted considerable attention to the generalized behaviour of the low-level flow over the Arctic. Using a polynomial specification technique that made inter-map correlation easy, Wilson (1958, 1967) was able to show that well-defined periods of stability of the large-scale motion occurred over the Arctic, especially between December and May. These stable periods averaged six days in length, and "were separated by periods of abrupt change, when the circulation seemed to flip from one pattern of stability to another" (1967, p. 55). No classification of the stable patterns of large-scale motion could be attempted, nor did she find any simple relationship between the 'flips' and reorganizations of the westerly belt. It is my impression that there is a distinct synoptic régime in high latitudes that cannot simply be treated as the shadow of events further south.

3. The Arctic and Sub-Arctic Land Areas

(a) *Vegetation zones* The limit of the Arctic over land areas is normally taken to be the Arctic tree-line. Historically this line has been assumed to follow the 10°C isotherm for the warmest month, or a similar line defined by Nordenskjöld to allow for variations in the severity of winter (Hare 1951). It is highly probable that the northern limits of forest growth and of individual tree growth are thermally determined, but ecologists cannot yet say precisely in what way. It is usual to define the vegetation zones shown in Table 1, working southward from the Arctic coasts of Eurasia and North America.

Figs. 8 to 11, taken in Labrador-Ungava, illustrate the nature of zones I, III and IV. Similar coniferous forest or woodland types encircle the hemisphere from Alaska to Labrador and from Norway to the Pacific coast of

Table 1
Northern Vegetation Zones

Zone	Limit	Character
I		Tundra (treeless, open landscape)
	(i)	...... *the Arctic tree-line*
II		Forest-Tundra (tundra intermingled with mainly coniferous forest stands)
	(ii)	...... *the northern forest-line*
III		Boreal Woodland zone (open mainly coniferous woodland; extensive lichen floors)
	(iii)	...... *forest/woodland boundary*
IV		Boreal Forest (closed-crown mainly coniferous forest)
	(iv)	...... *southern Boreal limit*
V		Mid-Latitude Forests and Grasslands

(Hare 1954; Hustich 1966)

Siberia. On Figs. 1, 2, 3 and 5, I have drawn in the Arctic tree-line and the southern limit of the forest-tundra (following Tikhomirov 1963; Hustich 1966); zone II between them is taken here as effectively delimiting the Arctic land areas.

In early papers on this problem (Hare 1950, 1954) I showed that the Arctic tree-line and the northern forest-line (limits (i) and (ii) in Table 1), defining zone II, followed isolines of certain values of Thornwaite's potential evapotranspiration, a function of summer temperature roughly correlated with net radiation for the growing season. This is reasonable, since survival of woody species along the tree-line depends (a) on an adequate energy supply for photo-synthesis and transpiration; (b) on radiation and humidity conditions during fruiting and seeding; (c) on the intensity of winter tissue-damage, especially by desiccation; and (d) on non-climatic factors not discussed here. The fact that at least one or other of the sub-Arctic species of *Abies, Larix, Picea* and *Pinus* reaches roughly the same thermal limit all round the pole indicates that the arctic tree-line and forest-line are in some kind of equilibrium with present-day climate. In several areas it has been proved that the forest-limit advanced into what is now the tundra or forest-tundra during the climatic optimum and again in the 'little optimum.' Thus Bryson, Irving and Larsen (1965) discussed evidence for two advances of the northern forest-line of the order of 100 to 150 km west of Hudson Bay in about 3500 B.P. and 900 B.P. Andreev (1956) demonstrated a comparable post-glacial advance of the Arctic tree-line between the White Sea and the Urals, and a subsequent retreat. It is thus reasonable to assume that these lines are genuinely in climatic equilibrium, in that they have clearly been able to follow the thermal isolines about the map during the post-glacial climatic variations.

What is striking about their present position is that they appear to follow the contours of the 500 mb chart for July (Fig. 2). Basing himself on the surface wind confluence, and on its correspondence with Reed and Kunkel's line of maximum frontal activity in summer, Bryson (1966) decided that "the northern edge of the boreal forest or the southern edge of the tundra is associated with the mean position or modal position of the Arctic front in summer" (Fig. 5). He showed further that the southern Boreal limit (limit (iv) in Table 1) is very close to the winter position of the same front. He thus arrived at the remarkable result, which I accept, that the Boreal Forest formation in North America corresponds to the zone over which the Arctic front oscillates in the mean during the year. Since the alignment of this front must on the average be parallel to the 500 mb contours, the correspondence of limits (i) and (ii) to the July 500 mb contours is comprehensible, as is their bipolar shape.

If this relationship is valid, post-glacial variations in the Arctic tree-line and northern forest-line, which can be established by palynological and other non-climatological techniques, offer a clue to the standing circulation of the corresponding epochs. Does the relationship depend, as Bryson suggests, on the fact that forest vegetation responds to a characteristic *assemblage* of climatic values, and that these are best defined by the traditional air mass concept? If so, the position of the Arctic front can be expected to correspond to some vegetational boundary. Or does the front lie in its chosen position because of differences in albedo, evaporating potential and aerodynamic roughness between the tundra and the forest? I cannot answer the question categorically, but incline to the view that both biotic limits and the circulation reflect overriding radiative controls. Both are effects, that is to say, rather than cause and effect.

(b) *Land ice* The residual land ice of the Northern Hemisphere is shown in Fig. 1, below the mean January topography of the 500-mb surface. The large ice-sheet covering Greenland was formerly matched by three other major continental glaciers—those of North America, centred over sub-Arctic Canada and the western Cordillera of North America, and that of Fennoscandia centred over the northern Baltic region. There were smaller ice-masses associated with the upland areas of central and eastern Siberia. All but a few relic plateau and alpine cirque glaciers have vanished, leaving Greenland unmatched in the hemisphere, though dwarfed by the Antarctic glacier.

Alaska, Mackenzie and Keewatin Districts (Canada) and much of Siberia, together with much of Fennoscandia, lie in similar latitudes to the very active southern half of the Greenland glacier. The problem thus arises: why are these areas unglaciated?

The centre of gravity of the remaining non-alpine glaciers is in about 45 deg W, 70 deg N. Broadly speaking they lie along the northern fringe of the Atlantic, and it is easy to relate them simply to the prevailing paths of mid-latitude cyclones involving moist airstreams from the Atlantic. The Greenland, Baffin, Ellesmere, Iceland and Svalbard glaciers also lie along and east of the trough-line of wave-number two in the westerly current (Fig. 12), in longitudes where there is strongly diffluent flow, and hence frequent cyclonic deepening. The unglaciated areas of Canada and Alaska (heavily glaciated in Wisconsin times, and until 8000 years B.P., after which retreat was rapid) lie in the confluent zone west of the major trough. From Hudson Bay westwards the principal source of precipitation is now the stream of weak cyclones travelling along the Arctic frontal zone.

The position of this long-wave trough might be thought to account for the survival of the existing glaciers. East of it, Greenland has kept

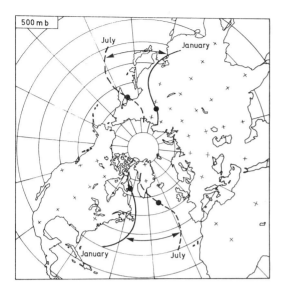

Fig. 12 Trough-line of wave number two, January and July, showing phase-shifts. Maximum amplitude is indicated by black circles on trough-lines.

its glaciers. Immediately west of it Baffin (which is physiographically similar) has been almost completely deglaciated in the past 7,000 years (Andrews and Barry 1967), and at the time of the post-glacial thermal maximum, 3500 to 5000 years B.P., may have been near to total deglaciation. Yet north-east Siberia and western Alaska, in a comparable relation to the complementary trough, are unglaciated, and have been little affected in the past. It is possible that the main effects of the trough in promoting snowfall are felt over the Bering Sea, and that glaciation is thus avoided. It is true, of course, that the eastern hemisphere trough lies over or very near the Asiatic land-mass, and that the lack of moisture in the zone of deepening may be the critical factor.

The glacial ice of today must be largely in dependent relationship to the radiative régime in high latitudes and the present topography of the circumpolar Arctic circulation. In Wisconsin times (before 8000 B.P.) the ice was so much more extensive that it must itself have heavily influenced, via radiative processes, the thermal régime of the entire hemisphere, and must also have affected the long-wave structure

of the westerlies — including those of mid-latitudes — by orographic barrier effects. The existing glaciers are probably too small to produce either effect: they are creatures of the present climate rather that its creators. It is hence difficult, and probably misleading, to argue back naively from present climate to Wisconsin conditions.

Equally passive in its relation to climate is the distribution of permanently frozen ground (permafrost). Fig. 1 shows the estimated southern limit of continuous permafrost (following Stearns 1966, Brown 1967, 1968; Ferrians 1965). In addition discontinuous or sporadic permafrost occurs in a belt 500 to 1,500 km south of this line. Permafrost also occurs in fresh-water filled rock beneath coastal seas, and probably below most glaciers. In high latitude land areas its thickness may exceed 500 metres. The southern limit of continuous permafrost is usually taken as corresponding to the isotherm of —5°C in the permafrost below the zone of annual temperature variation. According to Brown (1967) this temperature is roughly 3°C *above* mean annual air temperature. The limit shown in Fig. 1 is thus near a mean annual air temperature of —8°C.

Between them the areas of glacial ice and permafrost cover fully a quarter of the land area of the Earth and more than a third of that of the Northern Hemisphere (Stearns 1966). Both areas are slowly changing, since there has been some marginal glacial recession and probably melting of permafrost in the past century. But these are slow processes involving little energy. In any period of more rapid warming of the hemisphere the glaciers and permafrost must constitute a formidable heat sink retarding the process.

4. The Arctic Seas and The Energy Balance

(a) The general problem The present-day climates of the Arctic and sub-Arctic depend critically on the energy balance over the Arctic Ocean, which is largely an ice surface with attendant high albedos and other special properties. In winter this frozen area extends more deeply into the interior of Asia and America, and Greenland extends it at a high altitude

Fig. 8   Forest-tundra (zone II) near Schefferville, Quebec province, Canada (near 56 N), showing inter-mingled tundra (foreground) and woodland.

Fig. 9   Boreal woodland (zone III), showing scattered spruce (Picea) *trees in a carpet of lichen (chiefly* Cladonia *spp.) and a broken layer of shrubs (chiefly* Betula *spp.), Quebec province, Canada, about 58 N (photo by R. N. Drummond).*

throughout the year. If the Arctic Ocean lacked its permanent ice-cover — and this is conceivable within the present day radiative régime of the Earth as a whole — the entire hemispheric energy balance, and hence general circulation, would be drastically altered. I suspect also that the tropospheric Arctic core I have previously discussed is related in origin to the persistence of the permanent pack-ice of the Arctic Ocean and the Greenland glacier. Clearly then, these elements have a strong bearing on the problem of climatic change, and on the causes of the present day climate.

In many ways, given the data, the Arctic Ocean forms an ideal site for a thorough-going analysis of the energy balance, since the sea is largely enclosed, and it is possible to make some estimate of the role of the ocean circulation and heat transport. Until recently the necessary data were lacking, but the past fifteen years have provided long radiosonde runs; much oceanographic data; extensive observational runs and experimental research from drifting stations on the pack-ice or on ice-islands, and on coastal and island sites; and extensive reconnaissance over the pack-ice by aircraft. Even so, it has only just become possible to make estimates of the overall energy balance of the ocean-ice-atmosphere system.

(b) The seas and their ice cover The Polar Ocean proper, which is the Arctic Ocean excluding the Norwegian and Barents Seas, has a thick permanent pack-ice cover, with only infrequent and short-lived patches of open water misleadingly called *leads* in this paper. This pack-ice dominates the energy exchanges and latitudinal temperature gradient. In Untersteiner's words (1966), the ice cover "suppresses wind stress and wind mixing, reflects a large proportion of the incoming shortwave radiation, limits surface temperature, and impedes evaporation; its changes of phase are heat buffers, and the transfer of heat is essentially reduced to molecular conduction."

Fig. 13 shows the surface and subsurface currents schematically, and Fig. 14 shows how the various manned stations on the pack-ice or on ice islands have drifted (Sater *et al.* 1963). The slow anticyclonic surface motion is clearly indicated. Figs. 1-2 also show the

variation of normal ice-cover between January and July. The annual average extent of this ice is about 10^7 km², and the thickness (pressure-ridges apart) of the old polar ice of the Polar Ocean is 3 to 4 metres; pressure-ridged ice, covering perhaps 15 per cent of the Polar Ocean, may attain 18 to 30 metres (Wittman and Schule 1966). The extent of open water in the Polar Ocean is uncertain, estimates varying from less than 2 to 12 percent, with a seasonal variation still imperfectly understood. The new ice added annually to the permanent pack amounts to about 0.5 m (Untersteiner 1964), equivalent to 4×10^{18} g. A further 8×10^{18} g is formed annually in the marginal areas of the Arctic, including the enormous winter-ice masses of Hudson Bay, Foxe Basin and Baffin Bay (Hare and Montgomery 1949; Coachman 1966).

As Fig. 13 implies, a large volume of warm Atlantic water enters the Norwegian and Barents Seas, mainly from the Faroes-Shetland channel. Estimates of the volume vary, and have been reviewed by Mosby (1962) and Vowinckel and Orvig (1962); a figure between 95,000 and 152,000 km³ yr⁻¹ is probable. A comparable volume of water, still largely of Atlantic origin, flows from the Norwegian Sea under the Arctic water and its associated pack-ice, forming a distinct warm layer between 150 and 900 m depth. This layer is the main source of the annual net upward heat-flux entering the base of the pack-ice and the leads (though much of the seasonal flux is due to the release of locally absorbed radiation).

(c) The energy balance of the Arctic Ocean area The calculation of the energy balance is exceedingly complex: it involves the estimation of the mass budget of the sea, and of the heat transport across the Ocean's conventional boundaries, including warm stream run-off from land areas; the corresponding transport by the atmosphere; the characteristics of the ice-cover, and its extent; and the radiative balance at the surface, in the troposphere and at the effective upper limit of the atmosphere. The radiative balance in turn requires adequate temperatures and humidity soundings, cloudiness observations, and as much radiation data as possible. Russian initiatives in these areas

Fig. 10   Close-up of lichen floor of zone III, showing thick layer of non-vascular plants covering soil. The basal layer remains heavily waterlogged, and evaporation comes also entirely from this layer (photo by June S. Carroll).

Fig. 11   Boreal forest (zone IV) showing dense stand of well-developed spruce (Picea spp.) *balsam fir* (Abies balsamea) *and white birch* (Betula papyrifera), *with floor entirely shaded.*

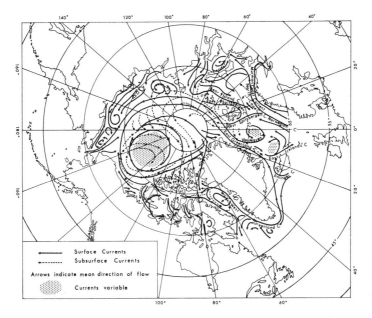

Fig. 13   Surface and subsurface currents of the Arctic Ocean and adjacent seas (after Sater et al. 1963).

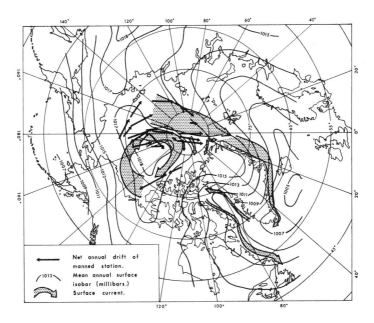

Fig. 14   Drift of manned pack-ice or ice-island stations since 1952. Shaded areas show heavily generalized lines of maximum ice movement. Sea-level mean annual atmospheric pressure is also shown (after Sater et al. 1963).

have been formidable: the best review of the radiative observations at the surface is by Marshunova and Chernigovskiy (1966), and of the nature of the heat exchange at ice surfaces by Doronin (1966). Untersteiner (1964) and Badgley (1966) have conducted the most significant experimental studies on the drifting ice, based on prolonged studies by the University of Washington group.

Estimates of the energy balance have been made by Fletcher (1965, 1966), by Gavrilova (1963) and by a group at McGill University under Svenn Orvig and Eberhart Vowinckel (Vowinckel 1962, 1964; Vowinckel and Orvig 1964, 1965a, 1965b, 1966; Vowinckel and Taylor 1965). The last series of studies was carried out by my own colleagues in the Arctic Meteorology Research Group, and I shall use it here, because it deserves to be more widely known, though I did none of the work involved. The results are comprehensive, covering monthly budgets over the central polar ocean and the marginal seas of the Arctic Ocean. There is some conflict between the calculated results and observations (though

the agreement is in general good). Considerable uncertainty is introduced by circumstances special to the Arctic, as yet unquantifiable:

(i) the prevalence of ice haze, thought to be important in the interception of solar radiation, and the absorption of terrestrial radiation (Fletcher 1966);

(ii) multiple diffuse reflection of the solar beam between snow and ice surfaces and stratus cloud leading to anomalously high isolation totals and to considerable absorption in the stratus (Marshunova and Chernigovskiy 1966);

(iii) the vital importance of open leads in the ice in releasing sensible and latent heat and infra-red radiation to the atmosphere in the cooler season at a rate believed to be two order of magnitudes greater than from old floes in the pack-ice (Badgley 1966) combined with uncertainty within one order of magnitude as to the extent of the open water.

I have summarized the Vowinckel-Orvig estimates for the annual balance in Table 2, converting to mechanical energy units.

The picture that emerges is one of near balance at the surface, summer gains offsetting

TABLE 2.   Annual mean heat exchanges over the Arctic Ocean

Sea	Layer	Heat exchange process (positive for gain) (kilojoule cm^{-2} yr^{-1})							
		1	2	3	4	5	6	7	8
Polar Ocean	Atmosphere	+ 79	+119	−527	+272	+ 27	+ 16	+ 13	—
	Surface	+130	−119	—	—	—	− 16	− 13	+ 19
	Net surface radiation: + 11 Net loss to space: − 318								
Norwegian and Barents Seas	Atmosphere	+ 90	+163	−579	+127	+ 15	+ 69	+115	—
	Surface	+211	−163	—	—	—	− 69	−115	+135
	Net surface radiation: + 48 Net loss to space: − 278								

Key to process

1. Absorbed solar radiation.
2. Net infra-red flux, surface to atmosphere.
3. Infra-red flux, atmosphere to space.
4. Atmosphere advection, sensible heat.
5. Atmosphere advection, latent heat.
6. Turbulent flux of sensible heat.
7. Turbulent flux of latent heat.
8. Transfer from ocean and ice to surface.

Adapted from Vowinckel and Orvig 1966

winter losses in what is basically a low-intensity exchange, thanks to the damping influence of the ice, less noticeable in the Norwegian and Barents sea areas than in the Polar Ocean. The small surplus of net annual radiation at the surface is not confirmed over the Polar Ocean by Marshunova and Chernigovskiy (1966) who find annual net radiation deficits ranging from 8 to 14 kj cm^{-2}yr^{-1}. Gavrilova (1963) gives a surplus of about 8 kj cm^{-2}yr^{-1} in the Siberian-Alaskan sector of the Ocean. Badgley (1966) reports near zero annual values over a mature ice-flow and a deficit of 23 kj cm^{-2}yr^{-1} over a lead, both in 80 deg N. All these values, however, are small by comparison with the infra-red exchanges between ocean, ice and atmosphere (concealed in column 2 of Table 2) which are fluxes of the order of 700 to 1,000 kj cm $^{-2}$yr^{-1}. And all are small by comparison with the flux of infra-red from atmosphere to space (including in Table 2, for simplicity's sake, direct fluxes from the surface to space). The dominant energy exchange in Arctic latitudes consists of transport towards the pole by the large-scale horizontal eddies of the atmosphere and loss to space from the troposphere. The surface — even the ocean — plays a much smaller rôle.

Obviously the exchanges shown in Table 2 would be quite different in the absence of the ice-cover. Can we reasonably speculate about its disappearance? Using the conservation equations for the ice-cover, Budyko (1966) estimated that a summer temperature anomaly of +4°C would destroy 4 m of pack-ice in four years, and that subsequently only winter ice would form. An anomaly of +2°C would achieve the same end, but would take much longer. Using a more sophisticated model incorporating meridional temperature gradients and the eddy heat-fluxes he concluded that under 'ice-out' conditions the temperatures of both water and air should remain above 0°C throughout the year. He found that in 70 deg N mid-summer temperature should rise by 2.5°C, and mid-winter temperature by 15°C. Clearly such anomalies, if realized, could readily account for the northward migration of the tree-line (discussed in Section 3 (a)) by 100 to 150 km at the time of post-glacial thermal maximum — when the Arctic Ocean may well have been ice-free on Budyko's arguments.

Similar calculations were made by Fletcher (1965, 1966) and Vowinckel and Orvig (1967) with broadly comparable results — though Fletcher pointed out that an ice-out régime should not diminish cyclonic activity: the decreased meridional heat transport required for balance would have to be achieved with a lower temperature gradient, and hence the intensity of meridional exchanges would be maintained. He concluded, as others have done, that this would favour the accumulation of continental glaciers, especially near onshore coasts.

It thus appears that quite small, and possibly random, fluctuations of the present-day temperature (as in 1959, for example) could conceivably destroy the Arctic pack-ice — and reform it as readily. My own opinion is that purely random combinations of enough successive cold or warm summers are sufficiently unlikely to make external factors far more probable as causes. The instability of the pack-ice does, however, guarantee that the effects of such external factors will be excessively amplified — and this, in my view, is the key rôle of the ice.

The frustrating thing about these speculations is the absence of a sufficiently sophisticated body of climatic theory against which they can be tested. The various attempts at numerical simulation of the general circulation are advancing in this direction, and in the Soviet Union Rakipova (1966) has been able to put the theory of zonal temperature distribution into such a shape that she can present pictures of the zonal distribution corresponding to the ice-out régime up to 20 km, and can offer quantitative estimates of the decrease in meridional heat-flux — 27 per cent in the colder half-year — and of the zonal index (about 22 per cent less in the warmer half-year, 42 per cent less in the colder). But much more is required. The entire inner Arctic — atmosphere, ice, ocean — is a single system, full of feedbacks and non-linearities (like the dependence of ocean and ice drifts and heat transports on wind, and of wind on temperature gradient). We are obviously a long way off a general theory for this system — so it still pays to speculate.

5. Conclusion

I have discussed a large number of different high-latitude distributions and processes. Some are largely the passive creations of present conditions, others (like the pack-ice) are active controls of those conditions. All are inter-connected, and all are related to climate. The Arctic means a climate, a region, a particular kind of vegetation, a particular fauna, or a particular aboriginal culture, according to one's discipline and interests. A great deal is lost, however, if we let specialist training and outlook blind us to the inter-connections.

The climatological view of the Arctic absolutely demands the interdisciplinary approach. The oceanography of the Arctic Ocean and its neighbouring seas and the glaciology of the land ice (including the geology and geophysics of permafrost) are central to the causes and stability of past and present climates. The ecology of the northern vegetation must also be considered, since the different zones have widely different albedos and aerodynamic roughness; Bryson's demonstration of the link between the Boreal Forest and the Arctic front is a striking instance of an interconnection. It has been the palaeoecology of the northern vegetation, moreover, that has given us most of our evidence of past climates.

Soviet scientists realized these interconnections many years ago, largely because of their early involvement in the problems of Arctic navigation along the Siberian coast, and in the economic development of Siberia. Such bodies as the Arctic and Antarctic Scientific Research Institute and the A. I. Voeikov Main Geophysical Observatory in Leningrad have pioneered in the broader approach to northern climatology. This has also been an article of faith for the Arctic Institute of North America and the Defence Research Board of Canada. And it is surely one of the ironies of the cold war that the most important single fusion of Russian and North American achievements in this field should have been sponsored by the Rand Corporation of Santa Monica (Rand Corporation 1966). Perhaps this implies that Arctic science is one of those areas, like meteorology itself, where national rivalries are made nonsense by the unity of nature.

Thus far, we have noted the greatly increased knowledge of the Arctic concerning macroclimatological research of a dynamic and physical nature. However, as stated by McKay (1970) "Required, therefore, is an understanding of the base line micro- and mesoclimates . . . and also an understanding of how and in what degree man and nature may upset these microclimates." Consequently, energy exchange processes on a micro, meso or macro scale must be studied to facilitate complete climatic analysis of the natural and disturbed tundra surface. Such data are limited at present but some ideas of the range of micro-studies is obtained from reference to publications by Vowinckel (1966), Ahrnsbrak (1968), Holmgren (1971), Hannell (1972), Brazel (1972), and Rouse and Stewart (1972).

Holmgren (1971) undertook very comprehensive research on the energy exchange on the Devon Island Ice Cap which will be referred to in the following snow and ice section. Ahrnsbrak (1968) has completed the most detailed energy budget studies of the Canadian tundra, at three sites west of Churchill at about 66° north, 88° to 101° west. The summertime climate of the region is shown to be one in which net radiation is nearly balanced by sensible heat transfer (apart from Snowbunting Lake) with the conduction of heat in the soil accounting for approximately 15 percent of the energy balance.

Energy Budget Values Summer 1966 (Source Ahrnsbrak 1968)

(Values in 1y day^{-1})

Location	Net Radiation	Sensible heat flux	Heat storage in soil	Latent heat flux
Pelly Lake	376.7	316.3	55.0	5.4
Snowbunting Lake	322.5	127.8[1]	100.0	104.7[2]
Curtis Lake	256.3	256.6	0.0	−0.3

[1]Low value due partly to soil being considerably wetter than other sites, with greater downward heat conduction.

[2]High value due to greater availability of water for evaporation than other sites.

The above table indicates the energy budget values and fluctuations over the three sites. Commenting on these flux patterns, Ahrnsbrak stated:

> In contrast, the July energy budget at Resolute Bay (Vowinckel 1966) is one in which the net radiation of 181 ly day^{-1} is nearly balanced by the heat storage in soil, with the transfer of sensible and latent heat accounting for only 10 percent of the energy budget. By August, however, the sign of the storage term reverses and heat exchange with the atmosphere is balanced by net radiation (96 ly day^{-1}) and heat from the soil (4 ly day^{-1}).*

Brazel (1972) has attempted to measure energy balance variations over the terrain "to suggest whether topoclimatic variables—such as slope, aspect, surface composition, subsurface properties and elevation—generate significant variations of individual components within the energy balance across alpine–periglacial terrain." The major results are as follows:

Phase–amplitude relations between global radiation and recorded surface temperatures across the terrain are highly correlated for clear sky conditions. Mean daily global radiation totals are not significantly different for slope angles ranging from 0° − 34°.

The major spatial variations of net radiation occur between bare scree and tundra surfaces.

There is very little heat storage in the active layer above the frozen ground, as the majority of heat is consumed in the thawing of frozen ground.

There is very little spatial variation in the subsurface heat flux even though the subsurface thermal properties are considerably different. This is because of the offsetting effect generated by variations in net global radiation.

Few cases of significant spatial variations in sensible heat exchange occurred, other than between bare scree and vegetated surfaces on rare occasions.

Spatial variations in the latent heat exchange were associated with differences in surface cover rather than slope characteristics.

There was a strong relationship between net radiation and evaporation, which consumed most of the net radiant energy available over the wet tundra interface.

To conclude, mean daily spatial variations in energy balance components at a high latitude Alpine pass appear to be generated by different surface covers, more than by changes in slope and aspect. Diurnal phase–amplitude relations of various components are governed, however, by site slope and aspect characteristics.

---

* Reproduced in part from *Tech. Rept. 37*, University of Wisconsin, 1968, "Summer Radiation Balance and Energy Budget of the Canadian Tundra" by W. Ahrnsbrak with the permission of the publisher.

Brazel's paper has been summarized at length since the program of work represents a valuable approach to tundra microclimatology, where natural and man-made surface variations are so abundant.

Rouse and Stewart (1972) applied energy budget calculations and equilibrium model estimates to the evaporation from a lichen-dominated upland site in the Hudson Bay lowlands. The energy budget data (using the solution presented by Bowen, 1926) indicated that the lichen surface is relatively resistant to evaporation with an average of only 54 percent of the daily net radiation being utilized in the evaporative process, compared to sensible heat and soil heat flux values of 39 percent and 7 percent of net radiation respectively. Equilibrium model estimates of evaporation (based on Slatyer and McIlroy 1961, and Monteith 1965) were derived from a comparison of actual and equilibrium evaporation, which required inputs of net radiation, soil heat flow, and screen temperatures. Tests of the model indicated that it will predict actual evaporation within 5 percent, for both hourly and daily totals, and that it can probably be applied to any high-latitude surface which exhibits a relatively large diffusive resistance to evaporation.

Future development of the Arctic will be influenced quite considerably by the existence of permafrost, the perennially frozen ground that covers up to 50 percent of Canada's total land surface of 6 million km^2. The economic exploitation of the permafrost zone is accelerating as the utilization of northern natural resources becomes increasingly profitable, e.g., oil and natural gas research in the Arctic in 1969 cost over $70 million, compared with a mere $5 million in 1958 (Crawford and Johnston 1971). Consequently the rate of surface disturbance will increase dramatically in the years ahead with the construction of buildings and transportation facilities.

Permafrost is particularly sensitive to the energy balance changes associated with man-induced disturbance in the area, in the form of clearing soil, disrupting drainage, or the erection of structures. For example, a gulley 7 m wide and 2.4 m deep developed over a four-year period following the bulldozing of a seismic line west of the Mackenzie delta (Crawford and Johnston 1971). Consequently, microclimatological research in the area is vital if we wish to minimize the detrimental effects of construction operations on the very fragile permafrost environment. For example, it has been estimated by Lachenbruch (1970) that a 1.2 m pipeline buried 1.8 m in permafrost and carrying oil at an operating temperature of 80°C would thaw the ground to a depth of more than 9 m within five years.

The distribution of permafrost in Canada was first mapped by Nikiforoff in 1928 (Brown 1960), where the southern limit (coinciding approximately with the −2°C isotherm) was represented by a single line extending from the central Yukon to James Bay to the Labrador Sea. The first Canadian map of permafrost distribution was produced by Jenness (1949), where the mapping was based on field observations of continuous and sporadic permafrost. For example, at Aishihik, Yukon Territory, the thickness of discontinuous permafrost varied between 15 and 30 m and the mean annual temperature equalled −4.2°C. In 1950, Black mapped the Canadian permafrost in terms of Russian ground temperature criteria, based on assumed temperatures which were below −5°C for continuous permafrost and between −5°C and −1°C for discontinuous permafrost at a depth of 10-15 m (the vicinity of the depth of zero annual amplitude.) In the light of present knowledge, the southern limit of the continuous zone was drawn too far north (particularly in the eastern Arctic) and the southernmost limit was located too far south (Brown 1960). However, despite the boundary inaccuracies, the

Black zonation was the first distribution of permafrost in Canada based on ground temperatures.

Since 1953, the Division of Building Research of the National Research Council of Canada has been investigating the distribution of permafrost in Canada, based on direct field observations. Since the mid 1950s, McGill University personnel have operated a research laboratory at Schefferville, Quebec, and have undertaken research on the permafrost of central Labrador–Ungava (Bird 1964).

The major problem in permafrost research in Canada has been associated with the climatic controls of perennially frozen ground. For example, in terms of air temperature the southern boundary of the discontinuous zone approximates the −1.1°C isotherm, although near the Great Slave Lake and Ungava the boundary is close to −4°C. Consequently, the poor correlation between mean annual temperature and permafrost forced Brown (1960) to relate permafrost formation to freezing and thawing indices. This partly accounted for the deviation of the boundary in the vicinity of the Great Slave Lake outlined above since continental extremes give the area a high freezing index, which is counteracted by a high thawing index and prohibits the formation of permafrost in northern Alberta (compared with its occurrence in northwestern Ontario, some 7° latitude further south).

The heat exchange mechanisms at the ground surface are too complex and variable to establish a simple practical relationship between ground and air temperatures. Snow cover, for example, insulates the ground from chilling, and the greater snow accumulation on the east side of Hudson Bay in October to December may help to explain the northward extension of both permafrost boundaries in crossing the bay (Brown 1960, Fig. 6). Other factors complicate the ground and air temperature relationships, for example, surface colour and albedo, vegetation, surface moisture, and aspect. This is best demonstrated by the field observation that permafrost may exist in patches where the mean annual temperature is only 0.5°C below freezing and that it may be absent where the value is 6.7°C below freezing (Crawford and Johnston 1971). Obviously the micrometeorological parameters are the major controlling factors of permafrost formation in terms of the components of the energy balance (radiative, sensible and latent heat fluxes) and the terrain factors (soil moisture content and conductivity, slope orientation, vegetation and snow cover) which influence the energy exchange mechanisms. Unfortunately, the above energy balance data are limited in Arctic Canada and the most comprehensive study is that of Ahrnsbrak (1968) described earlier which did not relate surface energy fluxes to active layer and permafrost table fluctuations. Perhaps the most relevant data are available in the reprint of the Brown (1963) article "The Influence of Vegetation on Permafrost." The title is rather misleading with regard to the climatological content since the analysis concerns the role of vegetation in the modification of the energy regime at the earth's surface, which controls the melt of the permafrost zone. Future programs of work must be devoted to modelling energy balance and permafrost relationships, particularly over man-induced surface disturbances, so that one can predict the thermal amelioration of the permafrost following changes in the aerial environment.

INFLUENCE OF VEGETATION ON PERMAFROST*

R. J. E. Brown

In permafrost regions numerous climatic and terrain features operate singly and in combination, determining the extent, thickness, and thermal regime of the perennially frozen ground. One of the terrain features is vegetation which forms a continuous or discontinuous mantle on the ground (soil) surface and exerts direct and indirect influences on the underlying permafrost.

Vegetation has a direct influence on the permafrost by its thermal properties which determine the quantity of heat that enters and leaves the underlying ground in which the permafrost exists. Components of the energy exchange regime at the ground surface and thermal contribution of each of them to permafrost are modified by surface vegetative cover. Vegetation also exerts an indirect influence on permafrost by affecting climatic and other terrain features, which in turn have a direct influence on the permafrost.

These direct and indirect influences vary with time and space. The environment in which permafrost exists is dynamic as are the individual components of this environment. Vegetation is one of these dynamic factors and varies with time. As one type of vegetation is succeeded by another, so the underlying permafrost is changed with time. As a result, permafrost existing today reflects direct and indirect influences of the past as well as the present. The effects of vegetation also vary in space, being greater, for example, in the taiga zone than in the tundra.

The sum of these influences with their variation in time and space is manifested in the variations which exist in the permafrost. The depth from ground surface to permafrost table, the temperature regime of the ground above the permafrost and of the permafrost, and the extent and thickness of permafrost are all conditioned by vegetation on the ground surface. Because vegetation is so closely interwoven with climatic and other terrain factors affecting permafrost, it is frequently difficult or impossible to single out the role of vegetation alone.

Even within the vegetation complex itself, some components have more influence than others. It is frequently difficult to delineate the boundary between living vegetation and underlying organic matter, litter, humus, peat, muck, and to separate the influence of these two layers. It is more convenient, therefore, to consider the combined living and dead material lying on the mineral soil as vegetation, although each may have a different effect.

Further complications are caused by the fact that the influences of vegetation may vary depending on conditions under which they occur. As a result, it is possible for a certain combination of vegetation and other factors to produce one set of permafrost conditions at one time or place, and a different set of conditions at another time or place.

Because vegetation is such a widespread factor of the permafrost environment, a large body of literature is devoted to this topic in North America and the USSR. Russian literature is particularly voluminous, the most prominent authors being B. A. Tikhomirov and A. P. Tyrtikov. (The latter has compiled two lengthy and comprehensive review papers [1, 2].) Work has also been carried out in Scandinavia on variations in near-surface ground temperatures under different types of plants and on use of plant types as snow depth indicators [3, 4]. Both have important implications in the relationship between vegetation and permafrost.

Present knowledge of the influence of vegetation on permafrost are reviewed. The com-

* Reprinted from *Permafrost: Proceedings of an International Conference.* Publication number 1287, National Academy of Sciences, National Research Council, Washington, D.C., 1966, by permission of the publishers. (Note: the publication cited is now out of print.)

plex nature of the problem makes it impossible to cover all facets. Canadian experience is cited when available with supplementary information derived mostly from Russian observations.

Surface Energy Exchange

The annual heat exchange equation [5, 6] at the earth's surface can be written as

$$R + LE + P + B = 0 \qquad (1)$$

where (R) is the annual radiation balance (net radiation), (LE) is heat involved in evaporation (including evapotranspiration)-condensation, (P) is heat involved in conduction-convection (turbulent heat exchange), and (B) is thermal exchange in the ground.

The contribution of each of these components to the heat transfer mechanism operating between soil and atmosphere is modified by vegetation properties at the interface of these two media which comprise the permafrost environment. Russian studies show that the heat exchange equation can be used in permafrost regions to relate the ground thermal regime to the surface energy exchange regime [7, 8].

Net Radiation and Albedo Studies undertaken in Labrador-Ungava [9], using a Kipp and Zonen solarimeter, confirm the higher albedo—solar radiation mostly 0.3-20.$_u$—of treeless lichen-covered surfaces being 13.55% in contrast to 11.67% for lichen-spruce woodland, 6.52% for spruce bog, 9.78% for muskeg, and 11.13% for a closed forest.

At Norman Wells, NWT, where permafrost is widespread, observations by the Division of Building Research, National Research Council, Canada, showed that lichen possessed higher reflectivity values than true mosses and *Sphagnum*. Nevertheless, these two plant types maintain the permafrost table at about the same level in a given area, and ground temperatures under both plant types are similar. Therefore, if lichen rejects a higher proportion of the net radiative flux than moss, this may be compensated by the moss rejecting a higher proportion than lichen of some other component of the energy exchange.

Another aspect of the radiation component of heat exchange is the influence of forest growth in reducing the intensity of solar radiation at the ground surface which decreases the heating of the ground. If, for example, radiation in an open area is 1.5 cal/sq cm/min it may be reduced to 0.01 cal/sq cm/min under a dense forest canopy. In winter the forest hinders the rate of radiation from the ground but the effect is not as pronounced because of reduced foliage [1, 2].

Conduction-Convection Direct measurement of the convective component is extremely difficult. In energy exchange studies, quantitative values can be obtained by two methods: (1) All of the other components of the heat exchange equation are measured and it is assumed that the remainder equals the convective component; (2) Bowen's ratio can be used which relates this component to the evaporative flux. Since the annual heat exchange equation at the earth's surface can be written as

$$R = LE + P \qquad (2)$$

it follows that an increase in heat transfer by evaporation decreases the amount of turbulent exchange and vice versa. Investigators in the USSR stated that the ratio of (P) to (LE) at the treeline is about 1 to 6 or 1 to 7, increasing to about 1 to 3 with increased surface roughness in the south-central part of the taiga where the permafrost boundary occurs [5]. The (P) to (LE) ratio at Point Barrow, Alaska, was 1 to 0.7 [10, 11]. Bowen's ratio for saturated *Sphagnum* in Ottawa, Canada, at the Division of Building Research, was about 1 to 8.

Variations in turbulent heat exchange with different plant species have not been examined in permafrost regions. On a microscale, the roughness factor provided by the vegetation is greatly reduced north of the treeline. On a microscale, *Sphagnum* tends to produce an uneven microrelief in the form of hummocks, mounds, and peat plateaus in contrast to areas covered with true mosses and lichen which have less microrelief. This could result in ground surface air turbulence being significantly greater over a *Sphagnum*-covered area.

Evaporation Evaporation (including evapo-transpiration) withdraws heat from the surrounding atmosphere and from incident solar radiation [12]. Vegetation draws water from the soil by transpiration, thus depleting the soil of the heat held by that water. There appear to be variations in these mechanisms from one plant species to another in permafrost regions, but the magnitude of variations and the contribution of this component to ground thermal regime are not clear.

The mechanism of moisture transfer in moss and lichen is not clearly understood, but both are nonvascular plants and cannot transpire in the sense that vascular plants do. *Sphagnum* especially, and to a lesser degree true mosses, tend to absorb and raise water in the manner of a lamp wick. Mosses and lichens have a large water-holding capacity and are strongly hygroscopic [13, 14, 15]. They absorb water not only from precipitation but also from atmospheric vapor, the latter being absorbed in direct proportion to the relative humidity of the air. Yet during a dry period they tend to lose moisture rapidly. Probably at some point in the humidity scale for the atmosphere, losses exceed the gains of moisture by the lichens.

Tyrtikov [2] postulates that as vegetation absorbs moisture from the soil there is a commensurate increase in the soil thermal conductivity. This occurs at the same time as the evaporation of the water lowers the air temperature, especially near the ground surface, and so reduces the warming of the soil. Immediately after a rainfall, the sun rapidly dries the tips of the moss but this drying extends only to a depth of about 5 cm, thus protecting the lower layer of moss from excessive loss of moisture.

When dry, lichens absorb water slowly from water vapor. This process is negligible, however, compared with the absorption of liquid water because, during a rainfall, the water content of the aerial parts may rise from 50 to 250% within a few hours [14]. The loss of water vapor to the air may occur rapidly if the difference in vapor pressure between the air and the lichen is great, so that on a drying day, the surface of the lichen becomes dry rapidly. Observations by the Division of Building

Research at Norman Wells, NWT, of the moisture content in a lichen cover during a dry period showed 8% moisture content in the top 1 in. of the lichen in contrast to nearly 200% at the bottom of the 2-in. lichen cover. Unlike *Sphagnum* and true mosses, lichen may not be acting as a wick in drawing moisture from beneath, but it appears to be protecting underlying moisture from evaporating influences of the air above. Rapid evaporation or diffusion exchange of water vapor from the wet basal layer to the atmosphere above the lichen may contribute, however, to low soil temperature and a high permafrost table.

Despite speculation that the high permafrost table under true moss, *Sphagnum,* and lichen may be caused in part by high evaporation rates that prevent large quantities of heat from entering the ground, observations at Norman Wells in the summer of 1959 showed that rates of water loss from moss and lichen were about equal to each other but lower than from a grass-like sedge, species unknown, or grass [16]. Observations in the summer of 1960 showed that rates for moss, lichen, and the sedge were about equal to or lower than for grass.

Meteorological factors play a prominent role in evaporation and transpiration rates where soil moisture is not a limiting factor. Physiological characteristics and radiation and thermal properties of plants such as moss and lichen which maintain high permafrost tables, probably contribute significantly by evaporation and transpiration to the energy exchange regime of permafrost. A discrepancy arose at Norman Wells where the sedge did not maintain a high permafrost table but displayed evapotranspiration rates comparable to rates of moss and lichen. This may have been caused by the lower insulating values of the sedge which permitted a greater depth of thaw during the summer.

Conductivity Vegetation has a marked insulating effect on underlying permafrost. This is true of mosses, lichens, and particularly, of peat. Increase in depth of thaw in permafrost areas where vegetation has been removed has been widely observed.

Variations in the thermal conductivity of peat with moisture content contribute to conditions conducive to formation of permafrost [2]. When dry, peat has a low thermal conductivity, equivalent to that of snow (about 0.00017 g cal/sec sq cm °C cm). When wet, its thermal conductivity is greatly increased (unsaturated peat is about 0.0007 g cal/sec sq cm °C cm; saturated peat—e.g., about 0.0011 g cal/sec sq cm °C cm); when frozen, its thermal conductivity is increased many times over that of dried peat and approaches the value for ice (e.g., saturated frozen peat about 0.0056 g cal/sec sq cm °C cm). During the summer a thin surface layer of dried peat, which has a low thermal conductivity, hinders heat transfer to underlying soil. During the cold part of the year, peat is saturated from the surface, and when it freezes its thermal conductivity greatly increases. Because of this, the amount of heat transferred in winter from ground to atmosphere through the frozen ice-saturated peat is greater than the amount transmitted in the opposite direction through the surface layer of dry peat and underlying wet peat in summer. A considerable portion of heat is also required during the warm period to melt the ice and to warm and evaporate the water. The net result is favorable to a permafrost condition.

The rate of organic and peat accumulation varies with the type of vegetation and influences thermal conductivity of the surface soil layer and thermal regime of the underlying permafrost. In a given climatic zone, less organic material accumulates from meadow and steppe vegetation than from forest and bog vegetation. Organic material accumulates more quickly in a coniferous forest than in a deciduous forest. Coniferous forests with their dense tree crowns and acidic litter tend to create conditions suitable for the development of mosses, particularly where a cool climate has reduced evaporation to a point where the forest floor will be moist despite a low rain fall. The rate of accumulation in a forest is related to many factors, including the presence or absence of moss and lichen cover, its species composition, and the degree of swampiness. In a forest with a surface cover of *Sphagnum,* a

peat horizon forms more quickly than in a forest with true moss.

Climatic and Terrain Features

Vegetation exerts both an indirect influence on permafrost by modifying climatic and other terrain features, which themselves influence permafrost, and a direct influence by its role in the heat transfer mechanism between ground and atmosphere. The influence of vegetation on various microclimatic features, drainage and the water regime, snow cover, and the influence of one type of vegetation on another, are important aspects. These features are so closely interrelated that it is difficult to assess their individual contributions without including some aspects of other features.

Microclimate Vegetation decreases air current velocities within its strata and therefore impedes heat radiation from the soil to the air when the latter is cooler as at night [15], and during periods of the year when soil temperatures are warmer than the air [13].

The density and height of trees influence wind velocities near the ground surface. Wind velocities are lower in areas of tall dense tree growth than where trees are stunted and scattered, or absent. Higher velocities, resulting possibly from fewer obstructions in areas of sparse tree growth cause more heat to move away from these areas per unit time than from the areas of dense growth. In the southern fringe of the permafrost region, permafrost is more commonly associated with areas of sparse or no tree growth for a number of reasons. Other factors being equal, the possibility of slightly lower air temperatures because of higher wind velocities in these areas may contribute to a ground temperature condition conducive to the existence of permafrost.

Air movement, such as the drainage of cold air at night from an elevated area downslope to a depression, is a microclimatic feature associated with relief, which may also be significant. Even microrelief features, such as peat plateaus, may produce sufficient differences in elevation to cause downslope air drainage at night.

Vegetation, especially tree growth, intercepts

a significant portion of atmospheric precipitation, both rain and snow, by as much as 10 to 40% [2]. Any rain that reaches and penetrates the ground carries heat with it toward permafrost so that interception of rain results in a reduction of heat entering the ground. On the other hand, interception of snow by trees results in a lower accumulation on the ground and the possibility of deeper seasonal frost penetration. Increased shading caused by snow on tree branches reduces the amount of solar radiation received at the ground surface, but this is counteracted partly, at least, by the reduction of radiation from ground surface into atmosphere.

Drainage   Ground that permits the greatest degree of water penetration usually thaws to the greatest depth [13]. There is evidence that extensive root systems impede downward percolation of water and therefore restrict soil thawing [13]. On the other hand, roots, especially dead and decaying ones, may provide channels for water penetration and sometimes loci for the growth of granules and small stringers of ice [13].

Vegetation impedes surface runoff. In forests, particularly where vegetation is not disturbed, runoff amounts to less than 3% of the total rainfall, whereas in open areas and on the plains, it exceeds 60% [1]. The rate of runoff to precipitation is probably also significant to permafrost. Subsurface drainage is slow in peat because of its filtration properties.

Snow Cover   The low thermal conductivity of snow and its double role as inhibitor of frost penetration during winter and soil thawing in spring has been noted by many authors [13]. The retention of snow in the crowns of trees has already been mentioned. In spring, the snow cover remains on the ground longer in forested areas than in the open. Where strong winds prevail, more snow accumulates under a vegetation cover than in open areas. Snow protects the ground from freezing in winter but it also increases the moisture content of the soil in summer, thus contributing to lower summer ground temperatures [1].

In Labrador-Ungava a good correlation was noted by Ives [17] and Annersten [18] between the vegetation and snow accumulation and the distribution of permafrost. On exposed ridge summits, where vegetation was virtually absent, snow accumulation was kept to a minimum by the wind, and permafrost 200 ft thick existed. In sheltered gullies, vegetation was better developed, snow accumulated in the winter, and permafrost was only a few feet thick or absent.

Vegetation Properties   Within the framework of the complex interrelation existing among various terrain features that effect the ground thermal regime, such as relief, drainage, soils, snow cover, and vegetation, special characteristics of some plant types may significantly influence permafrost and also indicate the existence of permafrost.

Tikhomirov [19] mentions several characteristics that influence the ground thermal regime of true mosses and *Sphagnum:* it possesses low thermal conductivity, high moisture holding capacity, and may shield roots of vascular plants from low air temperatures; it promotes uniform thawing and protects soil from runoff, solifluction, erosion, and thermokarst. Moss reduces the temperature amplitude of underlying soil. Tikhomirov postulates that heat from moss in late winter recrystallizes snow at the moss contact, that photosynthesis is possible under a thin snow cover, and that hollows form in the snow in which the air temperature is higher than the outside air temperature, producing favorable conditions for plant growth under the snow. Porsild states that solar radiation is the more likely source of heat [20]. He also questions photosynthetic activity beneath even a thin snow cover and the provision of favorable conditions for plant growth under the snow.

Robinson noted that in fall the melting of early snow fills the moss with moisture enabling it to conduct heat at a more rapid rate; this permits greater heat loss from the ground and deep penetration of seasonal frost [21]. In summer the top few inches dry to a point where they act as an effective insulating blanket. Therefore the presence of a deep layer of moss keeps the soil at a low temperature

Fig. 1 Variety of vegetation associations with related variations in permafrost occurrence.

continuously and favors development of perma-frost. Moss is very water absorbent. The lower portion of a moss layer is usually moist and this maintains the ground in a damp or wet condition.

Certain plant types provide rather reliable indicators of the existence of permafrost. At Thompson, Man., Canada, the presence of *Sphagnum,* and/or stands of spruce varying from small trees in open stands to moderately large trees in nearly closed stands, was found usually to be associated with permafrost, if the drainage was fairly good [22]. In northern Alberta, all but a few of the permafrost occur-rences were found in low flat depressions [23].

In these areas two associations of vegetation predominated. One was grass-like sedge with little or no tree growth and thin moss cover. These areas were almost always wet and no permafrost existed. The other consisted of *Sphagnum,* lichen patches, and scattered stunt-ed black spruce. Some of these areas were wet and contained no permafrost. The remainder were drier, and permafrost occurred at depths

ranging from about 2 to 4 ft. At the edges of these areas, the permafrost dropped off abruptly.

Three varieties of vegetation are shown in Fig. 1, taken Sept. 20, 1962 (57°47'N, 117°50'W), 3.2 miles west of Mackenzie Highway, Alberta, Canada, in the southern fringe of the discontinuous zone of the perma-frost region.

The light-toned vegetation in the foreground and middle ground consists of sedge (*Carex sp.*) and grass with a thin discontinuous cover of feather and other non-*Sphagnum* mosses in standing water. No permafrost was en-countered.

The dark toned island in foreground (man kneeling) is a slightly elevated peat plateau with ground cover of hummocky *Sphagnum* and Labrador tea. Depth of moss and peat is 4 ft. 2 in.; black organic silt 4 ft. 2 in. to 6 ft. 0 in.; dense bluish silt clay 6 ft. 0 in. to below 7 ft. Permafrost table extends from 2 ft. 9 in. to 7 ft. below ground surface. Permafrost occurs also in the dark almost treeless patch (same ground vegetation) at right, between sedge area in middle ground and forest in background. The forested area consists of spruce, poplar, jackpine, and birch; no permafrost was encountered here.

Vegetation Zones as a Permafrost Indicator

In Canada, Alaska, and the USSR, the influence of vegetation varies from one geobotanical zone to another. In Canada and Alaska, permafrost occurs in tundra and taiga zones. In the USSR, it occurs in these two zones and also extends southward into the steppe. The variety of vegetation associations, the quantity of vegetable matter, stand height, and density, and rate of peat accumulation are all greatest in the taiga, gradually diminishing northward into the tundra and southward into the steppe. The degree and variety of the influence of the vegetation on the permafrost changes in a parallel manner [2].

In the northern part of the tundra the vegetation has little influence on permafrost because it is sparse and the vegetative period lasts less than two months. It causes local variations in depth of thaw and helps impede erosion. The destruction of the vegetation accelerates thawing only slightly.

In the southern part of the tundra, the vegetation becomes more abundant, peat mantles part of the surface and attains thicknesses of several feet in some basins. The main influence of the vegetation is on the depth of thaw. If vegetation is removed, the depth of thaw will increase; erosion will increase and thermokarst will develop if thawing proceeds at different rates over an area or if there are local differences in the ice content of the frozen ground.

In forest tundra, vegetation mass is greater than in the tundra, and the rate of accumulation of organic material is higher. Extensive peat bogs form and water basins are encroached by vegetation and permafrost forms. Woody and brush vegetation grow which accumulate snow leading to higher permafrost temperatures than in the tundra. If the vegetation is removed, the depth of thaw increases but this is counteracted to some extent by

lower snow accumulation and a decrease in permafrost temperatures [1].

The maximum development of vegetation occurs in the taiga. Here vegetation has its greatest influence on permafrost even in very small localized areas causing variations in its extent, depth of thaw, and ground temperatures. Frequent forest fires cause variations in the occurrence and thickness of permafrost over short distances in the taiga.

Mass, density, height and influence of vegetation, and rate of accumulation of organic matter are less in the steppe than in the taiga, but depth to permafrost is greater.

Alteration of Permafrost Conditions

Changes take place in the permafrost as a result of the vegetation. The influences include the effect of vegetation on depth of thaw and depth to permafrost, the temperature regime in permafrost and the ground above, and extent and thickness of permafrost.

Depth of Thaw   The most easily observed and measured characteristic of permafrost is depth of thaw and variations in types of vegetation are often readily noticeable. Because this is so, more observations have been made on this aspect of the relation of vegetation to permafrost than any other. Despite the large number of observations reported in the literature, mechanisms operative in thawing of the active layer and permafrost and causes of variations from one type of vegetation to another are not clearly understood. One difficulty in comparing depth of thaw observations in various localities is caused by variations in climate, variations in other terrain factors closely associated with the vegetation, and by minute, but possibly significant, variations within a particular type of vegetation.

Removal of vegetation cover in a permafrost area causes an increase in the depth of thaw. At Inuvik, NWT, in the continuous permafrost zone, the maximum depth of thaw in an undisturbed moss-covered area was 2 ft. in contrast to depths of 5 to 8 ft. in areas stripped three years previously [24]. On land stripped for cultivation at Inuvik, the original maximum depth of thaw was about 2 ft. prior to disturbance (1956). By 1959 the ground thawed to a depth of 70 in. [25].

The surface cover and peat appear to have much greater influence on depth of thaw than the underbrush and trees. At Inuvik, in undisturbed areas and in areas where the underbrush and trees had been removed three years previously but with the moss cover left intact the depth of thaw was 2 ft.—similar to the depth prior to any disturbance [24]. At Norman Wells, depth of thaw measurements were recorded by the Division of Building Research in different types of vegetation from 1957 to 1959. The greatest depth of thaw occurred in the grass-like sedge area with no moss cover reaching a depth of 5 ft. 6 in. after about 3350 degree days of thawing. The next greatest depth of thaw was observed in a wooded area having a ground cover of 4 in. of moss overlying 3 in. of peat, reaching 4 ft. 6 in. after 3350 degree days of thawing. The next greatest thaw, 3 ft. 3 in., occurred in a treeless grass-like sedge and moss area with a 3-in. moss cover overlying 6 in. of peat. The shallowest depth of thaw, 2 ft., was observed in a sparsely treed area having 5 in. of moss overlying 18 in. of peat. Evidently, the depth of thaw decreased with an increase in the combined thickness of living moss and peat; the density and size of tree growth did not appear to make much difference.

Russian investigators found that of all the types of ground cover, *Sphagnum* appears to retard thawing most. In the Igarka region of the USSR, the depth of thaw in 1950 under this cover was 18 cm (7.1 in.) on 13 July, 22 cm (8.7 in.) on 3 August, and 26 cm (10.2 in.) on 2 September in contrast to 25, 31, and 35 cm (9.8, 12.2, 13.8 in.) on the same dates under true mosses consisting of *Hypnum* and other species [1, 2].

The relative influence of living ground cover and underlying peat has been investigated. It has been postulated that peat retards thawing even more than living mosses and lichens. In the arctic region of the Yenisey River in Siberia, removing the moss and lichen but leaving the peat layer resulted in an increase in the depth of thaw by 20 to 50%. After both living cover and peat were removed, depth of thaw increased by 1.5 to 2.5 times [1, 2].

Temperature Regime Just as the vegetation exerts a marked influence on the depth of thaw and the depth to permafrost, it also modified ground temperatures both in permafrost and the ground above. An increase in depth of thaw leads to an increase in ground temperature and degradation of permafrost. A decrease in depth of thaw leads eventually to an aggradation of the permafrost.

At Norman Wells, ground temperatures were measured in the thawed layer by the Division of Building Research in 1959 and 1960 to assess the effect of different types of vegetation on underlying soil temperatures during the summer. In September 1959 the mean air temperature was 41.2°F and the mean ground temperature at the 1-ft. depth for this period in various vegetation areas were: grass (no moss or peat) 40.0°F, sedge (no moss or peat) 36.5°F, grass-like sedge (3 in. true moss over 6 in. peat) 35.0°F, moss (5 in. *Sphagnum* over 18 in. peat) 32.5°F, lichen (2 in. lichen over 24 in. peat) 32.6°F.

There appeared to be a general decrease in temperature with increased combined moss and peat thickness. Temperatures were similar under moss and lichen although living moss was much thicker than lichen. The combined thickness of living cover and peat was, however, approximately equal. Ground temperatures taken in 1960 in the thawed layer showed that they were highest under the grass area, lower under the sedge, and lowest under the moss and lichen. Temperature amplitudes also decreased in the same order. Thermal resistance and damping effect of moss and lichen were shown by the fact that monthly mean air temperatures were about 10°F higher in 1960 than in 1959, but the difference in mean ground temperature at the 1-ft. depth was less than 1°F.

Ground temperature readings were also taken at Norman Wells down to the 20-ft. depth in permafrost under the sedge, moss, and lichen. Even below the 10-ft. depth, the mean temperature under the sedge for August and September 1960 was about 3°F higher than under the moss and lichen, which were about equal. Above this depth the difference was even greater. Similarly, the temperature amplitudes in the top 10 ft. were twice as much under the sedge.

There is no doubt that vegetation modifies the temperatures of the seasonally thawed layer and permafrost. In all cases, ground temperatures in summer will rise when the vegetation cover is removed. On the other hand, the effects of winter air temperatures will penetrate to greater depths than in undisturbed areas. The net effect on mean annual temperature will depend on other factors, such as snow cover.

It is difficult to compare the effects of different types of vegetation on permafrost because different types frequently grow in close association or in a mosaic, and the individual effect of each may not be readily apparent.

Extent and Thickness   A change in depth of thaw and a change in temperature of the active layer and the permafrost produced by a change in vegetation establishes a change in temperature gradient. This results in either aggradation or degradation of the permafrost. In the southern fringe of the permafrost, the removal of the vegetation may result in disappearance of the permafrost. The establishment of a moss cover may lead, perhaps, to the formation and accumulation of permafrost.

Conclusion

This paper reviews various ways in which vegetation affects permafrost. Some mechanisms add heat to the ground, others facilitate heat loss from the ground. Some add heat at one time and contribute to heat loss another time. Influences of vegetation are almost all reversible depending on the conditions under which they occur. The complexity and multifaceted effects of vegetation on permafrost often lead to a situation where under one set of conditions a plant community decreases the soil temperature and increases it under other conditions.

It is very difficult to attach absolute quantities to each facet of the vegetation influence—total them, and arrive at the resultant direction and magnitude of heat flow at a particular time, much less perform this operation for all the past influences and assess their effect on the present permafrost situation. Even if each factor could be measured quantitatively, the contribution of some is so minute that the cumulative error would be unrealistic. In addition, there are factors which are probably not even known or possible to measure.

The best solution appears to be to measure obvious effects of vegetation, such as depth of thaw, temperatures, extent and thickness of permafrost which are manifestations of the net heat gains and heat losses to the ground, and relate these to variations in environmental components. The same permafrost conditions may occur in two adjacent areas, but the combination of vegetation and other factors producing two similar sets of permafrost conditions may be quite different.

ACKNOWLEDGMENTS

The author wishes to acknowledge the helpful comments of A. E. Porsild and W. K. W. Baldwin, National Herbarium, Department of Northern Affairs and National Resources, Ottawa; the late J. A. Pihlainen, Ottawa, and W. S. Benninghoff, Department of Botany, University of Michigan. This is a contribution from the Division of Building Research, National Research Council, Canada, and is presented with permission of the director of the division.

REFERENCES

[1] A. P. Tyrtikov. "The Effect of Vegetation on Perennially Frozen Soil," Materialy K Osnovam Ucheniya O Merzlykh Zonakh Zemnoy Kory (Materials for the Basis of Study of the Frozen Zones of the Earth's Crust), Issue III, 1956, pp. 85-108.
[2] A. P. Tyrtikov. "Permafrost and Vegetation," *Fundamentals of Geocryology*, Vol. 1, Chap. 12, Acad. Sci. USSR, Moscow, 1959, pp. 399-421, (Text in Russian).
[3] E. R. Dahl. "Mountain Vegetation in South

Norway and Its Relation to the Environment," Skrift. utgitt av Vidensk.-Ak. Oslo, I, Mat. Natur. Kl., No. 3, 1956.

[4] R. Nordhagen. "Mountain Vegetation as a Record of Snow Cover Thickness and Duration," *Medd. fra Vegdirektøren,* Nos. 1, 2, 4, 1952, pp. 1-31.

[5] A. A. Grigor'ev. "Concerning Certain Geographical Relationships of Thermal and Water Exchange on the Surface of Dry Land and Means of Further Study of Materials and Energy in the Geographical Environment," *Izvest. Acad. Sci. USSR,* Geog. Ser., No. 3, 1958, pp. 17-21. (Text in Russian).

[6] M. I. Budyko. *The Heat Balance of the Earth's Surface,* U.S. Dept. of Commerce, Office of Climatology, 1958, (Transl. by N. A. Stepanova, from Gidrometeorologicheskoe izdatel'stvo, Leningrad, 1956).

[7] K. A. Sychev. "Thermal Balance of the Active Layer of Permafrost in Summer," *Problemy Arktiki I Antarktiki,* No. 1, 1959, pp. 87-93, (Text in Russian).

[8] N. A. Shpolyanskaya. "The Influence of the Heat Exchange Between the Soil and the Atmosphere on Permafrost in Transbaikalia," *Vestnik Moskovskogo Universiteta,* No. 2, 1962, pp. 37-42, (Text in Russian).

[9] I. C. Jackson. "Insolation and Albedo in Quebec-Labrador," *McGill Subarctic Res. Papers, No. 5,* Jan. 1959.

[10] J. R. Mather, C. W. Thornthwaite. "Microclimatic Investigations at Point Barrow, Alaska, 1956," *Climatology,* Drexel Inst. Tech., Lab. of Climatology, Centerton, N.J., Vol. 9, No. 1, 1956.

[11] J. R. Mather, C. W. Thornthwaite. "Microclimatic Investigations at Point Barrow, Alaska, 1957-1958," *ibid.* No. 2.

[12] R. F. Legget, H. B. Dickens, R. J. E. Brown. "Permafrost Investigations in Canada," Proc. 1st Intl. Symp. Arctic Geology, Vol. 2, *Geology of the Arctic,* Univ. Toronto Press, 1961, pp. 956-969.

[13] W. S. Benninghoff. "Interaction of Vegetation and Soil Frost Phenomena," *Arctic,* Vol. 5, No. 1, March 1952.

[14] E. M. Fraser. "The Lichen Woodlands of the Knob Lake Area of Quebec-Labrador," *McGill Subarctic Res. Papers, No. 1,* Dec. 1956.

[15] M. I. Sumbin, S. P. Kachurin, N. I. Tolstikhin, V. F. Tumel'. "General Permafrost Studies," Acad. Sci. USSR, Moscow, 1940, pp. 160-165, (Text in Russian).

[16] R. J. E. Brown. "Potential Evapotranspiration and Evaporation Observations at Norman Wells, NWT." *Proc. Hydrology Symp. No. 2 — Evaporation,* Natl. Res. Coun., March 1961, pp. 123-127.

[17] J. D. Ives. "A Pilot Project for Permafrost Investigations in Central Labrador-Ungava," *Dept. Mines and Tech. Surv., Geog. Paper No. 28,* Ottawa, Canada, 1961.

[18] L. Annersten. "Ground Temperature Measurements in the Schefferville Area," *P. Q. Proc., 1st Canadian Conf. Permafrost,* Natl. Res. Coun., Assn. Comm. Soil and Snow Mech., Tech. Memo. No. 76, Jan. 1963, pp. 215-217.

[19] B. A. Tikhomirov. "The Importance of Moss Cover in the Vegetation of the Far North," *Botanicheskiy Zhur.,* Sept. 1952, pp. 629-638, (Text in Russian).

[20] A. E. Porsild. Natl. Herbarium, Ottawa, Canada, priv. comm., 1963.

[21] J. M. Robinson. Dept. of Forestry, Canada, priv. comm., 1959.

[22] G. H. Johnston, R. J. E. Brown, D. N. Pickersgill. "Permafrost Investigations at Thompson, Manitoba, Terrain Studies," Natl. Res. Coun., *Tech. Paper No. 158,* Oct. 1963 (NRC 7568).

[23] R. J. E. Brown. "Permafrost Investigations on the Mackenzie Highway in Alberta and Mackenzie District," Natl. Res. Coun., *Tech. Paper No. 175,* June 1964 (NRC 7885).

[24] J. A. Pihlainen. "Inuvik, NWT, Engineering Site Information," Natl. Res. Coun., *Tech. Paper No. 135,* Aug. 1962 (NRC 6757).

[25] F. S. Nowosad, A. Leahey. "Soils of the Arctic and Subarctic Regions of Canada," *Agr. Inst. Rev.,* March 1960, pp. 48-50.

Chapter 9

The Climatology of
Snow and Ice

The value of climatological research in this field was admirably stated by the Department of Energy, Mines and Resources (1969) *viz,* "ice and snow form an integral part of the Canadian environment and knowledge of the occurrence, behaviour and characteristics of ice and snow is required for effective management of water resources and the exploitation of other resources."

It is obvious that glaciometeorological research is required in the future, since some 200,000 km² of the Arctic archipelago is permanently covered by glaciers, permafrost, and winter snow which have a profound influence on the area's heat balance and a great flood potential with spring melting. Also, the Canadian ice and snow cover has a considerable economic value in terms of a resource potential for irrigation in the western dry lands, hydro power, and fresh drinking water (especially in the future when the forecasted population "explosion" will place astronomical demands on such water supplies). Consequently, it is essential for the climatologist and glaciologist to measure the energy and mass balances of the ice and snow cover in order to understand fully the condition (and future trends) of this vital reserve. Finally, it should be noted that patterns of glacial regimes and sea ice conditions will dominate the future activities of mining enterprises and oil companies in the exploration and exploitation of the natural wealth of the north-lands. For example, we will require knowledge of glacier flow and conditions at the ice-rock interface since the data will be of particular importance in the Arctic where mineral exploitation could develop under or near glaciers.

Snow

Research on the Canadian snow cover has been undertaken by various departments of the federal and provincial governments.

One of these departments is the Division of Building Research of the National Research Council of Canada (N.R.C.), where the programs of work are mainly related to the effect of snow on construction (especially snow loads on roofs), problems of snow removal in cities and across highways, and the control of avalanches. This emphasis has necessitated surveys of various climatological controls of the snow pack, in terms of physical characteristics, densities, heat and water budgets. Klein (1950) initiated one of the first examinations of the physical characteristics of each layer in the winter snow pack, measuring specific gravity, hardness, and depth at a number of observation sites. However, since 1959 the Division has concentrated more on the micrometeorology of the snow cover and a considerable number of publications by Gold, Williams, and Boyd have referred to the vital energy balance controls of snow preservation. The study by Gold and Williams in 1961, "Energy balance during the snow melt period at an Ottawa site," represented pioneer work in Canada on the influence of radiative and heat fluxes on snow ablation. The net radiation was found to contribute 2305 cal/cm² and convection 785 cal/cm²; of this heat, 2290 cal/cm² were used for evaporation and 800 cal/cm² for melting. The water equivalent of the snow cover lost was 13.4 cm and of this, 3.4 cm was estimated to have evaporated. Additional micrometeorological studies over the snow pack were completed by Williams in 1959 (evaporation from snow covers); Boyd, Gold, and Williams in 1963 (radiation balance during the snow ablation period); Gold in 1963 (influence of snow cover on ground temperatures); Gold and Boyd in 1965 (annual heat and mass transfer in terms of Fourier components). The research findings of all the micrometeorological projects of the Division were concisely summarized by Gold (1968) *Micrometeorological observations of the snow and ice section Division of Building Research, National Research Council* and a reprint of the paper is included in this section to emphasize the importance of heat and water balance studies in the climatological assessment of the snow cover.

MICROMETEOROLOGICAL OBSERVATIONS OF THE SNOW AND ICE SECTION, DIVISION OF BUILDING RESEARCH, NATIONAL RESEARCH COUNCIL*

L. W. Gold

1. Introduction

Micrometeorological studies of the Snow and Ice Section of the Division of Building Research of the National Research Council have been concerned with (1) the determination of the size of the components of the heat and moisture exchange associated with snow and ice cover formation and ablation, (2) the evaluation of field techniques for measuring these quantities, and (3) the establishment of the dependence of significant characteristics of the ground thermal regime on easily observed elements of the weather. The studies have not been concerned directly with the clarification of the physical processes by which heat and moisture are exchanged between the atmosphere and snow, water or ground surfaces. Most of the measurements have been made at Ottawa for two principal reasons:

(1) to carry out the measurements close to adequate supporting services, and

(2) to determine the quantities that should be measured for given field problems and establish the techniques for making the measurements.

2. Heat Exchange at Snow and Grassed Surfaces

The first problem considered was that of determining the amount of moisture lost from a snow-cover each year by sublimation and evaporation. Observations at Ottawa using shallow pans indicated that the loss was about 0.02 cm ice per day (about 15 cal/cm² day) during the month of January (1, 2). These observations were in reasonable agreement with observations made by Sverdrup (3), de Quervain (4), and the Central Sierra Snow Laboratory (5), and indicated that an equation of the form

$$E = k_e\ U_a\ (e_a - e_s)$$

is probably adequate for practical estimates of sublimation loss from snow-covers in midwinter. In this equation,

E = rate of heat loss by sublimation,

U_a = average wind speed at height a above the ground,

e_a = vapour pressure in the air at the height a,

* Reproduced from N. R. C. 10018. "Micrometeorological observations of the Snow and Ice Section, DBR, NRC" in *Proceedings of the First Canadian Conference on Micrometeorology*, Part 1, 1967, pp. 136-144 by permission of the editor.

e_s = saturation vapour pressure over ice at the average temperature of the surface of the snow-cover.

For E in cal/cm²hr, U_a in cm/hr, e_a, e_s in mb, k_e was found to be about 9.75 x 10^{-7} cal/cm³mb.

Calculations of the evaporation loss during the spring melt period of 1959, using the energy balance equation and assuming Bowen's ratio to be valid, indicated a larger value for k_e than was obtained by pan measurements (6). This difference may be real because the snow-cover over which the observations were made was quite rough, due to the formation of penitent snow; whereas for the evaporation pan observations, melting often resulted in a surface of part snow and part water. It is quite possible, however, that the difference was due to observational errors, particularly in the value of the net radiation, which was measured with a Schulze radiometer. Observations in the spring of 1960 gave high values for evaporation as well (7). In this case, the value used for net radiation was the mean of that measured with the Schulze and with a Beckman and Whitley instrument. The observations indicate that during the spring melt period at Ottawa about 10 and 20 per cent of the water equivalent of the snow-cover is lost by evaporation, and the rate of heat loss by evaporation is about 110/cm²day.

The difficulties experienced in measuring net radiation led to a field comparison of two of the better net radiometers available; namely, a shielded type developed by the Commonwealth Scientific and Industrial Research Organization of Australia, and a ventilated Suomi type made by the Canadian Meteorological Service. A random difference was observed between the average daily values measured by the two instruments that was about equal to the average daily net radiation being observed in mid-winter (8). Calibration in the long-wave length region carried out about two years later showed the calibration constant for the CSIRO instrument to be within 1 per cent of that given by the manufacturer. The calibration constant for each side of the Suomi type instrument is dependent on the

degree of ventilation over each surface of the heat meter. When the ventilation was adjusted so that it was equal on both sides of the heat meter, the calibration constant was also found to be within 1 per cent of that supplied by the manufacturer.

The observations on the heat and water vapour exchange at the snow surface were extended into the summer over a grassed surface in order to obtain an appreciation of the size of the winter exchange relative to that in summer. The observations were conducted over a two-year period. Net radiation was measured directly; heat flow from the ground was estimated from ground temperature observations: evapotranspiration was estimated from Class A pan observations as well as from observations on the wind speed and vapour pressure difference between the surface and the two-meter level; and convection was determined as the difference in the heat balance equation. The maximum monthly average net radiation was observed to be about 280 cal/cm²day, and the heat loss by evapotranspiration about 225 cal/cm²day. During a six-month period, 1 April to 30 September, about 48 per cent of the observed net solar radiation was dissipated by evapotranspiration, 42 per cent by longwave radiation, 7 per cent by convection, and 3 per cent by conduction into the ground (9). The sublimation loss in mid-winter was about 10 per cent of that by evapotranspiration from the same site in mid-summer. The observations indicate that, on the average, there is a net loss of heat by radiation in mid-winter, and that the absolute value of this loss is about 10 per cent of the maximum monthly average daily gain in summer. As it is difficult to measure evaporation, evapotranspiration and net radiation under field conditions in mid-summer, with an accuracy of 10 per cent, it appears that the size of these components in mid-winter is about the same as the accuracy possible with current instruments and techniques.

Observations were made on the heat balance at the surface of moss contained in tanks about 4 ft. in diameter and 12 in. deep. It was found that about 90 per cent of the net radiation was dissipated by evaporation. These observations, as well as associated observations made at a bog near Ottawa, were presented to this symposium by G. P. Williams.

One of the conclusions of the heat exchange studies was that if the components of the heat and water vapour balance need to be measured with reasonable accuracy in mid-winter, it will be necessary to develop suitable equipment and techniques. Following are representative values for the Ottawa area of the size of quantities to be measured and comments pertinent to the development of instrumentation.

The average daily net radiative gain to the snow surface in the daytime in mid-winter was about 30 cal/cm², and the long-wave loss at night about 40 cal/cm². In order to measure the small difference between these two quantities with reasonable confidence it will be necessary to prove the reliability under field conditions of net radiometers available at present and, perhaps, to develop equipment with adequate accuracy. In addition, careful and continuous control of field installations and frequent checks of the calibration will be required particularly for measurements extending over long periods of time.

The vapour pressure difference between the surface and two-meter level was between 0 and 2.5 mmHg. It can probably be measured accurately enough for estimates of sublimation based on empirical formulae of the Penman type, but the accuracy and speed of response of instruments available at present for measuring water vapour density or vapour pressure impose severe limitations on the application of aerodynamic or eddy-correlation techniques for estimating mass transfer in mid-winter. This problem becomes more serious the lower the air temperature.

The rate at which heat was removed from the snow surface by convection during the daytime in mid-winter was about 3 cal/cm²hr, and the transfer rate at night to the surface was about 1.5 cal/cm²hr. Field observations indicate that the temperature gradient in the air at the two-meter level during the daytime under clear sky conditions was about 10^{-3} C deg/cm and considerably larger but of opposite sign at night. Measuring the convective heat transfer over a snow surface using the aerodynamic or eddy-correlation technique should not be any more difficult than making the corresponding observations in summer, except for the additional instrument problems introduced by the colder temperatures.

Observations made during January and February indicated that during the daytime the ratio between the convective and evaporative loss was between ½ and ⅔. It was observed also that the following correlation existed between the hourly average vapour pressure difference between the two-meter level and the surface, and the associated temperature difference

$$e_a - e_s = 1.0 + 0.2 \ (T_a - T_s).$$

(e_a and e_s in mmHg, T_a and T_s in °C). These observations were made when the average air temperature was about $-10°C$. It would be expected that the ratio between the convective and evaporative loss would increase with decreasing average air temperature.

Some observations were made of albedo using two Eppley pyrheliometers mounted back to back (9). The monthly average albedo of the snow surface was found to be between 55 and 65 per cent, that of the grassed surface between 10 and 20 per cent. There appeared to be a tendency for the albedo of the grassed surface to increase from a minimum in spring to a maximum in autumn.

Consideration was given to the possibility of advancing breakup of lakes and rivers by increasing the albedo of the cover through the application of dust (10). Calculations showed that it might be possible to advance break-up by about 2 weeks in southern Canada and by about 4 weeks in the Arctic. Field trials showed, however, that in much of the south of Canada the variable weather conditions that exist during the spring thaw would allow probably less than a 50 − 50 chance for dusting to be successful.

Although preliminary calculations showed that the smaller the grain size of the dust the smaller the amount that must be applied to obtain adequate coverage, the field trials indicated that economic considerations make grain size a secondary consideration. The optical properties of the dust appear to be a secondary factor, as well, as long as the dust is dark. Once the dust has initiated melting and is submerged in water, the albedo of the surface is determined primarily by the water. Observations made at Inuvik suggest that clearing away the snow so as to expose the ice may be almost as effective as dusting, from the practical point of view, for advancing melting. It is considered that sufficient study has now been made of this technique to allow a realistic appraisal of the advantages to be gained through applying it in given areas from a study of meteorological records.

3. Ground Thermal Regime

Concurrent with the foregoing observations on the heat and moisture exchange at grassed and snow surfaces, observations were made at the same site on ground temperature and the influence of snow-cover on heat flow from the ground. The heat flow from the ground was directly proportional to the difference between the average shielded air temperature and the ground surface temperature, and inversely proportional to the average depth of the snow-cover (11). The observations showed that the snow-cover could be assumed to have a thermal conductivity equal to that for snow of density equal to the average density of the cover. The maximum monthly average heat flow from the ground was about 20 cal/cm² day, and varied approximately sinusoidally throughout the year. For the temperature conditions that exist at Ottawa, a snow-cover 2 ft. deep would normally allow a depth of frost penetration into the ground of less than 6 in.

The ground temperature observations demonstrated that the clay in which the measurements were made satisfied quite well the conditions for simple heat transfer in a semi-infinite medium for temperature disturbances of period greater than about one-half year (12).

During the period 23 December 1956 to 31 March 1964, the annual average ground temperature at the observation site, which was grass-covered in summer, was between 1.25 and 3.25 C deg warmer than the annual average air temperature (13). Snow-cover was found to be the principal reason for this differ-

ence. The ground surface temperature under the snow-cover was quite constant during winter and between 5.5 and 11.5 C deg warmer than the minimum monthly average air temperature. On the other hand, the surface temperature at the base of the grass stems was only between 1 and 2 C deg warmer than the monthly average air temperature at the time of maximum in summer. It was observed that the monthly average surface temperature correlated well with the monthly average air temperature during spring, summer and fall.

Ground temperature was measured under two parking lots, one cleared of snow in winter, the other not, and was compared with those observed under the grassed site (14). This comparison demonstrated not only the influence of snow-cover on the ground thermal regime but also the influence of the size of the convective component. The surface of both parking lots was gravel-covered so that it was reasonably well drained and without vegetation. The average monthly surface temperature of the two lots at the time of maximum in mid-summer was about 7 C deg warmer than the monthly average air temperature. In mid-winter the parking lot cleared of snow had a monthly average surface temperature about equal to the monthly average air temperature. It was found that the difference between monthly average air and surface temperatures correlated well with incoming solar and net radiation. An approximate calculation indicated that the convective loss from the parking lots in summer was about 30 per cent of the net solar radiation, and the evaporation loss 15 to 20 per cent.

Average snow depth, average air temperature for given periods, and average rainfall vary from one year to the next and from one location to another. As such factors influence the exchange of heat and moisture between the atmosphere and the surface, it would be expected that fluctuations in them would induce fluctuations in the ground thermal regime as well. The observations made at Ottawa indicate that fluctuations in the annual average ground surface temperature had an amplitude of about 1 C deg, and correlated with changes from one year to the next in average snow depth and air temperature (14, 15). The amplitude of these variations decreased with depth, but were still significant 20 ft. below the surface. There was evidence of a gradual rise in temperature at the 20-ft. depth of about 0.2 C deg in 5 years. Snow-cover introduced a significant component with period one-half year into the ground temperature wave.

4. Heat Exchange at Water Surfaces

The Snow and Ice Section has given attention to the problems of predicting dates of freeze-up and break-up, ice growth rates, maximum ice thickness and formation of frazil ice. This has led to observations on the rate at which heat is transferred between the atmosphere and water or ice surfaces.

It was found in the studies on freeze-up that the cooling of a lake should be considered in two consecutive periods (16). The first is the period when the lake is cooled from the summer condition to the isothermal condition at 4°C. Cooling during this period involves mixing to the thermocline, and consequently the lake has a large "thermal inertia." Under these conditions the rate of heat loss from the lake was directly proportional to the difference between the average air and water surface temperature. The constant of proportionality depends in part on exposure and time of year, increasing with degree of exposure. Observations on a small sheltered lake in Ottawa gave a value for the average heat transfer coefficient of about 25 cal/cm²day°C (17, 18). During this fall cooling period, the evaporation loss appears to be about equal to the solar radiation absorbed by the water.

When the average temperature of the lake is lower than 4°C, only the near-surface water is involved in the cooling process, and the surface temperature is much more sensitive to changes in weather. Prediction of when ice will begin to form, therefore, becomes a problem of predicting weather. If the time for a given thickness of ice to form is used as the criterion for freeze-up, then it may be possible to determine this time with greater certainty than the time to formation of the first permanent ice. It was observed that for the

small sheltered lake the time to formation of 4 to 6 in. of ice correlated well with the number of degree-days since the water was isothermal at 4°C (16). It should be appreciated that the degree-days method is more a method of correlation than of prediction.

An analysis of available observations demonstrated the very significant importance of such factors as wind, size of lake, and speed of current in determining when break-up will occur. Studies on the small sheltered lake in Ottawa indicated, as well, that incoming solar radiation must be taken into consideration when estimating the heat gain to the ice-cover during this period (16). It was found that the heat gain to an ice-cover that melts in place was given, approximately, by the following formula:

$$\text{where } Q_m = 0.5 \, \Sigma R_{sw} + 18 \Sigma \Delta T \text{ cal/cm}^2$$

Q_m = heat available for melting ice,

ΣR_{sw} = accumulated short-wave radiation over break-up period (cal/cm²),

$\Sigma \Delta T$ = accumulated degree-days over break-up period °C).

Our ability to predict the rate of heat and moisture transfer between the atmosphere and surfaces by calculating or measuring each component of the transfer is still in a very imperfect state. For engineering purposes, therefore, it is probably more practical at this time to predict information, such as probable rate of cooling of water bodies, rate of ice growth and ice thickness at given times, from available meteorological and climatological records using statistical correlations or empirical or semi-empirical formulae, as above.

A study was undertaken of available ice thickness records to determine whether information of this nature could be obtained for rate of ice growth. It indicated that the average rate of growth of ice on lakes during the growth period was about ½ in./day and that this rate was largely independent of time of year or region (19). The reason for this is that conditions at the time of freeze-up are about the same at all locations, and the increase in heat loss that would be expected due to the drop in average air temperature as the winter progresses is largely compensated for by the increase in ice thickness. Snow-cover is the major factor that influences the rate of ice growth and, consequently, the thickness at a given time. Six inches of snow-cover reduces the rate of growth to ½ to ⅓ that for a snow-cover less than 2 in. thick. From the analysis it was possible to prepare probability charts giving rate of ice growth and ice thickness to be expected after a given number of days from the date of freeze-up.

The information on ice growth rates was converted to rate of heat loss in order to show the range in the net heat exchange for Canadian lakes in winter (20). It was observed that during the period of ice formation the rate of net heat loss from the ice surface exceeds 25 cal/cm²day 90 per cent of the time and 150 cal/cm²day 10 per cent of the time. The rate of heat loss appears to decrease gradually with increasing ice thickness and to increase slightly with latitude.

5. Concluding Remarks

Information on the exchange of heat and moisture between the atmosphere and ground surface is required for the development of satisfactory solutions for many engineering problems. This is particularly true where the problem involves freezing or thawing of water, in either lakes or rivers or the ground. It is of some importance, therefore, that we increase our knowledge of the factors that control the exchange of heat and moisture at the earth's surface, and develop instruments and techniques for measuring the size of the components under field conditions. It is hoped that the information described in this review will be a useful contribution not only to the solution of engineering problems but also to the required programs of instrument development and observations on the physical processes.

This paper is a contribution from the Division of Building Research, National Research Council, and is published with the approval of the Director of the Division.

REFERENCES

1. Williams, G. P. 1959: Evaporation from Snow Covers in Eastern Canada, Division of Building Research, National Research Council, NRC 5003.
2. Williams, G. P. 1961: Evaporation from Water, Snow and Ice. Proc. of Hydrology Symposium No. 2, p. 31-51, Assoc. Com. on Geodesy and Geophysics.
3. Sverdrup, H. U. 1936: The Eddy Conductivity of the Air Over a Smooth Snow Field. Geofysiske Publikasjoner, 11 (7).
4. de Quervain, M. 1952: Evaporation from the Snowpack. U.S. Army Corps of Eng., Translation, Res. Note No. 8.
5. 1956: Snow Hydrology. Summary Report of the Snow Investigations, North Pacific Division, U.S. Army Corps of Eng., Portland, Oregon.
6. Gold, L. W. and G. P. Williams, 1961: Energy Balance During the Snow Melt Period at an Ottawa Site. Int. Assoc. of Sc. Hyd., I.U.G.G., Pub. No. 54, p. 288-294.
7. Boyd, D. W., L. W. Gold and G. P. Williams. Radiation Balance During the Snow Melt Period at Ottawa, Canada. Proc. of the 1961 and 1962 Annual Meeting of the Eastern Snow Conf. (NRC 7152).
8. Boyd, D. W. 1962: A Field Comparison of Two Types of Net Radiometers. Symposium on the Heat Exchange at Snow and Ice Surfaces, Snow and Ice Subcommittee, Associate Committee on Soil and Snow Mechanics, 26 October 1962, TM 78.
9. Gold, L. W. and Boyd, D. W. 1965: Annual Heat and Mass Transfer at an Ottawa Site. Can. J. of Earth Sci., 2 (1), p. 1-10.
10. Williams, G. P. and Gold, L. W. The Use of Dust to Advance Break-up of Ice on Lakes and Rivers. Proc. of the 1963 Annual Meeting of the Eastern Snow Conf.
11. Gold, L. W. Influence of Snow Cover on Heat Flow from the Ground — Some observations made in the Ottawa area. Gentbrugge 1958, Toronto General Assembly 1957, Volume IV, Int. Assoc. of Sc. Hydrology, I.U.G.G.
12. Pearce, D. C. and Gold, L. W. 1959: Observations of Ground Temperature and Heat Flow at Ottawa, Canada. Geophysical Research, 64 (9).
13. Gold, L. W. 1963: Influence of the Snow Cover on the Average Annual Ground

Temperature at Ottawa, Canada. Int. Assoc. of Sci. Hydrology, I.U.G.G., Pub. No. 61.

14. Gold, L. W. Influence of Surface Conditions on Ground Temperature. Submitted for publication.

15. Gold, L. W. 1964: Analysis of Annual Variations in Ground Temperature at an Ottawa Site. Can. J. of Earth Sciences, 1 (2).

16. Williams, G. P. Correlating Freeze-up and Break-up with Weather Conditions. Submitted for publication.

17. Williams, G. P. 1963: Heat Transfer Coefficients for Natural Water Surfaces. Int. Assoc. of Sci. Hydrology, I.U.G.G., Pub. No. 62.

18. Williams, G. P. 1959: Two Notes Relating to

Frazil Ice Formation — 1. An empirical method of estimating total heat losses from open-water surfaces; 2. Some observations on super-cooling and frazil ice production. Seminar on Ice Problems in Hydraulic Structures, August 1959, Int. Assoc. of Sc. Hydrology, I.U.G.G.

19. Williams, G. P. 1963: Probability Charts for Predicting Ice Thickness. The Engineering Journal.

20. Williams, G. P. 1963: Ice Growth Rates and Heat Exchange. Symposium on Heat Exchange at Snow and Ice Surfaces, TM78, 1963. Associate Committee on Soil and Snow Mechanics, National Research Council, Ottawa.

The understanding of avalanche development in the Cordillera has been a major concern of the Division, particularly the study of climatological factors which stimulate the movement of the material to facilitate the execution of a successful defence plan. For example, Schaerer (1962) classified avalanches in terms of climatic causative factors, which proved to be most practical for the design of the active defence system at Rogers Pass (Alberta):

Avalanche Class	*Climatic Cause*	*Active Defence System for All Classes*
Dry snow direct action	snowfall 16" (406 mm) wind	1. *Permanent structures* retaining barriers; snow fences and wind baffles; braking barriers (earth mounds/ catching dams); diverting dams; snow sheds; reforestation
Wet snow direct action	snowfall 12" (305 mm) with rain or temperatures > 32°F (0°C)	
Dry snow delayed (climax avalanche)	unstable snow cover, snow fall and wind over a long period of time	
Wet snow delayed (spring thaw avalanche)	high temperatures over a few hot days	2. *Temporary measures* — explosives

The Atmospheric Environment Service (A.E.S.), Department of the Environment, has been publishing snow data for the past seventy years. Prior to 1940, the data were mostly in the form of snow-depth data recorded on the last day of the month at a limited number of climatological stations. However, since 1941, snow-depth data have become part of the 1230 G.M.T. synoptic messages providing daily data over the winter for all Canadian synoptic stations. These data are published in the Monthly Record of Meteorological Observations. The first snow cover data report appeared in the winter of 1954-55 and presented depth, water equivalent, and crust data for eastern Canadian stations. In the winter of 1962-63, the coverage was expanded to include data from all Canada.

The early data were used by Connor (1941) in the first attempt to map the distribution of snow cover in Canada, mainly to the south of latitude 55°N. In the last two decades, A.E.S. has been associated with numerous snow publications as departmental reports (e.g., Thomas 1964, Potter 1965, and Thompson 1969) and research papers by its personnel (e.g., Potter 1961, Richards and Derco 1963, McKay and Thompson 1967, and Findlay and McKay 1972). These publications mainly include average and extreme snowfall depths and durations (in tables, charts, and frequency diagrams) to represent an estimation of Canadian snow resources. The major exceptions were the publications of Richards and Derco (1963) and Thompson (1969) where snowfall distribution was analyzed in terms of winter lake-effect storms. This bias recognized the influence of vast water bodies on weather modification, particularly in the lee of the Great Lakes which represents the residential environment of about 60 percent of the total Canadian population.

The Georgian Bay snow-belt is a well-known snow distribution phenomenon in Ontario, and the Richards and Derco paper indicated that lake-effect storms are responsible for about 50 percent of the snowfall in the area. The lake-effect is mainly in the form of changes in the moisture, heat, and instability characteristics of the airflow and a thermal troughing which results in pronounced convergence, and all these modifications are compounded by orographic uplift. Also, the streamline analysis of Thompson (1969) revealed that a weak airflow did not encourage streamline confluence so that lake-effect snow was confined to areas *immediately* in the vicinity of the lake, compared to a snow band extension up to 160 km inland from strong airflow and associated streamline confluence.

Snow surveys by the Ontario Water Resources Commission in Wilmot and Oakville basins have provided detailed analyses of snow distribution (including depth and water equivalents) over intensive snow courses in small drainage basins. For example, Logan (1971) analyzed the snow depth, density, distribution, and storage along eight snow courses in the Oakville basin, with statistical evaluation of the data to establish the accuracy and reliability of the gravimetric method of sampling. The pattern of basin snow cover distribution and time trends in accumulation and ablation showed that the snow pack depth and water equivalent increased with increasing basin elevation. Also, whereas the major snow storms tended to be proportionally distributed over the basin, the minor storm distributions were affected by local terrain parameters, thereby increasing the areal variability of snow cover on the basin. Major research on the Canadian snow pack has also been undertaken by a number of individuals and universities. For example,

1) the McGill University snowmelt-runoff project at Schefferville (Jackson 1960) where the spatial distribution of snow cover on various terrain-vegetation units is explained in terms of spatial energy balance variations;

2) Pysklywec, Davar, and Bray (1968) have determined a reliable regional relationship between snowmelt (measured at a small experimental plot in New Brunswick) and meteorological parameters thermodynamically related to the ablation process. Analyses included the degree day method, the heat transfer index and multiple linear regression equations relating regional snowmelt to thermal budget indices. All three methods gave generally comparable results; the percent deviation of seasonal totals by the three methods varied from -15.6 to $+54.5$ percent whereas the root-mean squared values of the deviations of daily estimated snowmelt ranged from 0.17 to 0.22 inches (4.3 to 5.6 mm);

3) O'Neill and Gray (1971) studied the energy balance control of snowmelt in terms of solar radiation, net all-wave exchange, sensible and latent heat transfers, heat supplied by rainfall, and heat conduction from the underlying ground. The

1971 paper largely represents a valuable theoretical approach to the energy balance concept as applied to the snowmelt phenomenon, with discussion on the assumptions and problems underlying its application over a watershed in particular.

Land Ice

Climatological studies of glacial ice have been traditionally concerned with an assessment of the mass balance of selected glaciers and ice caps in terms of measuring the accumulation, ablation, and firnification (that is, the conversion of snow to glacial ice) of glacier material, recording meltwater discharge in englacial-peripheral streams and relating glacial fluctuations to recent and palaeo-climatic trends. In more recent years, with the availability of sophisticated micro-meteorological instruments (e.g., the net radiometer), glaciometeorologists have been able to undertake most valuable research on the energy balance controls of the ablation process which represents the very essence of mass balance trends.

Research on Canadian glaciers commenced as late as 1950 with a program of work initiated on Baffin Island by McGill University and the Arctic Institute of North America (Baird 1950, 1952; Orvig 1951, 1954; Ward and Orvig 1953; and Hale 1952). The project was associated mainly with the accumulation, ablation, and glacial-meteorology of the Barnes and Penny ice caps on Baffin and represented a pioneer mass balance study of Canadian land ice. Since 1950, and particularly in the last decade, glaciological research has accelerated in the Arctic and Cordillera and is associated with the activities of the following institutions:

McGill University has been most active in research on Arctic land ice during the past two decades, being partly responsible for the initiation of the pioneer glacio-meteorological research on Baffin Island in 1950 (Baird 1950, 1952; Hale 1952; Orvig 1951, 1954; Ward and Orvig 1953) and, since 1959, continuing research on Axel Heiberg Island (Andrews 1964; Havens 1964; Havens, Muller and Wilmot 1965; Muller 1966), Ellesmere Isand (Sagar 1960), and Devon Island (E.M.R., page 61, 1971). The studies have been associated mainly with the heat balance of the accumulation and ablation of selected glaciers and ice caps in the three research areas, and the relative importance of the energy fluxes will be indicated in a heat source summary table later in this section. Muller (1966) went a stage further and correlated mass balance data with meteorological observations over the period 1959-65, which reflected strongly negative and positive balances and abnormal macroclimatological conditions. Also stratigraphic analyses of a firn core indicated the balance changes over past decades and revealed a change to less favourable conditions for the glaciers in about 1930. Finally, details of the climatic changes and glacial response during the Pleistocene were obtained by palynological studies and radiocarbon dating of a peat deposit at the snout of the Thompson Glacier, which revealed a vegetation cover in the current snout area some 6000 years B.P.

The Arctic Institute of North America (A.I.N.A.) has sponsored glacio-meteorological research in the polar world of Canada since 1950, when the joint McGill-A.I.N.A. Baffin Island expedition was launched (Baird, Hale, Orvig, and Ward/Orvig 1950-54 as above). The Institute has also been involved with mass and energy balance studies on Ellesmere Island (Lister 1962, whose energy flux data will be indicated in a later summary table) and on the Devon Island Ice Cap during the summers of 1961-63 (Keeler 1965; Koerner 1966, 1970; and Holmgren 1971) and current studies in conjunction with McGill and the Department of the Environment. The present program is associated with the relationship between glacier variations and meso-scale synoptic weather patterns, particularly the contrasts between the southeast and northwest sides of the Devon Ice Cap (E.M.R., page 61, 1971). The program of work by Holmgren (1971) represents the most comprehensive study available of climate and energy exchange on a subpolar ice cap in summer, but it is impossible to summarize the 347 pages of data in this discussion. Consequently, a listing of the topics covered by the six parts of the publication will have to suffice to indicate the detail and value of the data:

Part A — Physical climatology.

Part B — Wind and temperature field in the low layer on the top plateau of the ice cap.

Part C — On the Katabatic winds over the northwest slope of the ice cap. Variations of the surface roughness.

Part D — On the vertical turbulent fluxes of water vapour at ice cap station.

Part E — Radiation climate.

Part F — On the energy exchange of the snow surface at ice cap station.

Reference to the above reports will surely encourage the adoption of similar "discipline" approaches to energy flux-ice ablation relationships in the next decade or so, to compare with the Sverdrup "doctrine" over the decades following his Spitzbergen expedition in 1934. As far as our current theme of the role of climate on glacier behaviour is concerned, the most relevant part of the Holmgren study is the section on energy exchange at the snow surface (Part F), which refers to the relationship between ablation and energy fluxes above the snow surface and to the variations of energy balance in relation to the synoptic weather situation. The relative importance of these energy fluxes is summarized in the table on pages 163-4 which includes data from numerous McGill-A.I.N.A. projects referred to above.

The relative heat source summaries reveal that generally the high elevation ice masses (above 800 m), remote from the oceanic reserves of sensible and latent heat, experience a radiation dominance with latent heat transfers especially providing a very small contribution (less than 10 percent). The low elevation Sverdrup Glacier and White Glacier (in 1964) experienced a greater dominance of latent heat transfers (more than 15 percent) although the 1965 anomaly on the latter glacier indicates that table-based generalizations may be rather subjective interpretations of data.

The lowest elevation of all at 15 m on the Ward Hunt Ice Shelf (Lister 1962) had a 100 percent dominance of radiation since the 83° north latitude suggests remoteness from the heat sources of the low-Arctic Ocean. The Devon Ice Cap data of Holmgren are most interesting since they represent contrasting synoptic conditions of 1) anticyclonic stability, and 2) cyclonic cloud and turbulence. In the latter condition, the weak radiation contribution is observed with sensible heat dominance (46 percent) associated with the southerly airflows (advanced warmth) of cyclonic circulations.

Department of the Environment* glaciometeorological research was undertaken prior to 1966 by the Geographical Branch on Baffin Island, during the period 1961-65, and in the Cordillera in 1965 (Ostrem 1966). The Baffin project reached a peak in 1965 when a field party of twenty-eight assistants was active in the area (Ives 1966). Results of the field studies on the Barnes Ice Cap for the summers of 1962-65 were reported by Sagar (1966). Mass balance data indicated that, in the relatively warm summer of 1962, the net accumulation equalled the net ablation

---

* Before 1972, this department was referred to as Energy, Mines and Resources.

Summary of relative importance of heat sources during the ablation season over selected land ice areas (Source Paterson 1969* and references listed in table)

Location	Position	Elevation m	Radiation	Sensible heat transfers	Latent heat	Reference	Institution Association
Ward Hunt Ice Shelf Ellesmere Is.	83 12N 74 00W	15	100	—	—	Lister 1962	A.I.N.A.
McGill Ice Cap Axel Heiberg Is.	79 41N 90 27W	1530	64	35	1	Havens 1964	McGill
White Gl. Axel Heiberg Is.	79 26N 90 39W	208	48	32	20	Andrews 1964	McGill
White Gl. Axel Heiberg Is.	79 26N 90 39W	208	63	30	7	Havens et al 1965	McGill
Sverdrup Gl. Devon Is.	75 40N 83 15W	300	51	34	15	Keeler 1964	A.I.N.A.
Barnes Ice Cap Baffin Is.	69 43N 72 13W	865	68	32	0	Ward & Orvig 1953	McGill A.I.N.A.

(continued overleaf)

Summary of relative importance of heat sources during the ablation season over selected land ice areas (Source Paterson 1969* and references listed in table)

Location	Position	Elevation m	Radiation	Sensible heat transfers	Latent heat transfers	Reference	Institution Association
Penny Ice Cap Baffin Is.	66 59N 65 28W	2050	61	30	9	Orvig 1954	McGill A.I.N.A.
Barnes Ice Cap Baffin Is.	70 14N 73 55W	1075	70	18	7 (5% from refeezing rain)	Sagar 1966	E.M.R. Geographical Branch
Devon Ice Cap Devon Is.	75 28N 83 10W	1320	70 44	22 46	8 (1) 10 (2)	Holmgren 1971	A.I.N.A.
(1) = clear skies, light wind; (2) Overcast skies, strong wind.							
Salmon Glacier B.C.	56 10N 130 07W	1700	75	15	10	Adkins 1958	U. of Toronto —N.R.C.

* By permission of the publishers, Pergamon of Canada Limited, based on pp. 58-61 of *The Physics of Glaciers* by W. S. B. Paterson (Toronto: Pergamon, 1969).

over the ice cap crest compared with a net accumulation surplus of 17 cm over the 1962-1963 mass balance year and 38 cm between 1963-64. The energy balance control of ablation was dominated by radiative energy in 1962, (70 percent contribution) with a reasonable supply of heat from sensible, latent, and refreezing rain transfers (data summarized in the relative heat source table above). However, in 1963 and 1964 the radiation value was close to a complete (100 percent) dominance of the reduction of subsurface "cold content" and melt promotion. Analyses of surface and 850 mb level synoptic charts showed that southerly warm air advection, associated with cyclonic activity, was relatively frequent during the 1962 summer season and accounted for the heat transfer differential (and the equilibrium condition of the 1961-62 mass balance year outlined above).

Since 1966, glaciometeorological research in the Department of the Environment has been undertaken by the Inland Waters Branch (Glaciological Division) on Arctic and Cordillera land ice, associated with mass balance and energy balance investigations of Loken and Sagar (1968), Derikx and Loijens (1971), Goodison (1971), Stanley (1971), and a glacial inventory for all Canada (Ommanney 1970, 1971). The research by Derikx and Loijens (1971) was related to mass balance measurements, meteorological observations, and stream discharge data from the Peyto Glacier, Alberta, and the Berendon Glacier, British Columbia. The physical relations, which exist between the important energy balance parameters and the daily melt in a glacierized basin, were synthesized into a distributed dynamic model to provide a sufficiently accurate estimate of the heat balance for the entire glacier. On the average, the daily discharge calculated for the model for Peyto and Berendon glaciers was 17 percent and 12 percent respectively higher or lower than the measured values in 1968. It appeared that the model underestimated the mass loss in the relatively inaccessible upper regions of the glaciers (Peyto, especially) where it is difficult to obtain accurate mass balance data. Also, the melt in the lower parts of the Peyto Glacier was overestimated, despite very accurate mass balance measurements, since the calculated ablation gradient was too high. Comparison of observed and calculated mass balance at Berendon Glacier showed that the measured ablation gradient is higher than the one calculated by the model, the reverse of that of Peyto Glacier.

Moreover, for most of the melt season, the daily basin runoff can be calculated with an average accuracy of 10 to 20 percent, if basic meteorological variables are available for *one point only*. Although there are deficiencies in the model, its overall satisfactory agreement suggests that the most significant processes are identified (Derikx and Loijens 1971). The adoption of a physically based, parametric simulation model to estimate melt and runoff over an entire glacial basin has numerous advantages, particularly since direct measurements of moisture and heat fluxes are too complicated and expensive to be duplicated over all the contrasting physical environments in the basin. To produce acceptable daily runoff calculations for an entire inaccessible basin, by modelling the data from *one* energy and moisture flux measuring location, is both realistic and expedient for remote land areas in Canada.

The current research of the Department (E.M.R. 1971) is associated with mass and energy balance studies on a large number of glaciers in British Columbia and Alberta (i.e., Place, Sentinel, Woolsey, Peyto, Berendon, and Ram River glaciers) and in the Arctic (Barnes Ice Cap and Decade Glacier on Baffin Island, the Per Ardua Glacier on Ellesmere Island, the Meighen Ice Cap, and the Devon Island Ice Cap). The latter two Arctic projects are being undertaken primarily by the Polar Continental Shelf Project in cooperation with McGill University, A.I.N.A. and A.E.S.

There are numerous other institutions involved in glaciological research in Canada, including the Defence Research Board in Ottawa and the Institute of Arctic and Alpine Research (Instaar) of the University of Colorado, U.S.A. The former institution has been undertaking research on Ellesmere Island since 1957 (Hattersley-Smith 1960; Sagar 1960; and Hattersley-Smith and Serson 1970), mostly related to the mass balance of the Lake Hazen and Ward Hunt Ice Rise Shelf. The University of Colorado program is associated with Baffin Island and is being conducted mainly by personnel who were previously operating on Baffin Island with the Department of Energy, Mines and Resources (Andrews, Barry, and Drapier 1970). It involves an assessment of present and past glacieriza-tion of east Baffin Island in terms of palaeoclimatic considerations and current mass-energy balance investigations of land and sea ice.

Canadian research in glaciometeorology has provided valuable data on mass balance and energy balance regimes for a large number of glaciers and ice caps in the Cordillera and Arctic regions. Reference to the heat source data tabulated earlier will reveal unique energy balance data for nine different land ice locations. This table represents a wide coverage of knowledge concerning these intricate glacier-climate relationships. Furthermore, the glacier inventory survey of the Department of the Environment (Ommanney 1970, 1971) has provided a reliable estimation of land ice as a water resource in Canada. For example, perennial ice and snow masses cover about 2 percent of the total land area (some 201,000 km^2), in the form of 70,000 glaciers representing some 40,000 km^3 water or almost double the volume of the polluted Great Lakes. Future research into the climatology of land ice will be essential to conserve this valuable reservoir of pure fresh water and to provide new information to produce more effective water management in Canada.

Floating Ice

Floating ice studies are essential in Canada since sea ice hinders shipping, influences the energy and water balance of the water body, and exerts considerable pressure (with resultant damage) on riverbanks, shorelines, and structures such as docks and oil rig towers. Also, sea ice provides landing strips for light aircraft and routeways for surface vehicles. River ice development changes the regime of the water since, during the formation and decay of ice, river stability and circulation patterns of the waterbody are modified considerably, thus influencing the distribu-tion of nutrients and pollutants.

Research covers the distribution, movement, and growth of floating ice which largely reflect aspects of the macroclimate, energy balance, and synoptic weather patterns of the area concerned. River-lake ice studies are undertaken primarily by the Division of Building Research of the National Research Council (Gold 1960; Ager 1962; Gold and Williams 1963, Williams and Gold 1963; Williams 1965; and Legget and Gold 1966) and are mainly related to the freeze-up and break-up of ice and the effect of ice pressures on structures, especially bridge piers. The bias reflects the desire to lengthen artificially the shipping season on inland waterways and to minimize the expensive seasonal repairs of damaged structures.

Sea ice investigations were initiated by the Geographical Branch between 1947 and 1952 and have been completed by McGill University in the Arctic since 1954, with an Ice Research Project supported by the Defence Research Board of Canada. The research findings largely formed the basis of the publication by the director of the project (Pounder 1965) where sections referred to the properties of sea ice (i.e., physical, crystallographic, mechanical, thermal, and electrical). The climatology of sea ice was represented by sections on drifting ice and ice growth and decay, in terms of Stefan's Law for ice growth (that ice thickness varies as the square root of the freezing exposure), diffusion equations, and the heat budget before and after freeze-up. For example, in a study of the ice on Hudson Bay, Schwerdtfeger and Pounder (1963) discovered that although the convective flux in the air (Q_C) was large at any moment, it usually varied cyclically with energy lost from the ice during daylight hours and returned to it from the temperature inversion during the night, so that the twenty-four-hour mean of Q_C is small. They also indicated that during the period of ice growth, the heat transfer by evaporation or condensation of water substances (Q_E) is negligible at 60° North once the ice cover is established. Consequently, the radiative flux (Q_R) is the dominant method of heat loss from ice. Pounder (1965) concludes that "at the present stage of these studies of the heat budget of ice, they do not present a practical method of forecasting ice growth and Stefan's Law will continue to be useful. However, the approach is more fundamental and should lead to a better understanding of the processes involved, with ultimate improvement in forecasting techniques."

Sea ice studies in the Arctic are vital with regards to the future mineral and oil exploitation in the area, since the distribution and movement of pack ice will determine the type of marine transport to be utilized (i.e., conventional oil tankers with ice breaker support or the revolutionary *Manhattan* tanker system). Little is known about the sea ice conditions within the Arctic archipelago and adjacent Arctic Ocean, and the need for this knowledge is rising rapidly with the increasing economic potential in the "high" Arctic. Research is also needed on the physical properties of sea ice, especially the expulsion of brine with the freezing of salt water and its influence on structures (E.M.R. 1969).

The demand for Arctic sea ice knowledge has focussed the attention of the Polar Continental Shelf Project on the floating ice of the Canadian North (Lewis 1966; Lindsay 1969). The current research by the project includes ice drift studies (by aerial surveys, sonar tracking, and remote sensing), the thermal imagery of sea ice by infrared aerial reconnaissance, and the microclimate of the sea ice-water interface (E.M.R. 1969). These programs of work are dominated by the climatological control of the floating ice in terms of wind-advection of the pack ice and the energy balance of the freezing and break-up processes.

Numerous individuals have completed research on the Arctic sea ice, including Swithinbank (1969, Defence Research Board Contract research in the form of an ice atlas of Canada), Untersteiner (1961, 1964, the most comprehensive study of mass-energy budgets of the Arctic sea ice) and Koerner (1970). The latter program represented unique distribution and melt-freeze climatic data from Point Barrow to the North Pole, on the Canadian "leg" of the trans-Arctic crossing 1968-69.

Chapter 10

Agroclimatology

The traditional approach of the physical geographer to agricultural climatology was mapping the spatial distribution of crops (especially wheat) based on easily measured macroclimatic parameters to determine broad crop-climate relationships (Unstead 1912; and Zacks 1945). More recently, the approach has emphasized microscale aspects and has been concerned with the analysis of physical processes including air temperature, soil temperature, and moisture, water, and radiation balances and the effects on plants (Williams 1972). It is now possible to measure vital heat and water budgets from the soil and plant surfaces by energy balance, aerodynamic, and eddy correlation models. This change in research emphasis over the past sixty years has practically eliminated geographical involvement in current agroclimatological studies in Canada, which is primarily undertaken by the Plant Research Institute in Ottawa (Canada Department of Agriculture), the Atmospheric Environment Service (Department of the Environment), and the University of Guelph agrometeorology group in the Department of Land Resource Science (formerly Soil Science).

The earliest agroclimatological research in Canada was mainly associated with aspects of the water balance, with particular reference to the "dry belt" problems of the western Prairie provinces. The first recognition of the unfavourable water balance in southern Alberta and Saskatchewan was associated with the field work of Captain John Palliser during the period 1857-1860, when the dry belt known as Palliser's Triangle became a regional geographic reference (Villimow 1956). The actual climatic characteristics of the area were determined later by Stupart (1905) and Connor (1938), and the first successful climatic differentiation of the dry belt was related to the adaptation of the Thornthwaite (1948) classification of climate to the area by Sanderson (1948b). Research by Villimow (1956) analyzed the nature and origin of the dry belt in terms of regional features of precipitation (including moisture indices, precipitation deficiencies, and potential evapotranspiration or water need ratios after the Thornthwaite 1948 classification) and temperature, together with explanations for the dry belt related to cyclonic-anticyclonic frequencies, air mass trajectories, and upper pressure and wind patterns.

The assessment of water balance in Canada has changed dramatically over the past two decades, from the Thornthwaite type of approach to water need (i.e., potential evapotranspiration from the mathematical integration of several macroclimatic factors) of the late 1940s and early 1950s to atmometer-based evaporation data of the late 1950s and soil moisture-heat budget modelling of the 1960s and 1970s.

The Thornthwaite approach is well illustrated by the pioneer work of Sanderson (1948a) when she analyzed the climate of Canada according to the "new" classification, mapping the country in zones of seasonal variations of effective moisture and summer concentration of thermal efficiency. The value of this approach to agroclimatology at that time is evident from the observations in the summary and conclusion of the Sanderson paper (1948a) which follow:

The relationship of water surplus and water deficiency to water need, defined by the Thornthwaite classification, provides a new attack on the complex problems of climate in scientific agriculture as well as physical geography.

The scientific regional geographer for years has sought a rational quantitative delimitation of the climatic regions, since on it is based the explanation of vegetation and soil zones, the geomorphic nature of the landscape and the patterns of land use. Although Koeppen's classification has served as a standard for geographers, it cannot explain the distributions of these phenomena because it is derived from them.

The hydrologist can make use of the map of water surplus for Canada in gaining an approximate knowledge of average run-off where no actual measurement is made. In addition the method of computing current water surplus for a given season from the meteorological data alone is a useful starting point in predicting available run-off.

Crop testers and plant breeders have always been hindered in their work by the lack of fundamental climatic data and quantitatively defined climatic regions. The maps of potential water need, and water deficiency presented here for Canada partially fill this gap. The map of climatic regions indicates homoclimes for the benefit of the crop tester. For example, there is an interesting similarity of moisture relations between Harrow in southwestern Ontario, and Winnipeg in Manitoba. In a crop testing program worthwhile correlations could be made between the yield and quality of the crop and actual evapotranspiration and water deficiency computed from current meteorological data.

The plant pathologist could gain a clearer knowledge of the climate advantageous to insect pests and disease organisms through a correlation of the times of infestation and the life cycle of the pathogen with related water surplus and deficiency.

Soil scientists recognize the fact that water surplus and water deficiency both play important roles in the development of soil. Thornthwaite's 0 line indicates the theoretical boundary between the pedalfer and pedocal soils. Actual soil relationships along this line remain to be investigated.

The Thornthwaite method of using meteorological statistics to arrive at a knowledge of the real factors in climate, although not perfected, represents an invaluable addition to Canadian climatic research.*

Sanderson also made a detailed pioneer study of the characteristics of drought in the Canadian Northwest (1948b) when the agricultural potential and limitations of the area were defined in terms of water need, moisture deficiency, and water surplus according to the Thornthwaite (1948) formula.

Drought patterns in the Canadian Prairies were also mapped by Laycock (1960) when he applied the Thornthwaite (1948) formula and other deficit procedures to the data from 575 climatological stations for the period 1921-1950. The distribution patterns included average precipitation, average potential evapotranspiration, and moisture deficits over time, to illustrate varying regional patterns of drought. The frequencies of intermediate drought intensities, of great importance

* Reprinted from *Scientific Agriculture*, Vol. 28, 1948, by permission of the Agricultural Institute of Canada.

to the agriculturalist, were mapped in terms of lower quartiles (i.e., wettest quarters of the years) and upper quartiles (driest quarters) and indicated valuable irrigation and moisture conservation practices. The most pertinent agricultural part of the study was the approximate determination of local moisture deficiency patterns associated with varying soil moisture storage capacities, for example, light loam soils (used by cereal crops with moderate rootage), coarse textured sandy soils (plants have to withstand severe droughts here), and fine textured clays and clay loams (used by alfalfa and other crops with deep, extensive roots). Laycock also produced numerous maps to show drought conditions for specific crops by varying the length of the growing season and moisture storage capacities in the calculations. For example, wheat in July has a well-developed root system and droughts may be calculated using 6 in. (152 mm) soil moisture capacity, compared with drought based on 0.5 in. (13 mm) capacity in the germinating season (May). It should be emphasized that there was an acceptable similarity between the moisture deficits calculated by the three procedures used, viz, Thornthwaite (1948), Lowry-Johnson (1941), and Blaney-Criddle (1952).

It is apparent that empirical methods of evapotranspiration calculation were extensively used in Canada in the 1940s and 1950s with particular reference to the water need of the Prairie provinces. However, the 1950s also experienced the pioneering of new techniques for assessing water balance patterns with the measurement of evaporation by the Black Bellani plate atmometer, a new inexpensive design of the original 1820 version developed by the Canadian Department of Agriculture at the Central Experimental Farm in Ottawa (Robertson and Holmes 1958).

Tests conducted in Ottawa during the summer of 1953 indicated that the Black Bellani plate was more responsive to day-to-day variations in the meteorological environment than were other atmometers in common use in Canada at that time. This is shown in the following table where daily evaporation is correlated (using linear correlation coefficients) with data from observations taken during the period May 1 to September 30.

Meteorological Factor	Evaporimeter		
	Black Bellani Plate	Summerland Tank	Four-foot Tank
Mean daily temperature	0.46	0.42	0.30
Average wind speed	0.27	−0.01	0.23
Total solar energy	0.77	0.60	0.60
Average vapour pressure deficit	0.72	0.70	0.46

Significance (N = 153 cases):
5 percent level r = 0.16; 1 percent level r = 0.21
(Source: Robertson and Holmes, 1958).

Evaporation from different atmometers was correlated with evaporation calculated by means of the Penman 1948 equation (which makes use of air temperature, vapour pressure, sunshine, and wind data). Correlation coefficients using daily evaporation values for September 1953 revealed a highly significant correlation (r = 0.90, significant at the 1 percent level) for the Black Bellani plate compared with the Summerland Tank (r = 0.76) and Four Foot tank (r = 0.48). This indicated that the Black Bellani plate was a satisfactory integrator of the meteorological factors affecting evaporation as well as being superior to the others tested.

Further comparisons between evaporation from open water pans and latent-evaporation recorded by the Black Bellani plate in 1955-56 produced a correlation coefficient (r) of 0.93 which was significant at the 1 percent level. Also, measured potential evapotranspiration from the Thornthwaite evaporimeter was found to be related to latent evaporation from the Black Bellani plate by the ratio 0.0034. The good relationships outlined above were responsible for the extensive use of the Black Bellani plate by the Department of Agriculture, despite the fact that evaporimeters do not represent the natural-crop environment.

Direct measurements of soil moisture presented many practical difficulties although, in the 1960s, useful studies were made with lysimeters and absorption blocks. For example, Carder and Hennig (1966) measured the soil moisture balance under different cropping procedures, using the electrical resistance block to obtain an almost daily record of soil moisture content in fallowed land and under crops of spring wheat and red fescue over a five year period. The purpose of the experiment was to determine:

1) the efficiency of summer fallow for the conservation and storage of water;
2) the amount of water used by a cereal and a grass crop;
3) the water consumption of these crops during all stages of their development, collated with soil moisture changes;
4) the soil depths from which this water is obtained;
5) water source, i.e., soil reserves, rainfall, or snowmelt.

Observations revealed that, during the fallow year, 11.2 in. (285 mm) of water were lost by evaporation compared with a spring wheat loss of 13.2 in. (335 mm) and fescue loss of 12.9 in. (328 mm) over the same period. A feature of the study was the small amount of precipitation that was stored in the soil during the fallow period, for example, at Beaverlodge in Alberta, only 6 percent of the precipitation was stored compared with a 21 percent storage at Swift Current, Saskatchewan. Wheat stubble stored only 5 percent of the winter precipitation and fescue proved superior with a 10 percent storage. Wheat, with its extensive root system, withdrew moisture from at least 3 feet (1 m) depth, while fescue did not withdraw it from much beyond 1.5 feet (0.5 m). However, established fescue withdrew water more rapidly and thoroughly than wheat, particularly in the spring before wheat had a fully developed root system. The above moisture regime data for different cropping procedures were of practical value for the optimum adoption of irrigation and water conservation techniques.

Other direct measuring techniques for soil moisture were employed, for example, the neutron moisture meter and the tensiometer, which were too expensive and not sensitive enough to follow daily soil moisture changes (Baier 1965). Also, all the above methods provided only point measurements and failed to represent the soil moisture variations associated with actual soil changes. The review paper by Baier (1965) examined alternative moisture budgeting methods for estimating the moisture content in the soil, together with the interrelationship of meteorological factors and plant growth. It attemped to outline the problem of water budgeting from meteorological data and discussed research contributions directed towards a realistic yet simple crop-moisture budget model, over the seventeen years following the introduction of the potential evapotranspiration concept by Thornthwaite (1948). For example, the modulated budget approach to Canadian soil moisture problems was adopted by Holmes and Robertson (1959, 1964). Since 1966 there have been numerous applications of modulated studies to Canadian soil moisture budgets, for example, Baier and Robertson (1966, 1967), Baier (1967, 1969), Mack and Ferguson (1968), and Selirio and Brown (1971).

Mack and Ferguson (1968) applied the modulated soil moisture budget (Holmes and Robertson 1959) to a wheat crop over experimental plots at Brandon, Manitoba. Wheat yields were related more closely to the moisture stress function (potential evapotranspiration–actual evapotranspiration where the correlation coefficient, r, equalled –0.83) than to the seasonal precipitation and evapotranspiration ratio. Regression analyses showed that moisture deficiency affected the growth of cereals at certain stages of the crop development. The analyses indicated an average reduction of yield of 156 ∓ 40 kg/ha per cm of water deficit from emergence to harvest, but a reduction in yield of 311 kg/ha per cm of water deficit from the fifth leaf to the soft-dough stage and 69.1 kg/ha per cm of deficit from the soft-dough stage to maturity. Consequently, the moisture stress function may be used to characterize the soil-plant-atmosphere environment for the growing season of a crop.

Selirio and Brown (1971) applied the versatile moisture budgeting technique (Baier and Robertson 1966) to fallow soil in spring at Guelph, Ontario. This technique accounted for the transfer of moisture at different potentials within the root zone in relation to moisture content and recognized, indirectly, the moisture conductivity of the soil which was not represented in the Holmes-Robertson budget (1959). The budget data were compared with actual soil moisture–gravimetric measurements, where the correlation coefficient (r) was greater than 0.85 and highly significant for most soil-depth zones. Results in this study indicated that moisture budgeting techniques could be used to estimate soil moisture in the plough-layer in springtime. The technique has potential application in studies related to farm management, such as determining the suitability of soil for tillage, seeding, or other farm operations.

The role of heat budget studies in Canadian agroclimatological research was emphasized in a pioneer publication by King (1961) when he stated "that evapotranspiration from a land surface depends primarily on the radiant energy supply to the surface, but may be limited by the rate of movement of water to the evaporation surface." The paper outlined the principles which govern evaporation from land surfaces, particularly in terms of the energy balance method $E = \dfrac{Rn - S}{1 + \beta}$

(where E = latent heat flux of evaporation, Rn = net radiation flux, S = soil heat flux and β = Bowen ratio). The method was tested over a well-watered corn crop at Guelph, where the calculated evaporation showed close agreement with recorded evaporation from floating lysimeters. The energy for evaporation came mainly from the net radiation, since the ratio E/Rn on a day-time basis was 0.81 and about 1.0 for a twenty-four-hour day. When the cornfield was irrigated, with the surroundings relatively dry, a ratio of E/Rn above one was obtained, and there was a greater transfer of sensible heat to the crop than when both the field and surroundings were moist. The evaporation in irrigated fields may be 30 percent or more greater than the net radiation with advected energy from dry and warmer surroundings.

The paper also described the dynamics of vapour transport with reference to the application of vapour flux-direct methods and aerodynamic equations to crop surfaces. A simple aero-heat budget procedure to estimate daily or hourly evaporation was reviewed, and the evaporation values calculated by this method agreed closely with values calculated from the detailed energy balance as in the above equation. The paper concluded with a discussion of the effect of plants and soil on evaporation, which represents a dynamic flow of water occurring simultaneously with heat and vapour fluxes.

Water balance studies over the past decade have featured the aerodynamic and energy balance techniques of estimating evapotranspiration. For example, Mukammal, King, and Cork (1966) compared the two techniques in estimating evapotranspiration from a cornfield at Guelph, and the results were further compared with measurements from a floating lysimeter. Energy balance estimates were based on measurements of net radiation, wind speeds, and wet-dry bulb temperatures at seven heights between 25 and 540 cm. The flux of water vapour from 240 to 540 cm was computed using three aerodynamic formulae.

The aerodynamic methods yielded results which were close to the lysimeter measurements when the corn was short at the beginning of the season but fell to only 40 percent of the measured values at the end of the season. It is concluded that aerodynamic methods, when applied to calculating evapotranspiration from tall crops, cannot be expected to yield useful results. The authors recommended that the energy budget method be used instead since estimations using this approach were found to be within 11 percent of measured evapotranspiration values.

It is interesting to note that in moisture flux studies over the last five years, the role of plant water has become more significant than that of soil water in crop production. For example, the research of Gillespie and King (1971) determined night-time sink strength profiles for moisture in a corn canopy by blotting dew from leaves. Air temperatures and specific humidity in the canopy air layer were also measured, and from the flux gradient $E = \rho kw \ (dq/dz)$ (where E = water vapour flux density; ρ = air density; kw = diffusivity for water vapour, and dq/dz vertical specific humidity gradient) some apparent night-time diffusivity profiles for water vapour were deduced. These curves reached a minimum near the middle of the air layer occupied by the foliage, in contrast to corresponding day-time profiles which monotonically decrease with depth. Sensible heat was a stronger contributor of energy to the crop at night than was latent heat, but moisture sinks were more evenly distributed in the foliage.

The actual amount of turbulent transport within crop canopy layers (as was outlined by Gillespie and King, 1971, and Gillespie, 1971) was originally determined by the flux gradient equation $E = -\rho kw \ (dq/dz)$ explained earlier. However, problems with empirical models have forced the agroclimatologist to consider new and sophisticated techniques for accurately measuring the flux of water vapour and carbon dioxide, for example, instruments like the fluxatron and the eddy correlation measurement of shear stress. Shaw and Thurtell (1971) described the development and construction of a hot film anemometer, employing a new concept in angular detection suitable for the recording of horizontal and vertical components of the wind required by the eddy correlation measurement of fluxes within crop canopy layers. Shaw and Thurtell stated that:

> Since turbulent transport processes within the crop volume directly affect the supply of carbon dioxide to leaf surfaces and indirectly affect growth processes via the considerable effects of air movement on heat exchange, transpiration and leaf flutter, it is believed that a characterization of the ventilation processes within the plant layers will, in the future, be an important part of modelling plant community activity. The development of more complete models of crop growth should lead ultimately to the aim, expressed by Lemon

(1969) of improving the architecture of crop stands not only for light capture but for optimum ventilation as well.

The aims of the above project will clearly indicate the incredible progress made in agroclimatological research over the past twenty years. In the Thornthwaite era of the early 1950s, accurate and representative regional water balance data were the researchers' goals, with no hope of measuring moisture flux within the crop stand (which is relatively commonplace today).

Even though water balance studies have dominated agroclimatological research in Canada since the very early days, the agroclimatologist has always been aware of other important parameters and problems concerning agricultural production. For example, the influence of temperature on crop growth is well documented by Carder (1957), Holmes and Robertson (1959), Hopkins (1968), Robertson (1968), and Williams (1971), to name only a few. Furthermore, modern research on the turbulent transfer of carbon dioxide in the plant canopy (Gillespie (1971) represents a study of a much neglected atmosphere variable, which is one of the concerns of the current International Biological Program. In addition the complex and intricate relationships between climate and crop types have always interested the agroclimatologist, apart from the moisture regime patterns outlined earlier (Laycock 1960; King 1961; and Mukammal et al 1966). Examples include: oats (Zacks 1945), the soybean (Brown and Chapman 1960, 1961; Brown 1960), wheat (Williams 1969a, 1969b, 1971) and corn (Graham and King 1961; Rahn and Brown 1971; Gillespie 1971; Gillespie and King 1971). Future agroclimatological research will be concerned with the development of new crop varieties such as green corn which, as forage grain, could receive a great deal of attention particularly with regard to improving the ease of penetrating the canopy without damaging the crop. The agroclimatological techniques have been updated from the traditional approach, using climatic averages, to specific climatological descriptions (at all scales) for groups of crops, together with analysis of their response to the climatic environment.

The examination of development and progress in agroclimatological research in Canada has concentrated on the water balance of crops since moisture regime and water flux data, at every scale, will facilitate more successful crop management and more economic food production in the future. These improvements in agricultural practice are vital for a country such as Canada, which relies heavily on agricultural exports, and in which two-thirds of the territory is too cold and much of the remainder has either unsuitable surfaces or very restricted climatic resources for agriculture (Williams 1972). Crop production from the limited agricultural land in Canada must become increasingly efficient, especially since urban sprawl is devouring land in the areas of some of Canada's best agroclimates, such as southwest Ontario. This efficiency rise can only come from continued agroclimatological research so that, in the future, "farming the land" will be practised on physical environments which have been carefully controlled for optimum agricultural production.

Chapter 11

Urban Climatology

The current interest in environmental protection and conservation has focussed attention on the significant climatic implications of urbanization in Canada, especially the atmospheric pollution problem which has received so much publicity in the last few years from all forms of mass media. However, it must be emphasized that pollution and man-made climatological changes associated with urban sprawl have been a conspicuous part of the Canadian environment since the earliest days of town construction and industrialization. Unfortunately, the concentration of both these features in southwest Ontario has accentuated atmospheric and climatic changes in the area. This alarming trend is obvious to the naked eye on calm, stable anticyclonic days over the lake shore "belt" from Toronto to Hamilton and is also observed over all "pockets" of urbanization throughout Canada.

The qualitative awareness of the serious alteration of meso- and microclimate characteristics, associated with urban sprawl, is traced back into the last century and a peak stimulation of interest in the problem of atmospheric pollution (the most obvious, visual climatic implication) occurred in Canada during the first decade of this century. The first quantitative assessment of the problem appeared in 1922 when the Melton Institute of Industrial Research in Pittsburgh (engaged in a comprehensive survey of air pollution in that city between 1911 and 1914) published its findings and estimated the economic loss to be $10,000,000 per annum — the cost being made up of waste fuel, increased cleaning bills, ruined merchandise, and corrosion of structures. The alarming findings startled Canadians, mainly because it was now known that air pollution was actually costing money, damaging health, and causing the death of some urban dwellers, and forced Toronto to adopt a by-law which limited the emission of dense smoke. However, casual observation showed that the condition of Toronto's atmosphere left much to be desired, and it became obvious, over the decade following the Pittsburgh findings, that effective smoke abatement and pollution control must rest on a basis of accurate quantitative information regarding the degree of the existing pollution. For this reason, the first investigation of air pollution in Canada was initiated in Toronto as late as 1932 (Barrett 1938). Gauges were installed at four sites in the area, with locations ranging from a rural setting at the Connaught Laboratories Farm to the corner of King Street and University Avenue, in the business district near a railway terminal. The gauges measured impurities deposited during the period 1933-36 over an area of approximately 4 square feet, in the form of insoluble tar, carbon, and ash and soluble matter such as ash, sulphates, and ammonia. Total solids deposited at the three city locations averaged 341, 358, and 610 tons per square mile annually (with the greatest deposition recorded at the gauge sited at the corner of King Street and University Avenue) compared with an annual deposit of 133 tons at the rural farm location.

Suspended impurities in Toronto air were measured with an automatic air filter, similar to the type used in Great Britain between 1916 and 1920 which facilitated comparisons with other polluted centres. For example, in winter, the concentration of average hourly suspended impurities over Toronto was less than that recorded in London, England, but was almost identical to the level measured in the five most polluted American cities. In fact, during summer there were more impurities suspended in Toronto air than were found in the same five United States cities. Comparison of pollution present in the air on Sundays and week-days showed the greater part of Toronto's pollution to be of non-industrial origin. However, it should be noted that these measurements were made before the impact of the automobile and its effect on current week-day pollution concentration. The results of the first-ever pollution probe in Canada have shown that the amount of atmospheric pollution in Toronto was as great as had been found in badly polluted areas in the United States and Great Britain. The findings suggested that more adequate efforts should be made to bring about some abatement of this serious Toronto condition. Barrett's paper has been summarized at length since it represents a pioneer quantitative analysis of pollution in Canada and indicates the actual number of decades that pollution has been a serious problem in Toronto.

Atmospheric pollution studies were neglected in Canada during the fifteen years following the publication of the Barrett report until, in the early 1950s, the Detroit-Windsor condition was investigated by Baynton (1953, 1956) who correlated air pollution with meteorological factors. Further research, by Munn and Katz (1959, 1960) and Champ (1962), revealed the daily and seasonal cycles of pollution in the area, with particular reference to inversion intensities. Pollution in Toronto was analyzed by Mateer(1961) and Emslie (1964) in terms of its effect on solar radiation over the metropolitan area and represented the first significant pollution study in the city since the valuable pioneer findings of 1938. In the last decade, the increased public awareness of pollution has stimulated research in a considerable number of urban and industrial centres throughout Canada. The significant research papers have provided vital data on the degree and causes of pollution for Ottawa (Munn 1961; Ferland 1964), Winnipeg (Leahey 1962), Montreal (Summers 1962, 1966; Oke 1968), St. John, New Brunswick (Wilson 1964), Vancouver (Emslie and Satterthwaite 1966), and Hamilton (Oke and Hannel 1968; Rouse and McCutcheon 1972). The documentation of pollution data for Canadian municipalities, referred to above and in unpublished reports of numerous city health authorities, is quite extensive at this time and forms the basis of successful control measures. The daily broadcast of index levels of pollution by radio and newspapers emphasizes the current awareness of pollution and the amount of data available on the concentration of impurities in the air over the major Canadian cities. Industrial concerns have to limit smoke and impurity emission with the approach of dangerous pollution levels but it is ironic to record that, even under hazardous conditions, the automobile (which is responsible for more than 50 percent of the pollutants in the atmosphere) is allowed freedom of movement within the problem area. We have available adequate data on the causes and degree of pollution of the atmosphere over Canada, and the time has come to severely limit the emission of impurities. This applies particularly to the production of pollutants by the automobile since the current emissions could be well above acceptable provincial standards in the near future (which possibly could be modelled on the California level for 1974). Research by the Consumers' Gas Company in Toronto (1971) indicated a reduction in carbon monoxide from 33.0 gm per mile (using gasoline) to a permissable 3.07 gm per mile using natural gas (which is actually 20 gm *below* the California standards). The cost of 1 therm of natural gas was 33 cents compared to 50 cents average for the equivalent gallon of gasoline. Natural gas might not be the most practical way to provide power for automobiles at present but its usage does significantly reduce the emission of carbon monoxide, hydrocarbons, and oxides of nitrogen into the atmosphere. Research into new, clean fuel systems represents a positive approach to the pollution problem and, coupled with micrometeorological studies, will provide extremely valuable data for the future. The current program of work at McMaster University (Rouse and McCutcheon 1972) represents a modern micro-approach to the regional pollution of the Hamilton area, including the development of strong atmospheric heating and temperature inversions associated with the pollutant-modification of the incoming radiation fluxes over the city. Together with the Oke and Fuggle program at Montreal (1972), which is reprinted at the end of this section, it represents pioneer urban climatological research on the effect of atmosphere pollutants on the incoming long wave radiation.

The Hamilton data revealed an average long wave radiation increase of 11 percent at the industrial site compared with the control station on campus. The infrared excess peak of 33 percent was recorded for the average solar noon and reached maxima of 66 and 48 percent during the high-sun periods of May 6 and June 13 respectively. Rouse and McCutcheon state that:

> One of the most significant results in this study is the balance recorded between decreased solar radiation and increased infrared radiation. This balance was maintained for the whole study period and for individual days and part days. Thus there is no notable difference between the net incoming radiation load experienced by downtown areas and the suburbs; however, the former is subjected to substantially less sunlight and substantially more infrared radiation.

Increased pollution represents only one of many climatic implications of urbanization, although it does modify other meteorological variables with its direct influence on turbidity, visibility, and condensation nuclei. Consequently, urban climatological research in Canada has been concerned with other urban-induced parameter changes, particularly temperature patterns and the heat island development, which was the topic for the first-ever significant research paper related to Canadian urban climates written by Middleton and Miller in 1936 and associated with temperature profiles in Toronto. Since 1950, the amount of research into urban temperatures has increased considerably; for example: Thomas (1953, Toronto); Einarsson and Lowe (1955, Winnipeg); Crowe, Kuchinka, and Ross (1962, Winnipeg); Oke and Hannell (1970, Hamilton); Munn, Hirt, and Findlay (1969, Toronto); and Oke and Fuggle (1971, Montreal). Studies of other urban parameters include Robertson 1955 (Edmonton, fog); Summers 1963 (Montreal, wind); Potter 1961 (snowfall/temperature relationships) and Wright 1966 (Vancouver, precipitation).

The most recent studies in urban heat island development are by Oke and Hannell (1970) in Hamilton, Munn, Hirt, and Findlay (1969) in Toronto and Oke and East (1971) in Montreal.

The Hamilton survey (1970) examined heat island development under "ideal" conditions, associated with clear skies and winds from the south or southwest at $0 - 1.3$ m sec^{-1}, when the temperature patterns revealed four recurrent features. For example:

1) The heat island core had a magnitude of at least 6°C, but definite values were difficult to obtain due to topographic and other complicating factors. There appeared to be two separate cells to the heat island, firstly the large dominant cell of warm air centred on the heavy industrial zone to the north and attributed to the release of vast amounts of heat generated in the steel-making

process together with counter-radiation from the pollution haze. A secondary cell was located over the central business district, directly related to building density and latent heat of the urban fabric. The temperatures in this cell were often the same as those in the industrial zone and the cell was only relegated to a secondary role on the criterion of size.

2) The upper level of the city on the Niagara escarpment was dominated by two small distinct cells of warmer air related to commercial establishments and the concentration of tall apartment buildings in this area.

3) The re-entrant valleys encouraged katabatic (cold air) drainage down to the coastal plain, with the incursion of cold tongues of air on the western and eastern flanks of the heat island.

4) During the months of October and November the harbour and lakefront water did not exhibit a very strong influence on the heat island.

The survey also considered the effects of other meteorological conditions on the heat island development. For example, a northerly airflow drifted pollutants into the city (a situation producing the greatest pollution hazard) and displaced the two major heat island cells southwards into the same location as the highest particulate concentrations. The cells now became more clearly defined and lost their connection with air over the harbour. Analysis of all the surveys conducted suggested that the heat island effect of Hamilton was reduced to insignificant proportions when wind speeds were greater than 6–8 m sec^{-1}, with the resultant temperature differences between the urban and rural parts less than 1°C.

The Toronto survey (1969) examined the day-time heat island in the metropolitan area and its interaction with lake breeze patterns (including the modification of the lake breeze as it moves across the built-up area). The summary of results indicated that, when winds are off-water, the heat island was located to the northwest of the Bloor Street station in both summer and winter and over individual days. The displacement was greater in summer than in winter when, at the latter time, rural-urban snow cover and albedo differences became important. When the winds were off-land, the heat island was located slightly to the north-northeast of the Bloor Street station in summer. However, in winter with strong winds, the heat island was displaced southeastwards to the lake shore and, with lighter winds, the average position of the centre was 13 km to the northwest, although this latter pattern may be due to the small number of observations.

In a number of cases, the displacement of the urban heat island could be explained by the prevailing wind but other factors, such as surface roughness, intensity of heat sources, gravity winds, and sheltered topography, were also important considerations.

The Montreal survey (1971) indicated that, like Hamilton, Ontario, the city's thermal regime was complicated by controls such as topography and proximity to a water body. The heat island intensity varied between 7°C in March and 12°C in January (at times of anticyclonic stagnation) when the urban and rural boundaries of the city produced the familiar strong temperature gradient or cliff at the edge of the heat island. The extreme gradients of 4.0°C km^{-1} compared favourably with 3.8°C km^{-1} at Hamilton (1970) under similar experimental conditions. The interior of the heat island was relatively complex unlike that at Toronto (1969), but it was identical to the Hamilton (1970) pattern of haphazard organized cells of both warmer and cooler air. The cooler air was located in open areas within the city, for example, at Mt. Royal Park (similar to the Gage Park cold air-pool in

Hamilton) whereas the warm spots corresponded to the location of community shopping areas and the main downtown high-rise commercial centre.

Warming and cooling rates in urban, suburban, and rural areas were obtained to provide a new dimension to the normal static heat island studies. The analyses revealed that cooling rates in different parts of the city and surrounding countryside vary greatly and that the heat island intensity is closest to its maximum at 2100 hours EST.

The Montreal heat island was also studied with respect to both its horizontal and vertical extents to provide the first integrated picture of an urban heat island. Potential temperatures were used to determine the stability or instability of an air layer, which is fundamental to the diffusion of aerosols. The integrated surveys indicated that, under conditions of strong rural stability, the lowest layers of the urban atmosphere became progressively modified as air moved towards the centre of the city. Studies of the temperature excess at different elevations revealed a generalized trend (for all wind speeds) of a maximum heat island at the surface decreasing up to 200 m at which point there was a distinct break and a layer of constant excess up to 400 m. Above this height, the heat island effect was considered negligible. The existence of two internal boundary layers reflected the influence of two source levels of heat input. The main heat source was the city fabric which produced the lower boundary layer (and surface temperature excess) whereas stack heat-releases produced an additional but less effective heat input above the ground (the upper boundary layer of constant temperature excess, between 200–400m).

The most neglected aspect of urban climatology is the effect of urbanization on the surface heat balance and atmospheric environment of man's major habitat (Oke, Yap, and Fuggle 1972). Much research is presently underway on this topic, e.g., Oke, Yap, and Fuggle (1972), who are investigating "the notoriously complex turbulent exchanges of energy and mass. In particular . . . with the field determination of convective sensible heat transfer between the atmosphere and the urban interface." The heat transfer approaches include the energy balance, aerodynamic and eddy correlation techniques and the survey concludes that the eddy correlation method is most likely to achieve success in the context of the city. "It involves the least assumptions and has the great advantage of measuring the flux directly without need to resort to flux-gradient relations. In addition, this method performs without specific reference to the nature of the surface." Observations in Montreal provided preliminary data for the three heat flux techniques. The energy balance value of sensible heat transfer (H) indicated stable, reasonable values at midday, with H typically $= 20$mW cm^{-2} and in general $= 0.20 - 0.40$ net radiation (Rn). The most interesting part of the energy balance data is the Bowen ratio (which partitions energy between sensible and latent transfers), where the diurnal variations for an urban site ranged from 0.3 (between 1800-2000 hours) to a peak about 1.6 at midday. This diurnal course appeared quite unique, with positive values at all times reflecting the heat island effect which makes inversions rare in the lower urban atmosphere. It also appeared that a greater proportion of energy was used for the latent heat transfers to indicate that the city is less of a desert than has been suggested (the vapour source being possibly related to automobile, industrial, and domestic combustion). No meaningful estimates of H could be derived from the aerodynamic approach, whereas the H values obtained by eddy correlation techniques were encouraging in comparison with energy balance fluxes, but lacked consistency.

Canadian research in urban climatology has progressed remarkably since the early investigations of Barrett, Middleton, and Millar in the 1930s. Future studies of the micrometeorology of air pollution and the heat island will provide valuable information on the climatic implications of the urban sprawl, particularly if the energy balance data recorded at a limited number of study sites can actually represent the general urban conditions.

COMPARISON OF URBAN/RURAL COUNTER AND NET RADIATION AT NIGHT*

T. R. Oke and *R. F. Fuggle*

Abstract. Comparison of incoming long-wave radiation at night in urban and rural locations has been achieved via direct measurements. Counter radiation and air temperature were continuously monitored during automobile traverses across the Island of Montreal on clear nights. Results show that urban counter radiation values are consistently higher than rural values. The cross-Montreal radiation profiles closely parallel the form of the urban heat island.

Comparison of measured counter radiation with that computed from empirical equations shows a fair absolute agreement, but important urban/rural differences. Analysis of these differences suggests that the urban counter radiation anomaly is explainable on the basis of the vertical temperature structure of the city alone, without need to consider emissivity changes brought about by air pollution. This suggests that the increased counter radiation is a result, not a cause of the urban heat island.

Estimates of net long-wave radiation show only small urban rural differences. Thus the increased counter radiation appears to be nullified by an increased emission from the warmer city surface. The study indicates that radiative exchanges probably play a minor role in nocturnal urban/rural surface energy balance differences.

1. Introduction

Our understanding of the most basic energy exchanges necessary for a sound climatology is almost totally lacking for the urban environment. Even the workings of the fundamental radiative components remain largely speculative. At present there is a considerable drive to produce workable atmospheric dispersion models which take account of the complexities of the urban interface. There is, however, evidence that our ability to model has outstripped the physical data base (e.g., Stern, 1970). This leads to the use of unfounded assumptions regarding input data and boundary conditions, and little chance of 'validating' the output against the real-world situation. The present study is part of a larger project designed to provide some further insights into the energy exchanges and conversions at the atmosphere/city interface, and thereby to provide more realistic input criteria for modelling.

The radiation balance for any surface is

$$Rn = (Q + q)(1 - x) + I\downarrow - I\uparrow \qquad (1)$$

where, Rn – net surface all-wave radiation; Q, q – direct, and diffuse short-wave radiation; x – surface albedo; $I\downarrow$ – long-wave radiation emitted by the atmosphere (counter radiation); $I\uparrow$ – long-wave radiation emitted by the surface.

Some work is available regarding urban/rural $(Q + q)$ differences due to attenuation by urban pollutants (De Boer, 1966 for Rotterdam; East, 1968 for Montreal). The other term in Equation (1) which is determined by the nature of the atmosphere is $I\downarrow$. It is widely regarded that urban areas experience increased $I\downarrow$ due to atmospheric pollution, and that this

* Reprinted with permission from *Boundary-Layer Meteorology*, Vol. 2, 1972, pp. 290-307. Copyright © by D. Reidel Publishing Company, Dordrecht-Holland.

is one of the causes of the urban heat island effect (Kratzer, 1956). On closer inspection this statement is found to be almost totally unsupported by observational evidence. Bach and Patterson (1969) report increased $I\downarrow$ over Cincinnati compared to a rural station. Unfortunately their results are from daytime observations using economical net radiometers, and the reported differences lie within the accuracy of these instruments. Terjung (1970) gives some intra-urban variations in radiant sky temperature over a small portion of Los Angeles, but no rural values are available for comparison, and the data are complicated by cloud. Yamamoto (1957), using a radiation chart, concluded that aerosols contribute $<1\%$ extra to $I\downarrow$ from unpolluted skies. In summary, therefore, we may conclude that there is a dearth of information regarding $I\downarrow$ for urban areas, and that available data are crude or in conflict.

The objective of this paper is to establish the reality and size of any urban increase in counter radiation, based on a direct measurement approach, employing accurate instrumentation. In addition, it is hoped to provide some information on the role of counter radiation in the production of the urban heat island, and in the urban surface energy balance. Observations are restricted to nights because then the radiation balance equation is abbreviated to the simple form:

$$Rn = I = I\downarrow - I\uparrow \qquad (2)$$

where, I — net long-wave radiation. Because cloud would considerably confuse the $I\downarrow$ picture, only clear nights were used. In addition, the urban heat island effect attains its maximum intensity at night in Montreal (Oke and East, 1971) and this should maximize chances of establishing any links with $I\downarrow$.

2. Instrumentation

The observational programme involved the measurement of air temperature and atmospheric counter radiation, during automobile traverses across the Island of Montreal. The instrumentation was designed to yield high accuracy, even though combined with portability and fast response.

The radiation sensor was a polyethylene-shielded net radiometer (Swissteco Model S1) equipped with a uni-directional adaptor over the lower thermopile surface. In the absence of short-wave radiation at night, this arrangement provides on effective infra-red pyrgeometer. The radiometer time constant was 23 s, and the thermopile output was $\simeq 0.39$ mV/mW cm^{-2}. The upper thermopile surface was covered by a moulded polyethylene dome, 0.5 mm in thickness. The dome was kept rigid by a controlled flow of dry-nitrogen at $\simeq 35$ ml min^{-1}, and a pressure of 0.30 kg cm^{-2}. This is stronger than the normal purging rate when the sensor is maintained at a static site. The extra strength was deemed necessary to prevent deformation of the dome when the automobile was in motion. Laboratory tests showed the sensor output to be insensitive to flow rate. The instrument is provided with ports for external ventilation, but this feature was not utilized due to the problems of operating a compressor in an automobile. In addition, it was felt that adequate ventilation to offset condensation or the settling of aerosols was provided by the motion of the vehicle.

The adaptor was a hemispheric black-body cavity whose inner surface temperature was continuously monitored with a copper-constantan thermocouple. The thermocouple was referenced to a thermostatically-controlled constant temperature bath. The adaptor responded to ambient temperature changes. This was acceptable since a continuous temperature record was kept, but it introduces problems in reducing the data. To offset this, the adaptor was embedded in a foam-rubber block and short-period fluctuations (<15 min) were eliminated. Counter radiation values were obtained by computing the cavity output via Stefan-Boltzmann's Law (assuming a cavity emissivity of unity), and subtracting this value from the instrument's total output.

The signals from both the thermopile and the cavity were continuously monitored on a potentiometric recorder (Honeywell Electronik 194). A portion of the thermocouple signal was bucked-out to allow use of sensitive recorder ranges.

Air temperature was measured with a small thermistor probe mounted in a shield made of

Fig. 1   Radiation and temperature sensors mounted on the automobile.

polyvinyl-chloride tubing covered with reflective tape. The time constant of the probe was 0.5 s, and artificial aspiration was provided at a rate of 3.5 m s^{-1} by a small blower fan. A recording was registered every 0.25 s.

The net radiometer and temperature sensor were mounted on the roof of an automobile at $\simeq$1.5 m (Figure 1). Since observations were taken with the car in motion, levelling of the radiometer could only be made before a traverse. However, the route was level, and since physical rocking of the car whilst stationary caused no discernible fluctuations in the trace, this error was assumed negligible. In general a constant speed of 50 km hr^{-1} was maintained during a traverse, but Table 1 shows that counter-radiation ($/\downarrow$) values are not seriously affected by the car's speed between 30 and 100 km hr^{-1}. Similarly, tests showed the air temperature measurements to be unaffected if speeds were <70 km hr^{-1}.

Considering all the probable sources of error, it is considered that the absolute accuracy of the $I\downarrow$ observations is within 2.0 mW cm^{-2}, and that the relative accuracy within a particular traverse is $\leq$0.2 mW cm^{-2}. Similarly, the air temperature data possess an absolute accuracy of 0.5°C. and a relative accuracy of 0.2°C.

TABLE I
Effect of speed of automobile on $I\downarrow$ measurements

Speed (km hr^{-1})	<30	30–65	65–100	>100
$I\downarrow$* (mW cm^{-2})	0.998	1.000	1.004	1.006

$I\downarrow$* is the ratio of the value of $I\downarrow$ for the indicated speed range to the value for the speed range used in traverses (i.e., 30–65 km hr^{-1}).

3. Field Operations

The observation traverse route was arranged so as to completely cross the city of Montreal. The route, and the generalized land-uses of the Montreal region are shown in Figure 2. The main concentration of population (> 1 million) is on the Island of Montreal, which is essentially flat except for Mount Royal (200 m), which is an isolated hill to the south-east of the traverse route. The downtown core lies between the St. Lawrence River and Mount Royal. The rest of the city is characterized by mixed commercial, light industrial and residential land-uses. The ribbon of industrial uses lies south and east of the traverse route, except in the north-east where it runs through a large complex of oil refineries.

The traverse route (Figure 2) followed highway 40, completely crossing the Island of Montreal along a south-west/north-east transect. Moving from the south-west, the traverse

passed successively through rural, suburban, urban (commercial), suburban, industrial and back to rural land-uses. The portion of the route between Decarie Boulevard and Pie IX Boulevard is elevated to about 10 m. This is considered beneficial because this section contained the tallest buildings, which might otherwise have been in the radiometer's field of view. The route is a major 4–6 lane controlled-access freeway and hence buildings are set well back from the road, reducing view factor problems.

Each traverse consisted of two complete transects of the route in opposite directions. The recorder traces were annotated *en route* and additional descriptive comments were tape-recorded. (As an interesting practical note, the radiation trace provided its own event marks as sudden 'spikes' in the signal, due to travelling under overpasses which radiate strongly.) By maintaining a constant speed, and assuming that both air temperature and $I\downarrow$ varied linearly

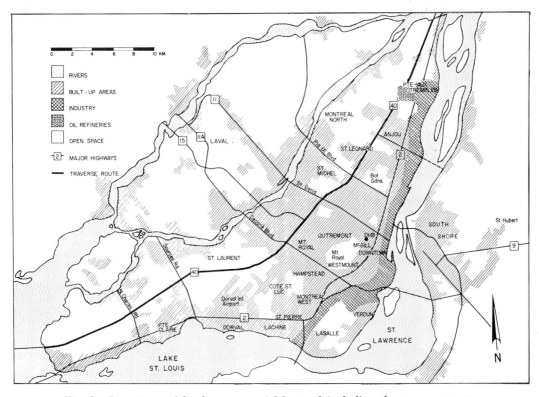

Fig. 2   Location and land-use map of Montreal including the traverse route.

with time, it was possible to average the data for any location and obtain results which were standardized to the mid-time of the traverse period. A typical traverse period was about 90 min. Only those traverses which showed <5% variation in $I\downarrow$ values between the outward and inward legs are presented here.

4. Traverse Observations

Counter-radiation $(I\downarrow)$ and temperature (T) traverses were conducted simultaneously during the period August 1969 to July 1970; two were during the winter and the rest in the spring and summer months. Twelve traverses satisfied the steady-state conditions mentioned above, and a summary of the data and the prevailing meteorological and pollution conditions is given in Table II.

Figure 3(a–c) shows a comparison of traverse results from three nights during June 1970. In this and following figures, the traverse route has been projected onto a straight line. This slightly exaggerates gradients of the measured parameters between St. Charles Road and Decarie Boulevard. With respect to cloud, wind direction and water vapour pressure, these nights are almost identical, but it will be noted (Table II) that wind speeds vary from 0.4 to 5.0 m s⁻¹.

The $I\downarrow$ and T profiles across the city parallel each other very closely on each occasion. Both parameters show a marked increase from a rural background level immediately upon encountering the built-up area, at the so-called heat-island 'cliff'. Both $I\downarrow$ and T peak at approximately the same location which is always within the urban area. Finally, both parameters decrease in value downwind of the urban centre, but the $I\downarrow$ values do not always appear to assume immediately their upwind rural levels, especially with good ventilation (e.g., Figure 3(a)).

In the following analyses, we will refer to urban/rural comparisons of a number of meteorological parameters. These will be indicated as a finite difference of the variable's highest urban, and lowest upwind rural values (e.g., ΔT_{u-r} is the urban heat-island intensity). In

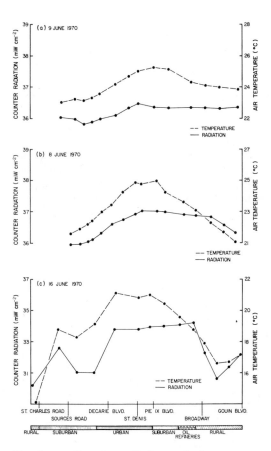

Fig. 3a-c   Counter radiation and air temperature profiles across Montreal for (a) June 9, 1970, wind SW 5.0 m s⁻¹, (b) June 8, 1970, wind S 2.0 m s⁻¹, and (c) June 16, 1970, wind W 0.4 m s⁻¹.

Figure 3 (a–c) and Table II, it is evident that for the illustrated three June days, as the wind speed $(\bar{u})$ decreases from 5.0 to 0.4 m s⁻¹, the values of ΔT_{u-r} increase from 2.2 to 7.0°C, and $\Delta I\downarrow_{u-r}$ increases from 0.7 to 4.0 mW cm⁻².

Figure 4 shows the traverse for February 15, 1970, one of the winter runs. Winds were SE 0.8 m s⁻¹, skies were clear, air temperatures very low ($\simeq -20$ C), and water vapour pressures correspondingly small ($\simeq 1$ mb). The urban heat island was impressively displayed, with $\Delta T_{u-r} = 9.5$°C (further information on this case, including urban/rural cooling rates and

TABLE II
Summary of traverse and general meteorological data

Date	Time[a] (EST)	$\bar{u}$[b] (m s^{-1})	Cloud[b]	e[c] (mb)	COH[d]	$\Delta T_{u\text{-}r}$[e] (°C)	$\Delta I\!\downarrow_{u\text{-}r}$[e] (mW cm^{-2})	$\Delta I\!\downarrow_{u\text{-}r}$[f] % increase
Aug. 12 1969	2140	SW 2.7	0	16.9	0.7	3.6	1.5	4.5
Oct. 30 1969	0100	N 1.8	0	3.9	1.2	4.5	1.2	5.4
Feb. 15 1970	0030	SE 0.8	0	1.3	1.4	9.5	4.0	25.3
May 23 1970	2200	W 3.0	AC 1	9.9	0.6	1.6	0.7	2.4
May 28 1970	2110	SW 1.8	0	7.9	0.9	3.7	1.2	4.0
June 8 1970	2140	S 2.0	Ci 1	16.0	0.7	3.4	1.1	3.1
June 9 1970	2045	SW 5.0	0	16.0	0.6	2.2	0.7	2.0
June 14 1970	2140	SSW 3.5	Ci 1	9.8	0.7	4.0	1.7	5.3
June 16 1970	2315	W 0.4	0	16.1	1.2	7.0	4.0	13.2
June 22 1970	2100	WSW 6.2	AC 1	12.9	0.4	1.8	0.6	1.8
June 25 1970	2100	W 2.6	SC 1	9.5	0.5	2.1	1.3	4.5
June 28 1970	2150	SW 4.5	0	11.0	0.5	1.1	0.6	1.9

[a] Mid-time of traverse.
[b] Average wind at Dorval.
[c] Vapour pressure at McGill Observatory.
[d] Average coefficient of haze of 3 urban stations along the traverse.
[e] Using highest urban and lowest upwind rural values.
[f] $\Delta I\!\downarrow_{u\text{-}r}/I\!\downarrow_r \times 100$.

the temporal variations of $\Delta T_{u\text{-}r}$, is given in Oke and East, 1971). On this occasion the $I\!\downarrow$ profile parallels the T profile across the route until it reaches the downwind urban/rural boundary where there is an even larger peak in $I\!\downarrow$ than that found in the city centre. It is felt that this apparent anomaly in the general pattern described above is due to the oil refinery complex. The main part of this complex lies to the south-east of the traverse route, and so a wind from this sector carries considerable effluent across the observation route. During the winter, this plume effect is likely to be most marked on $I\!\downarrow$ because the saturation vapour pressure is so low that water vapour almost immediately condenses into dense cloud after leaving the stacks, thus producing a very effective infra-red black-body absorption and emission layer at about 200 m. In addition, the temperature difference between the ground and the stack gases is at a maximum in winter, and hence the radiation exchange between them is also at a maximum. Similarly, the other winter run with low temperatures and humidities

showed a peak in the same area. In summer with light winds (Figure 3(c)), $I\!\downarrow$ is again seen to show an increase in the oil refinery area, but the peak is less pronounced.

From the foregoing it seems certain that an urban area such as Montreal does receive an increase in counter radiation, at least at night. The urban/rural increases, $(\Delta I\!\downarrow_{u\text{-}r})$, may be as great as 4.0 mW cm^{-2} ($\simeq 0.057$ cal cm^{-2} min^{-1}), and this could represent as much as a 25% increase over rural values. However, these are extremes, and an increase of 1 to 2 mW cm^{-2} ($\simeq 0.014$ to 0.028 cal cm^{-2} min^{-1}) appears more representative (Table II). The following section discusses the relation between this increase in $I\!\downarrow$ and atmospheric parameters which might be expected to be related.

5. Urban/Rural Counter Radiation Differences

Having established that $I\!\downarrow$ is greater in the urban area, a physical cause must be sought. This study was not originally designed to an-

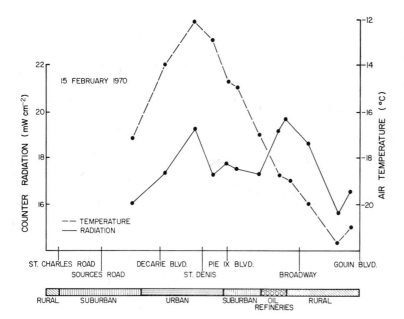

Fig. 4 Counter radiation and air temperature profile across Montreal for February 15, 1970, wind SE 0.8 m s⁻¹.

swer this question, but the following discussion may provide some insight into the problem.

Essentially we are concerned with the question whether the observed urban increase in $I\downarrow$ is due to an increase in the emissivity of the urban atmosphere, or whether it is simply due to the increase in urban air temperature. The former could be the result of a change in the atmospheric constituents which affect the exchange of long-wave radiation (e.g., water vapour, aerosols or gases). The latter could be due to non-radiative warming of the atmosphere through the turbulent sensible heat flux drawing upon artificial heat generation and stored heat, in the city.

First of all we may consider simple scatter plots of $\Delta I\downarrow_{u-r}$ versus pollution (COH), heat-island intensity (ΔT_{u-r}) and wind speed ($\bar{u}$) as shown in Figure 5(a–c). All three parameters show recognizable relationships to $\Delta I\downarrow_{u-r}$, but statistical correlations were not performed due to the small numbers of data points, and because the sample population is not normally distributed with respect to $\Delta I\downarrow_{u-r}$.

It appears that as the COH value increases, so does $\Delta I\downarrow_{u-r}$. The same type of response is even more clearly defined for ΔT_{u-r}, and as the heat island becomes negligible (< 1 °C), so do counter-radiation differences. These scatter

plots do not, however, help resolve the question of causation, and Figure 5(c) helps explain why. Wind speed exerts a strong influence on both COH and ΔT_{u-r} (see Table II), the former through its effect on diffusion, and the latter via advection.

The question of causation may, however, be studied by comparing the observed $I\downarrow$ values with those computed from empirical relations. The equations used were those of Ångström (1916), Brunt (1932), Swinbank (1963) and Idso and Jackson (1969). The equations and their constants are given in Table III. It will be noted that the Ångström and Brunt equations require both air temperature and humidity at screen-level, whereas the Swinbank and Idso and Jackson equations require only temperature measurements. In this study the 1.5-m traverse temperatures, and the McGill Observatory (Figure 2) vapour pressure observations were used in the absence of traverse humidities. Analysis of the Dorval (rural) and McGill (urban) vapour pressure readings showed that between 1900 and 0100 EST, urban/rural differences were <2 mb. The maximum error in computed $I\downarrow$ using *only* McGill data was found to be 0.24 mW cm⁻² at 25°C and 20 mb, and 0.25 mW cm⁻² at 0°C and 5 mb. These errors are approximately equivalent to instrumental errors and were considered acceptable. **The empirical coefficients** were not adjusted to provide a 'best-fit' between observed and calculated values for the Montreal area. Similarly no corrections were applied for the small amounts of Cirrus clouds observed on some nights (Table II).

Figure 6 shows a typical plot of the measured and calculated $I{\downarrow}$ values. From this and from data obtained on other nights, it was concluded that the empirical methods gave results usually within 5% of the measured values. The Angström equation consistently underestimated in both urban and rural areas, whereas the other three equations tended to underestimate in rural areas but approximate or overestimate in the urban area.

The real significance, however, lies not in the agreement between the measured and calculated values of $I{\downarrow}$, but in the trend of their differences. Figure 7(a–c) shows the differences between measured and calculated values for the three nights illustrated in Figure 3, using one empirical method requiring both temperature and vapour pressure (Brunt), and one requiring temperature alone (Idso and Jackson). (There is no special significance in this choice.) It should be noted that the *absolute* error in obtaining both calculated and observed values of $I{\downarrow}$ exceeds their difference and therefore the sign has no significance. However, the *relative* error between values on the same traverse is an order of magnitude smaller than their differences, therefore urban/rural variations are real and significant. Again it is important to stress that the absolute agreement between the two methods, or the agreement between the predicted and observed values is not particularly important. What is important is that each difference curve shows a dip which coincides with the urban area. This feature was found on almost all occasions. This dip indicates that the urban area receives less counter radiation

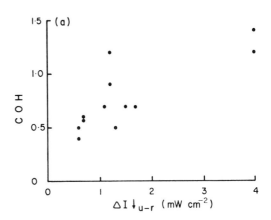

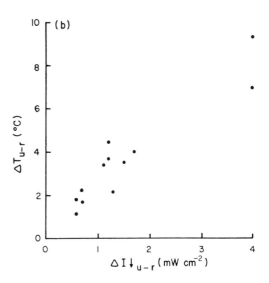

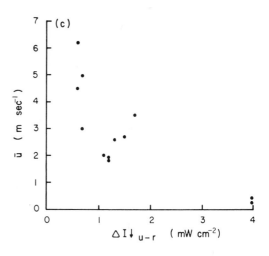

Fig. 5a-c Urban/rural counter radiation differences versus (a) COH, (b) urban heat island intensity and (c) wind speed in the Montreal region.

TABLE III
Empirical equations used to calculate $I\downarrow$ for clear skies (T in °K, e in mb, $I\downarrow$ in mW cm^{-2})

Author	Equation	Constants
Angström (1916)	$I\downarrow = \sigma\epsilon T^4[a-b\exp(-2.3\,ce)]$	$a = 0.82$ $b = 0.25, c = 0.94$[a]
Brunt (1932)	$I\downarrow = \sigma\epsilon T^4(a+b\sqrt{e})$	$a = 0.604$ $b = 0.048$[b]
Swinbank (1963)	$I\downarrow = 9.35\times10^{-6}\,\sigma\epsilon T^6$	$c = 0.261$
Idso and Jackson (1969)	$I\downarrow = \sigma\epsilon T^4\{1-c\,exp[-d(273-T)^2]\}$	$d = 7.77\times10^{-4}$

[a] Following Morgan et al. (1970)
[b] Following Sellers (1965)

relative to the rural area than is to be expected on the basis of calculations from surface temperature variations. This decrease is of the order of 1 to 2 mW cm^{-2}.

Considering the nature of the empirical equations, one, or both of two features of the atmosphere could account for this apparent decrease in calculated $I\downarrow$ for the urban area. Either the urban atmosphere has a lower emissivity than that in the country, or the vertical temperature structure is considerably different between the two areas. These two possibilities are explored further below.

It seems highly improbable that the addition of pollutants to the urban atmosphere would decrease its emissivity. Theoretical analyses by Yamamoto (1957) and Zdunkowski et al., (1966) indicate that the effects of smoke and haze are small, but that they tend to increase rather than decrease sky emissivities. Similarly Möller (1951), Monteith (1961) and Atwater (1971) have all argued that air pollution will produce an increase in $I\downarrow$. Inversely it might be argued that a reduction of water vapour in the city could produce a lowered atmospheric emissivity. This, however, is contrary to the observations of Chandler (1962, 1967) in London and Leicester, and Ackerman (1970) in Chicago, which all show slightly higher amounts of water vapour in the urban areas. In addition, it can be quantitatively shown that a 10 mb urban/rural vapour pressure difference is required to produce the 2 mW cm^{-2} difference

in $I\downarrow$. This grossly exceeds any reported urban/rural humidity differences, and does not agree with the Dorval-McGill analysis reported above.

On the other hand, the reported urban/rural differences in vertical temperature structure provide a plausible and sufficient explanation for the calculated decrease in sky emission over the urban area. The recent studies of Bornstein (1968) for New York, Clarke (1969) for Cincinnati and of special interest in this study, Yap et al. (1969), and Oke and East (1971) for Montreal have shown a similarity in urban/rural profile forms on clear nights. Characteristically the rural areas show surface-based inversions, whereas the urban area exhibits adiabatic or weak lapse profiles within a shallow (200-300 m) boundary layer. In Montreal this urban mixing layer often exhibits a lapse rate of −6°C km^{-1}. The effect of a rural inversion is to provide a warmer radiating atmosphere aloft which will give greater $I\downarrow$ than predicted from surface temperatures. Conversely, the urban lapse profile provides a cooler atmosphere aloft, thus decreasing $I\downarrow$ compared to that expected from surface temperature observations.

Möller (1951) has shown that variations in temperature gradient are capable of producing marked differences in $I\downarrow$ from clear skies. Table IV shows that the $I\downarrow$ difference between an urban lapse profile and a rural inversion profile is almost exactly the same as that needed to

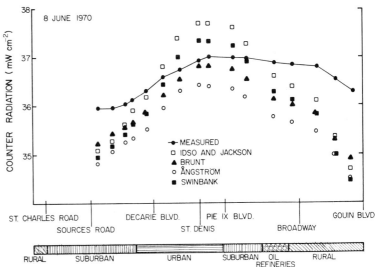

Fig. 6 Comparison of measured counter radiation across Montreal with that calculated from the Angstrom, Brunt, Swinbank and Idso and Jackson empirical methods, for June 8, 1970.

Fig. 7a-c Profiles of the difference between measured counter radiation and that computed to Brunt, and Idso and Jackson for (a) June 9, 1970 (b) June 8, 1970, and (c) June 16, 1970.

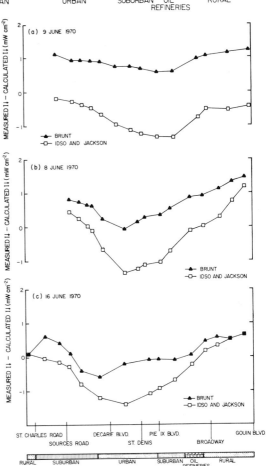

match the measured and computed $I{\downarrow}$ values in Montreal, and in its upwind rural area. This reasoning also explains why downwind rural differences between measured and calculated $I{\downarrow}$ were often greater than upwind rural differences (e.g., Figure 7(b) and (c)). In the downwind area when the surface temperatures drop and a surface-based inversion begins to develop, an 'urban plume' lies overhead (Clarke 1969). This elevated warm layer will augment the value of $I{\downarrow}$ and hence the underestimation based on cool surface temperatures.

The above analysis therefore shows that the role of pollutants and water vapour is probably not as important as the modification of the vertical temperature structure in contributing to an increase in $I{\downarrow}$ over the urban area. Indeed the numerical analysis of Atwater (1971), and the infra-red flux divergence measurements of Fuggle (1971), both indicate that the role of

TABLE IV

The effect of the vertical temperature distribution on $I\downarrow$ (after Möller, 1951)

Inversion thickness (m)	Temperature increase (°C)	$I\downarrow$[a] (mW cm^{-2})	Differences from lapse of 6°C km^{-1} (mW cm^{-2})
100	9.4	25.11	1.81
200	8.8	24.83	1.53
300	8.2	24.62	1.33
400	7.6	24.48	1.18
500	7.0	24.41	1.11
1000	0.0	23.99	0.70
	6°C km^{-1}	23.29	0.00
Lapse rate	Adiabatic	23.23	−0.06

[a] Based on a surface temperature of 0°C, and a total water vapour path of 1.12 g cm^{-2}.

pollutants is to cool rather than to warm the atmosphere, and would thus serve to decrease rather than increase the value of $I\downarrow$ in an urban atmosphere. It would appear, therefore, that counter radiation does not result from an increase in emissivity in polluted air and therefore does not represent a source of energy preferentially found in the city. As such, the observed increase in $I\downarrow$ is probably an effect rather than a cause of the urban heat island. Certainly, based on the calculations of Oke and East (1971), the energy necessary to support the modification of the urban boundary layer is at least 10 times greater than the observed 1 to 2 mW cm^{-2} increase in $I\downarrow$, even if it were considered to be a uniquely urban phenomenon.

6. Urban/Rural Net Radiation Differences

Using the measured $I\downarrow$ values as a base, it is possible to arrive at some preliminary estimates of the net long-wave radiative exchanges in Montreal and the upwind rural areas. Since at night the net long-wave radiation (I), represents the net all-wave radiative exchange (Rn), (Equation 2), this allows some insight into the possibility of using urban/rural differences to infer the basic radiative component of the surface energy balance.

Table V presents the components of the long-wave radiation balance for both the urban

and rural areas in and around Montreal for the nights on which traverses were conducted. The chosen points represent the maximum or minimum values in each case (e.g., the lowest upwind $I\downarrow_r$, and the highest $I\downarrow_u$). The $I\downarrow_u$, $I\downarrow_r$ and ΔI_{u-r} values were measured. Using the traverse temperature data, $I\uparrow_u$, $I\uparrow_r$ and $\Delta\uparrow_{u-r}$ values were computed using the simple Stefan-Boltzmann relation. In doing so it was necessary to assume that urban and rural emissivities were the same and equal in unity, and that the 1.5-m air temperature approximated that at the surface. The advisability of employing these assumptions is considered later. Having the above values, I_u, I_r and ΔI_{u-r} were easily obtained.

The individual $I\downarrow_u$ and $I\downarrow_r$ values show a considerable range due to ambient temperatures, but as noted previously their differences consistently show an increased receipt of radiation in the city. Similarly $I\uparrow$ values show a variation due to absolute temperatures, and because $\Delta I\uparrow_{u-r}$ is proportional to ΔT_{u-r} $I\uparrow$ always shows a greater loss for the warmer urban area at night. The difference lies between 0.6 and 3.9 mW cm^{-2}. The net radiative loss terms for both environments show little variability ranging only from 6.9 to 10.0 mW cm^{-2}.

The most important result of this analysis is the very small ΔI_{u-r} difference exhibited between the urban and rural areas. On two occasions the rural area shows a very slightly higher

TABLE V
Nocturnal urban/rural long-wave radiation components (mW cm^{-2})

	$I\downarrow_u$	$I\downarrow_r$	$\Delta I\downarrow_{u-r}^a$	$I\uparrow_u$	$I\uparrow_r$	$\Delta I\uparrow_{u-r}^a$	I_u $=Rn_u$	I_r $=Rn_r$	ΔI_{u-r}^a $=\Delta Rn_{u-r}^a$
Aug. 12 1969	34.9	33.4	+1.5	−43.9	−41.8	+2.1	− 9.0	−8.4	+0.6
Oct. 30 1969	23.4	22.2	+1.2	−32.0	−30.1	+1.9	− 8.6	−7.9	+0.7
Feb. 15 1970	19.8	15.8	+4.0	−26.7	−22.8	+3.9	− 6.9	−7.0	−0.1
May 23 1970	29.6	28.9	+0.7	−39.6	−38.8	+0.8	−10.0	−9.9	+0.1
May 28 1970	31.1	29.9	+1.2	−39.2	−37.2	+2.0	− 8.1	−7.3	+0.8
June 8 1970	37.0	35.9	+1.1	−45.0	−43.0	+2.0	− 8.0	−7.1	+0.9
June 9 1970	36.5	35.8	+0.7	−45.2	−43.9	+1.3	− 8.7	−8.1	+0.6
June 14 1970	33.5	31.8	+1.7	−42.7	−40.4	+2.3	− 9.2	−8.6	+0.6
June 16 1970	34.2	30.2	+4.0	−42.7	−38.8	+3.9	− 8.5	−8.6	−0.1
June 22 1970	33.7	33.1	+0.6	−42.6	−41.6	+1.0	− 8.9	−8.5	+0.4
June 25 1970	30.1	28.8	+1.3	−39.7	−38.6	+1.1	− 9.6	−9.8	+0.2
June 28 1970	31.9	31.3	+0.6	−41.4	−40.8	+0.6	− 9.5	−9.5	0.0
Average	31.3	29.8	+1.5	−40.1	−38.2	+1.9	− 8.8	−8.4	+0.4

[a] Positive values indicate that the urban value (loss or gain) is greater than the rural value.

radiative loss; the remaining cases show a very slightly larger urban deficit. The greatest urban/rural difference is only 0.9 mW cm^{-2} ($\approx$0.01 cal cm^{-2} min^{-1}), and the average is 0.4 mW cm^{-2} (0.006 cal cm^{-2} min^{-1}).

It is important at this juncture to consider further the nature of the assumptions used in this development. The similarity of urban/rural emissivities remains largely unproven. However, on the basis of previous assessments for concrete and other building materials, and for typical rural surfaces (Sellers, 1965), the indications are that emissivities are not radically dissimilar. This is supported by measurements in Columbia, Maryland (Landsberg, 1971, personal communication). The city may possess a slightly lower emissivity and this would serve to bring urban/rural values even closer, rather than accentuate differences. Considering the assumption that $T_{1.5} = T_0$, if one assumes a reasonable rural temperature difference between the surface and 1.5 m, on a clear night with weak winds, to be 2°C (e.g., Oke 1970), then $I\uparrow_r$ will be overestimated by <1 mW cm^{-2}. In the city with the weak lapse conditions already noted, $I\uparrow_u$ may be very slightly increased. Both of these errors, however, will be reduced at higher wind speeds due to mixing, and it therefore seems unlikely that the figures in

Table V will be radically affected. The analysis therefore indicates that the increase in $I\downarrow$ in the urban area is real, but is completely offset by the increased $I\uparrow$ from the warmer urban surface. This tends to substantiate the view that urban/rural radiation differences are unlikely to alter surface energy balance comparisons radically at night. This view, however, must await vindication through direct measurement of urban/rural net radiation differences. When taking such measurements, one must be careful to choose 'representative' sites, and one must include temporal variations in ΔI_{u-r} since all the traverses presented here were conducted at the time when Montreal's urban heat island is usually at its maximum (Oke and East, 1971). Measurements during the early evening, when ΔT_{u-r} is developing, would therefore be particularly helpful.

7. Conclusion

The results of this study indicate that a large urban area does receive a greater amount of long-wave radiation from the atmosphere at night in comparison with that received by a rural area upwind. This increase is still prevalent downwind of the city. It appears, however, that the increase is due to the increased warmth of the city atmosphere rather than to any change in the radiative properties of the atmosphere such as is often attributed to the pollution 'blanket'. Increased counter radiation cannot therefore be construed to be a cause of the heat island; rather it is the result of an atmosphere which is warmed by other means.

Although the results indicate only small changes in counter radiation due to urban effects, they do indicate that man-made, thermal, gaseous, and particulate pollution of the atmosphere can influence the infra-red radiation balance at the surface. It would be interesting, for example, to determine how many kilometres downwind the 'plume' effect continues. As urbanization proceeds, these ramifications will increasingly assume global importance.

The consideration of urban/rural net radiation differences shows a remarkably small difference in available radiant energy at the surface at night. Indeed, instead of the urban area showing a preferential radiative gain, on most occasions the city exhibits a deficit. Presumably therefore, our investigations regarding heat-island causation should concentrate more deeply upon eddy heat flux, storage and space-heating considerations.

These results tend to support the assumption that long-wave radiation can be omitted in the development of certain heat-island models (e.g., Summers, 1964; Myrup, 1969). However, the conclusions should not be construed as relegating long-wave exchanges to a minor role in the *atmospheric* energy balance, nor should the small *net* radiation differences be allowed to mask important changes in the size and nature of each *component* flux.

Ackowledgements

This research was supported by grants from the Canadian Meteorological Service, and the National Research Council of Canada, to whom grateful acknowledgements are extended.

Thanks are also due to Mr P. Hitschfeld for help in conducting traverses; to the National Atmospheric Radiation Centre, Canadian Meteorological Service, Toronto, for radiometer calibrations; and to the Health Department of the City of Montreal for providing air pollution data.

WEATHER AND CLIMATE BIBLIOGRAPHY

Ackerman, B.: 1970, "Urban-rural differences in the temperature and dew point in the Chicago area," *Argonne National Laboratory, Radiology Physics Division, Annual Report.* ANL 7760.

Adams, W. P. and F. M. Helleiner (Eds.): 1972, *International Geography, 1972,* 2 vols., University of Toronto Press, Toronto, 1354 pp.

Adkins, C. J.: 1958, "The summer climate in the accumulation area of the Salmon Glacier," *Jl. Glac.* 3, pp. 195-206.

Ager, B.: 1962, "Studies on density of naturally and artificially formed fresh-water ice," *Jl. Glac.* 4, pp. 207-14.

Ahrnsbrak, W.: 1968: "Summertime radiation balance and energy budget of the Canadian tundra," *Technical Report 37,* University of Wisconsin.

Andreev, V. N.: 1956, "The vegetation of the far north of the U.S.S.R. and its utilization," quoted in Tikhomirov 1963.

Andrews, R. H.: 1964, "Meteorology and heat balance of the ablation area White Glacier," *Axel Heiberg Island Research Report,* McGill University.

Andrews, J. T. and R. G. Barry: 1967, "Remarks on the retreat of the North American ice sheet," *Izu. Vsesoy. Geogr. Obsch.* (Leningrad), 99, pp. 230-231.

Andrews, J. T., R. G. Barry and L. Drapier: 1970: "An inventory of the present and past glacierization of Home Bay and Okoa Bay, East Baffin Island, NWT, Canada and some climatic and palaeoclimatic considerations," *Jl. Glac.* 9, pp. 337-62.

Angstrom, A.: 1916, "Uber der gegenstrahlung der atmosphare," *Met. Z. 33,* pp. 529-38.

Annersten, L.: 1963, "Ground temperature measurements in the Schefferville area, P.Q." 1st Canadian Conference on Permafrost, Association of Soil and Snow Mechanics, National Research Council, *Technical Memorandum 76,* pp. 215-317.

Arnold, K. C.: 1965, "Aspects of the glaciology of Meighen Is. N.W.T. Canada," *Jl. Glac. 5,* pp. 399-410.

Atwater, M. A.: 1971, "The radiation budget for polluted layers of the urban environment," *Journal of Applied Meteorology (Jl. Appl. Met.)* 10, pp. 205-14.

Bach, W. and W. Patterson: 1969, "Heat budget studies in greater Cincinnati," *Proceedings of the Association of American Geographers* 1, pp. 7-11.

Badgley, F. I.: 1966, "Heat budget at the surface of the Arctic Ocean," in *Proceedings of the Symposium on the Arctic Heat Budget and Atmospheric Circulation.* J. O. Fletcher (ed.) Rand Corporation, Santa Monica, California, pp. 269-277.

Baier, W.: 1964, "The interrelationship of meteorological factors, soil moisture and plant growth," *Agricultural Meteorology (Agr. Met.)* Technical Bulletin 5, Canadian Dept. of Agriculture, Ottawa.

Baier, W.: 1965, "The interrelationship of meteorological factors, soil, moisture and plant growth," *International Journal of Biometeorology (Intl. Jl. Biomet.)* 9, pp. 5-20.

Baier, W.: 1967, "Recent advancements in the use of standard climatic data for estimating soil moisture," *Annals of Arid Zone 6,* 21 pp.

Baier, W.: 1969, "Concepts of soil moisture availability and their effect on soil moisture estimates from a meteorological budget," *Agr. Met.* 6, pp. 165-178.

Baier, W., and G. W. Robertson: 1966, "A new versatile soil moisture budget," *Canadian Journal of Plant Science (Can. Jl. Pl. Sci.)* 46, pp. 299-315.

Baier, W. and G. W. Robertson: 1967, "Estimating yield components of wheat from calculated soil moisture," *Can. Jl. Pl. Sci.* 47, pp. 617-630.

Baird, P. D.: 1950, "Baffin Island Expedition 1950: a preliminary report," *Arctic* 3, pp. 131-149.

Baird, P. D.: 1952, "Method of nourishment of the Barnes Ice Cap." *Jl. Glac.* 2, pp. 2-9.

Barrett, H. M.: 1938, "Atmospheric pollution in Toronto, Canada." *Canadian Public Health Journal* 29, pp. 1-12.

Barry, R. G.: 1960, "The application of synoptic studies in palaeoclimatology: a case study for Labrador-Ungava," *Geog. Annal.* 42, pp. 36-44.

Barry, R. G.: 1967, "Seasonal location of the Arctic front over North America," *Geog. Bull.* 9, pp. 79-95.

Barry, R. G. and S. Fogarasi: 1968, "Climatological studies of Baffin Is. N.W.T.," *EMR IWB Technical Bulletin 13.*

Baynton, H. W.: 1953, "The role of meteorology in the Detroit-Windsor pollution study." *Royal Meteorological Society,* Canadian Branch, p. 49.

Benninghoff, W. S.: 1952, "Interaction of vegetation and soil frost phenomena," *Arctic* 5, pp. 34-44.

Bird, .J B.: 1964, "Permafrost studies in Central Labrador-Ungava," *Subarctic Research Paper 16,* McGill University.

Bornstein, R. D.; 1968, "Observations of the urban heat island effect in New York City," *Jl. Appl. Met.* 7, pp. 575-82.

Boville, B. W.: 1961, "A dynamical study of the 1958-59 stratospheric polar vortex," *Pub. Met.,* McGill University 36.

Boyd, D. W.: 1962, "A field comparison of two types of net radiometers," *Symposium on the Heat Exchange at Snow and Ice Surfaces.* Snow and Ice Subcommittee, Associate Committee on Soil and Snow Mechanics, TM 78.

Boyd, D. W., L. W. A. Gold and G. Williams: 1962, *Radiation balance during the snow melt period at Ottawa,* NRC 7152.

Brazel, A. J.: 1972, "Micro and topo-climatology: the case of an alpine pass: Chitistone Pass Alaska," *International Geography* (Adams and Helleiner, eds.) 1, pp. 133-143.

Brown, D. M.: 1960, "Soybean ecology I, development-temperature relationships from controlled environmental studies," *Agronomy Journal (Agr. Jl.)* 52, pp. 493-96.

Brown, D. M. and L. Chapman: "Soybean ecology II, development-temperature-moisture relationships from field studies," *Agr. Jl.* 52, pp. 496-99.

Brown, D. M. and L. Chapman: 1961, "Soybean ecology III, soybean development units for zones and varieties in the Great Lakes region," *Agr. Jl.* 53, pp. 306-308.

Brown, R. J. E.: 1960, "The distribution of permafrost and its relation to air temperature in Canada and the U.S.S.R.," *Arctic* 13, pp. 163-177.

Brown, R. J. E.: 1961, "Potential evapotranspiration and evaporation observations at Norman Wells, NWT." *Proceedings of Hydrology Symposium No. 2—Evaporation,* National Research Council, March 1961, pp. 123-127.

Brown, R. J. E.: 1964, "Permafrost investigations on the Mackenzie Highway in Alberta and Mackenzie District," *National Research Council Technical Paper No. 175,* (NRC 7885).

Brown, R. J. E.: 1965, "Some observations on the influence of climatic and terrain features on permafrost at Norman Wells N.W.T., Canada," *Can. Jl. Earth Sci.* 2, pp. 15-31.

Brown, R. J. E.: 1966, "Influence of vegetation on permafrost," *Proceedings, Permafrost International Conference,* Nov. 1963, pp. 20-25 (Res. Paper No. 298 DBR).

Brown, R. J. E.: 1967, "Permafrost in Canada," Map 124A (with text), Geological Survey of Canada, Ottawa.

Brown, R. J. E.: 1968, "Permafrost map of Canada," *Can. Geog. Jl.* 74, pp. 56-63.

Brunt, D.: 1932, "Notes on radiation in the atmosphere," *Quarterly Journal of the Royal Meteorological Society (Q. Jl. Roy. Met. Soc.)* 58, pp. 389-420.

Bryson, R. A.: 1966, "Air masses, streamlines, and the boreal forest," *Geog. Bull.* 8, pp. 228-296.

Bryson, R. A.: 1968, "Radiocarbon isochrones of the Laurentide Ice Sheet," Map in *Technical Report* 35, University of Wisconsin, Madison.

Bryson, R. A., W. N. Irving and J. A. Larson: 1965, "Radiocarbon and soil evidence of former forest in the Southern Canadian tundra," *Science* 147, pp. 46-48.

Budyko, M. I.: "The heat balance of the earth's surface," *U.S. Department of Commerce, Office of Climatology,* 1958, (Transl. by N. A. Stepanova, from Gidrometeorologicheskoe izdatel'stvo. Leningrad. 1956).

Budyko, M. I.: 1966, "Polar ice and climate," in *Proceedings of the Symposium on the Arctic Heat Budget and Atmospheric Circulation.* J. V. Fletcher (editor). Rand Corporation 1966, pp. 5-21.

Burbridge, F.: 1951, "The modification of continental air over Hudson Bay," *Q. Jl. Roy. Met. Soc.* 77, pp. 365-74.

Carder, A. C.: 1957, "Growth and development of some field crops as influenced by climatic phenomena at two diverse latitudes," *Can. J. Pl. Sci.* 37, pp. 392-406.

Carder, A. C. and A. M. Hennig: 1966, "Soil moisture regimes under summer fallow, wheat and red fescue in the Upper Peace River Region," *Agr. Met.* 3, pp. 311-31.

Champ, H.: 1962, "A comparison of the diurnal relationships between the mean smokiness and the mean inversion intensity at Detroit-Windsor during June and December," Department of Transport, Meteorological Branch, CIR-3162, TEC-399. Feb. 23, 1962. 34 pp.

Chandler, T. J.: 1962, "Temperature and humidity traverses across London," *Weather* 17, pp. 235-42.

Chandler, T. J.: 1967, "Absolute and relative humidities in towns," *Bulletin of the American Meteorological Society (Bull. Am. Met. Soc.)* 48, pp. 394-439.

Clarke, J. F.: 1969, "Nocturnal urban boundary layer over Cincinnati," *Monthly Weather Review (Mon. Weath. Rev.)* 97, pp. 582-589.

Coachman, L. K.: 1966, "Production of supercooled water during sea ice formation," Rand Corporation 1966, pp. 497-529.

Connor, A. J.: 1938, "The climate of Canada" in Koppen and Geiger *Handbuch der Klimatologie* Bd. 11, pp. 332-424.

Connor, A. J.: 1941, "Snowfall maps of Canada," *Proceedings, Central Snow Conference* (East Lansing) 1, pp. 153-159.

Crawford, C. and G. Johnston: 1971, "Construction on permafrost," *Canadian Geotechnical Journal (Can. Geotech. Jl.)* 8, pp. 236-251.

Crowe, B. W., C. L. Kuchinka and W. A. Ross: 1962, *Modification of extreme winter temperatures at Winnipeg International Airport due to urban development,* Department of Transport, Meteorological Branch, CIR-3754, TEC-432 4 p.

Dahl, E. R.: 1956, "Mountain vegetation in South Norway and its relation to the environment," Skr. utgitt av Vidensk. Ak. Oslo, 1, Mat. Natur. Kl., 3.

De Boer, H. J.: 1966, "Attenuation of solar radiation due to air pollution in Rotterdam and its surroundings," Koninklijk Nederlands Meteorologisch Instituut, *Wetenschappelijk Rapport W. R. 66-1,* 36 pp.

de Quervain, M.: 1952, "Evaporation from the snowpack," U.S. Army Corps of Engineers, Translation, *Res. Note No. 8.*

Derikx, L., H. Loijens.: 1971, "Model of runoff from glaciers," *Hydrology Symposium No. 8,* vol. 1, pp. 153-199. NRC Association Commission on Geodesy and Geophysics. Subcommittee on Hydrology.

Doronin, Yu. P.: 1966, "Characteristics of the heat exchange," Rand Corporation 1966, pp. 247-266.

Dorsey, H. G., Jr.: 1951, "Arctic meteorology," *Compendium of Meteorology,* pp. 942-951.

Doughty, J., W. Staple, J. Lehane, R. Warder and F. Bisal: 1949, "Soil moisture, wind erosion and fertility of some Canadian prairie soils," *Canadian Department of Agriculture Publication 819,* 78 pp.

Dzerdzeyevakii, B. L.: 1945, "On the distribution of atmospheric pressure over the central Arctic," *Met. i Hydrol.* 1, pp. 33-38.

East, C.: 1968, "Comparaison du rayonnement solaire en ville et à la campagne," *Cahiers de géographie de Québec* 12, pp. 81-89.

Einarsson, E. and A. B. Lowe: 1955, *A study of horizontal temperature variations in the Winnipeg area on nights favouring radiational cooling,* Department of Transport, Meteorological Branch, CIR-2647, TEC-214, 19 pp.

Emslie, J. H.: 1964, *The reduction of solar radiation by atmospheric pollution at Toronto, Canada,* Dept. of Transport, Met. Br., CIR-4094, TEC-535, 8 pp.

Emslie, J. H. and J. Satterthwaite: 1966, *Air pollution: meteorological relationships at Vancouver, British Columbia.* Dept. of Transport, Met. Br., CIR-4396, TEC-605, 9 pp.

Ferland, M.: 1964, *Pollution de l'air à Ottawa, en rapport avec les facteurs météorologiques locaux.* Dept. of Transport, Met. Br., CIR-3970, TEC-502, 6 pp.

Ferguson, H., A. O'Neill and H. Cork: 1970, "Mean evaporation over Canada," *Water Resources Research (Wat. Res. Res.)* 6, pp. 1618-1633.

Ferrians, O. J., Jr.: 1965, "Permafrost map of Alaska." Map 1,445, Miscellaneous Geologic Investigation, U.S. Geological Survey, Washington.

Findlay, B. and G. McKay: 1972, "Climatological estimation of Canadian snow resource." *International Geography* (Adams and Helleiner Eds.), 1, pp. 139-141.

Fletcher, J.: 1965, *The influence of the Arctic pack ice on climate,* Paper -P- 3208, Rand Corp. Santa Monica, California.

Fletcher, J.: 1966, *The Arctic heat budget and atmospheric circulation,* Rand Corporation 1966, pp. 23-43.

Flohn, H.: 1952, "Zur aerologie der polargebiete," *Met. Rund.* 5, pp. 81-87.

Fraser, E. M.: 1956, "The lichen woodlands of the Knob Lake Area of Quebec-Labrador," *Subarctic Research Paper* 1, McGill University.

Frith, R.: 1968, "The earth's higher atmosphere," *Weather* 23, pp. 142-155.

Fuggle, R. E.: 1971, "Nocturnal atmospheric infrared radiation in Montreal." Unpub. Ph.D. thesis, McGill University, 237 pp.

Garnier, B.: 1972, "A viewpoint on the evaluation of potential evapotranspiration." *International Geography* (Adams and Helleiner Eds.), 1, pp. 143-145.

Gavrilova, M. K.: 1963, *Radiatsionnyi klimat Arktiki, Leningrad,* 1963. Translated by Israel Program for Scientific Translation, Jerusalem, 1966, 178 pp. (Title in translation: Radiation climate of the Arctic.)

Gillespie, T.: 1971, "Carbon dioxide profiles and apparent diffusivities in cornfields at night," *Agr. Met.* 8, pp. 51-57.

Gillespie, T. and K. King: 1971, "Night-time sink strength and apparent diffusivities within a corn canopy," *Agr. Met.* 8, pp, 59-67.

Godson, W. L.: 1950, "The structure of North American weather systems," *Cent. Proc. Royal Meteorological Society,* pp. 89-106.

Gold, L. W.: 1958, "Influence of snow cover on heat flow from the ground: some observations made in the Ottawa area." Toronto General Assembly 1957, Vol. IV, *International Association of Scientific Hydrology,* International Union in Geodesy and Geophysics (I.U.G.G.).

Gold, L. W.: 1960, "Field study on the load bearing capacity of ice covers," *Woodland Research (Wood. Res.)* 61, p. 11.

Gold, L. W.: 1963, "Influence of the snow cover on the average annual ground temperature at Ottawa, Canada," *International Association of Scientific Hydrology.* I.U.G.G., Pub. No. 61.

Gold, L. W.: 1964, "Analysis of annual variations in ground temperature at an Ottawa site," *Can. Jl. Earth Sci.* 1, pp. 146-157.

Gold, L. W.: 1967, "Influence of surface conditions on ground temperature," *Can. Jl. Earth Sci.* 4, pp. 199-208.

Gold, L. W.: 1968, "Micrometeorological observations of the snow and ice section D.B.R. NRC." *Proceedings, 1st Canadian Conference on Micrometeorology.* Part 1, pp. 136-144. (Research Paper No. 351 D.B.R.).

Gold, L. W. and D. W. Boyd: 1965, "Annual heat and mass transfer at an Ottawa site," *Can. Jl. Earth Sci.* 2, pp. 1-10.

Gold, L. W. and G. P. Williams: 1961, "Energy balance during the snow melt period at an Ottawa site." *International Association of Scientific Hydrology,* I.U.G.G., Pub. No. 54, pp. 288-294.

Gold, L. W. and G. P. Williams: 1963, "An unusual ice formation on the Ottawa River," *Jl. Glac.* 4, pp. 569-73.

Goodison, B.: 1969, I.W.B. Reports a) *Distribution of global radiation over Peyto Glacier, Alberta* b) *An analysis of climate-runoff events for Peyto Glacier* (in print).

Goodison, B.: 1971, "The relationship between ablation and global radiation over Peyto Glacier Alberta," *Glaciers; Proc. of Workshop Seminar 1970* Can. Nat. Comm. for I.H.D., pp. 38-42.

Graham, W. and K. King: 1961, "Fraction of net radiation utilized in evapotranspiration from a corn crop." *Proceedings of the Soil Science Society of America* 25, pp. 158-160.

Grigor'ev, A. A.: 1958, "Concerning certain geographical relationships of thermal and water exchange on the surface of dry land and means of further study of materials and energy in the geographical environment," *Izvest. Acad. Sci. U.S.S.R., Geog. Ser.* 3, pp. 17-21. (Text in Russian).

Hale, M. E.: 1952, "Glaciological studies of the Baffin Island Expedition 1950," *Jl. Glac.* Part II Appendix 2, pp. 22-23.

Hannell, F. G.: 1972, "Subsurface temperatures on Arctic slopes." *International Geography* (Adams and Helleiner Eds.) 1, pp. 145-147.

Hare, F. K.: 1950, "Climate and zonal divisions of the boreal forest formation in Eastern Canada," *Geog. Rev.* 40, pp. 615-635.

Hare, F. K.: 1951, "Some climatological problems of the Arctic and Sub-Arctic," *Compendium of Meteorology,* American Meteorological Society, Boston, pp. 917-941.

Hare, F. K.: 1954, "The boreal conifer zone," *Geographical Studies* 1, pp. 4-18.

Hare, F. K.: 1955, "Dynamic and synoptic climatology," *Annals of the Association of American Geographers* 45, pp. 152-162.

Hare, F. K.: 1959, "A photo-reconnaissance survey of Labrador-Ungava," *Memoir* 6, Geographical Branch, Dept. of Mines and Technical Surveys, Ottawa. 83 pp.

Hare, F. K.: 1968, "The Arctic," *Q. Jl. Roy. Met. Soc.* 94, pp. 439-459.

Hare, F. K. and M. R. Montgomery: 1949, "Ice, open water, and winter climate in the eastern Arctic of North America," *Arctic* 2, pp. 79-89 and 149-164.

Hare, F. K. and S. Orvig: 1958, "The Arctic circulation," *Arctic Meteorological Research Group Publication 12,* McGill University.

Hare, F. K. and J. Hay: 1971, "Anomalies in the large-scale annual water balance over northern North America," *Can. Geog.* 15, pp. 79-94.

Hattersley-Smith, G.: 1960, "Studies of englacial profiles in the Lake Hazen area of North Ellesmere Island," *Jl. Glac.* 3, pp. 610-625.

Hattersley-Smith, G. and H. Serson: 1970, "Mass balance of the Ward Hunt Ice Rise and Ice Shelf, a 10 year record," *Jl. Glac.* 9, pp. 247-52.

Havens, J. M.: 1964, "Meteorology and heat balance of the accumulation area White Glacier," *Axel Heiberg Island Research Paper,* McGill University.

Havens, J. M., F. Muller and G. C. Wilmot: 1968, "Complete meteorological survey and a short-term heat balance study of the White Glacier," *Axel Heiberg Island Research Paper,* McGill University.

Holmes, R. and G. Robertson: 1959, *Heat units and crop growth,* Canada Department of Agriculture Publication 1042.

Holmes, R. and G. Robertson: 1959, "A modulated soil moisture budget," *Mon. Weath. Rev.* 87, pp. 101-106.

Holmes, R. and G. Robertson: 1964, "The calculation of the soil moisture profile under various conditions using the modulated soil moisture budget," *International Association of Scientific Hydrology (I.A.S.H.), I. U. G. G. Pub. No. 65,* pp. 454-461.

Holmgren, B.: 1971, "Climate and energy exchange on a sub-polar ice cap in summer," Parts A-F, University of Uppsala Pub. No. 107-112 (inc.).

Hopkins, J. W.: 1968, "Correlation of air temperature normals for the Canadian Great Plains with latitude, longitude and altitude," *Can. Jl. Earth Sci.* 5, pp. 199-210.

Hustich, I.: 1966, "On the forest-tundra and the northern tree-lines," *Annals of the University of Turku,* A.II: 36, pp. 7-47.

Idso, S. B. and R. D. Jackson: 1969, "Thermal radiation from the atmosphere," *Journal of Geophysical Research (Jl. Geophys. Res.)* 74, pp. 5397-403.

Ives, J. D.: 1961, *A pilot project for permafrost investigations in central Labrador-Ungava,* Department of Mines and Technical Surveys, Geog. Paper No. 28, Ottawa, Canada.

Ives, J. D.: 1966, "La glaciologie à la direction de la géographie 1961-65," *Geog. Bull.* 8, p. 2.

Jackson, I. C.: 1959, "Insolation and albedo in Quebec-Labrador," *Subarctic Research Paper* 5, McGill University.

Jackson, I. C.: 1960, "Snowfall measurement in Northern Canada," *Q. Jl. Roy. Met. Soc.* 86, pp. 273-5.

Jenness, J.: 1949, "Permafrost in Canada," *Arctic* 2, pp. 13-27.

Johnston, G. H., R. J. E. Brown and D. N. Pickersgill: 1963, *Permafrost investigations at Thompson, Manitoba, terrain studies,* National Research Council, Tech. Paper No. 158, (NRC 7568).

Keegan, J. T.: 1958, "Arctic synoptic activity in winter," *Jl. Met.* 15, pp. 513-521.

Keeler, C. W.: 1964, *Relationship between climate, ablation and runoff on the Sverdrup Glacier 1963, Devon Is. N.W.T.,* Arctic Institute of North America Research Paper 27.

King, K.: 1961, "Evaporation from land surfaces-evaporation," *Proceedings of the Hydrology Symposium No. 2, Toronto,* pp. 55-73. Dept. of Northern Affairs and Natural Resources, Water Resources Branch.

Klein, G. J.: 1950, *Canadian survey of physical characteristics of snow covers,* NRC, DBR Tech. Mem. 15.

Koerner, R.: 1966, "Accumulation on the Devon Island Ice Cap N.W.T. Canada," *Jl. Glac.* 6, pp. 383-92.

Koerner, R.: 1970, "Weather and ice observations of the British Trans-Arctic Expedition 1968-69." *Weather* 25, pp. 218-28.

Koerner, R.: 1970, "The mass balance of the Devon Island Ice Cap N.W.T. Canada, 1961-66," *Jl. Glac.* 9, pp. 235-36.

Kratzer, P.: 1956, *Das Stadtklima,* F. Vieweg and Sohn, Braunschweig, 221 pp.

Lachenbruch, A.: 1970, "Some estimates of the thermal effects of a heated pipeline in Permafrost," *Wash. U.S. Geol. Surv. Circ.* 623, pp. 23.

Lamb, H. H., R. P. U. Lewis and A. Woodroffe: "Atmospheric circulation and the main climatic variables between 8000 and 0 B.C.: meteorological evidence," *World Climate from 8000 to 0 B.C.,* Royal Meteorological Society, pp. 174-217.

Laycock, A.: 1960, "Drought Patterns in the Canadian Prairies," Symposium of *Int. Assoc. for Sci. Hydrol. I.U.G.G. Helsinki,* 51, pp. 34-47.

Leahey, D. M.: 1962, *An analysis of smoke observations in Winnipeg,* Dept. of Transport, Met. Br., CIR-3769, TEC-439, 6 pp.

Lee, R.: 1960, "The circulation of the Arctic," *Can. Geog.* 16, pp. 1-13.

Legget, R. F., H. B. Dickens, and R. J. E. Brown: 1961, "Permafrost investigations in Canada," Proc. 1st International Symposium on Arctic Geology Vol. 2, *Geology of the Arctic,* pp. 956-969.

Legget, R. F. and L. W. Gold: 1966, *Ice pressures on structures: a Canadian problem,* D.B.R. Tech. Paper No. 282.

Lettau, H.: 1969, "Evapotranspiration climatonomy," *Mon. Weath. Rev.* 97, pp. 691-699.

Lewis, E. L.: 1966, "Heat flow through winter ice," *Proceedings of the International Conference on Low Temperature Science, Japan,* 1, pp. 611-637.

Lindsay, D. G.: 1969, "Ice distribution in the Queen Elizabeth Islands," *Bulletin of the Canadian Institute of Mining and Metallurgy* 10, pp. 45-60.

Lister, H.: 1962, *Heat and mass balance at the surface of the Ward Hunt Ice Shelf 1960,* Arctic Institute of North America Paper 19.

Logan, L.: *Snow Survey Report East and Middle Oakville Creeks Drainage Basin.* O.W.R.C. Bulletin 4-1 (Climatic series) 86 pp.

Løken, O. and R. Sagar: 1968, "Mass balance observations on the Barnes Ice Cap Baffin Island, Canada." *I.A.S.H. Pub. No. 79,* pp. 282-291.

Mack, A. R. and W. S. Ferguson: 1968, "A moisture stress index for wheat by means of a modulated soil moisture budget," *Can. Jl. Pl. Sci.* 48, pp. 535-543.

Malone, T. F. (Ed.): 1951, *Compendium of Meteorology* 1, American Meteorological Society (Am. Met. Soc.), Boston, 344 pp.

Marshunova, M. S. and N. T. Chernigovskiy: 1966, "Numerical characteristics of the radiation regime in the Soviet Arctic," in Rand Corporation 1966, pp. 279-297.

Mateer, C. L.: 1961, "Note on the effect of the weekly cycles of air pollution on solar radiation at Toronto," *International Journal of Air and Water Pollution (Int. Jl. Air and Water Poll.)* 4, pp. 52-54.

Mather, J. R. and C. W. Thornthwaite: 1956, "Microclimatic Investigations at Point Barrow, Alaska, 1956," Drexel Inst. Tech., Lab. of Climatology, Centerton, N.J., Vol. 9, No. 1.

Mather, J. R. and C. W. Thornthwaite: 1958, "Microclimatic investigations at Point Barrow, Alaska, 1957-1958." Ibid. No. 2.

McIntyre, D. P.: 1950, "On the air-mass temperature distribution in the middle and high troposphere in winter," *Jl. Met.* 7, pp. 101-107.

McIntyre, D. P.: 1955, "On the barocline structure of the westerlies," *Jl. Met.* 12, pp. 201-210.

McIntyre, D. P.: 1959, "The Canadian 3-front, 3-jet stream model," *Geophysica.* 6, pp. 309-324.

McKay, G. and H. Thompson: 1967, "Snow cover in the Prairie Provinces," *60th Annual Meeting, American Society of Agricultural Engineers, Saskatoon, transcript.* No. 67-206.

McKay, G.: 1970, "Climate: a critical factor in the tundra," *Trans. R. Soc. Can.* 11, pp. 43-50.

Middleton, W. E. K. and F. G. Millar: 1936, "Temperature profiles in Toronto," *Journal of the Royal Astronomical Society of Canada* 30, pp. 265-272.

Moller, F.: 1951, "Long-wave radiation" in *Compendium of Meteorology,* pp. 34-49.

Monteith, J. L.: 1961, "An empirical model for estimating long-wave radiational exchanges in the British Isles," *Q. Jl. Roy. Met. Soc.* 87, pp. 171-179.

Morgan, D. L., W. O. Pruitt, F. J. Lourence: 1970, *Radiation data and analyses for the 1966 and 1967 micrometeorological field runs at Davis, California.* Tech. Report ECOM 68-G10-2, Dept. Water Sci. and Eng., Davis, Calif., 45 pp.

Mosby, H.: 1962, "Water, salt and heat balance of the North Polar Sea and of the Norwegian Sea," *Geogys. Publr.* 24, pp. 289-313.

Mukammal, E., K. King, and H. Cork: 1966, "Comparison of aerodynamic and energy budget techniques in estimating evapotranspiration from a cornfield," *Arch. Met. Geophys. Bioklim.* 14, pp. 384-95.

Muller, F.: 1966, "Evidence of Climatic Fluctuations on Axel Heiberg Island," Rand Corporation 1966, pp. 137-156.

Munn, R. E.: 1961, "Analysis of smoke observations at Ottawa, Canada," *J. Air Pollut. Contr. Assoc.* 11, pp. 410-416.

Munn, R. E., M. Hirt and B. Findlay: 1969, "A climatological study of the urban temperature anomaly in the lakeshore environment at Toronto," *J. Appl. Met.* 8, pp. 411-422.

Munn, R. E. and M. Katz: 1959, "Daily and seasonal pollution cycles in the Detroit-Windsor area," *Int. Jl. Air and Water Poll.* 2, pp. 51-76.

Munn, R. E. and M. Katz: 1960, "Air pollution levels associated with a 49-hr. inversion at Detroit-Windsor," *Bull. Am Met. Soc.* 41, pp. 245-259.

Murgatroyd, R. J., F. K. Hare, B. W. Boville, S. Teweles and A. Kochanski: 1965, "The circulation in the stratosphere, mesosphere and lower thermosphere," *Technical Note 70,* World Meteorological Organization, Geneva, 206 pp.

Myrup, L. O.: 1969, "A Numerical Model of the Urban Heat Island." *Jl. App. Met.* 8, pp. 908-918.

Namais, J.: 1958, "Synoptic and climatological problems associated with the general circulation at the Arctic," Trans. *Am. Geophys. Un.* 39, pp. 40-51.

Nordhagen, R.: 1952, "Mountain vegetation as a record of snow cover thickness and duration," *Medd. fra Vegdirektoren,* Nos. 1, 2, 4, pp. 1-31.

Nowosad, F. A. and A. Leahey: 1960, "Soils of the Arctic and Sub-arctic regions of Canada," *Agricultural Institute Review (Agric. Inst. Rev.),* pp. 48-50.

Oke, T. R.: 1968, "Some results of a pilot study of the urban climate of Montreal," *Clim. Bull.* 3, pp. 36-41.

Oke, T. R.: 1970, "The temperature profile near the ground on calm clear nights," *Q. Jl. Roy. Met. Soc.* 96, pp. 14-23.

Oke, T. R. and C. East: 1971, "The urban boundary layer in Montreal," *Boundary-Layer Met. 1,* 411-437 pp.

Oke, T. R. and R. Fuggle: 1972, "Comparison of urban/rural counter and net radiation at night," *Boundary-Layer Met.* 2, pp. 290-308.

Oke, T. R. and F. G. Hannell: 1970, *The Form of the Urban Heat Island in Hamilton, Canada.* W.M.O. Tech. Note No. 108, W.M.O. No. 254 TP. 141, pp. 113-126.

Oke, T. R., D. Yap, and R. Fuggle: 1972, "Determination of urban sensible heat fluxes," *International Geography* (Adams and Helleiner Eds.), 1, pp. 176-178.

Ommanney, C.: 1970, "A pilot study for an inventory of the High Arctic Axel Heiberg Is. N.W.T.," *Perennial Ice and Snow Masses,* Unesco/IASH. Tech. Pap. in Hydrol. 1, A 2486, pp. 25-35.

Ommanney, C.: 1971, "The Canadian Glacier Inventory," *Glaciers, Proc. of Workshop Seminar 1970,* Can. Nat. Comm. for I.H.D.

O'Neill, A. and D. Gray: 1971, "Energy balance and melt theories," *Runoff from Snow and Ice.* Vol. 1, Symp. No. 8, Quebec City. pp. 31-58. E.M.R., I.W.B.

Orvig, S.: 1951, "The climate of the ablation period on the Barnes Ice Cap in 1950," *Geog. Annal.* 33, pp. 166-209.

Orvig, S.: 1954, "Glaciological-meteorological observations on ice caps in Baffin Island," *Geog. Annal.* 36, pp. 197-318.

Orvig, S.: 1965, "Surface heat balance studies at McGill University," *Proc. 1st Canadian Conference on Micrometeorology, Toronto,* pp. 315-322.

Østrem, G.: 1966, "Mass balance studies on glaciers in Western Canada," *Geogr. Bull.* 8, pp. 81-107.

Østrem, G. and A. D. Stanley: 1969, *Glacier mass balance measurements: a manual for field work and office work,* E.M.R., I.W.B. Reprint Series No. 66.

Paterson, W. S. B.: 1969, *The physics of glaciers.* Pergamon Press, Toronto, pp. 58-61.

Pearce, D. C. and L. W. Gold: 1959, "Observations of ground temperature and heat flow at Ottawa, Canada," *Jl. Geophys. Res.* 64, p. 1293.

Penner, C. M.: 1955, "A three-front model for synoptic analysis," *Q. Jl. Roy. Met. Soc.* 81, pp. 80-91.

Petterssen, S.: 1940, *Weather analysis and forecasting,* McGraw-Hill, New York, pp. 169-174.

Pihlainen, J. A.: *Inuvik, N.W.T., engineering site information,* Natl. Res. Coun., Tech. Paper No. 135, Aug. 1962 (NRC 6757).

Potter, J.: 1961, "Changes in seasonal snowfall in cities," *Can. Geog.* 5, pp. 37-42.

Potter, J.: 1965, "Snow cover Canada," *Climatological Studies No. 3,* Dept. of Transport, Met. Br.

Pounder, E. R.: 1965, *Physics of ice,* Pergamon Press, Toronto, 151 pp.

Putnam, D. F. and L. J. Chapman: 1938, "The climate of Southern Ontario," *Scientific Agriculture (Scient. Agric.)* 18, pp. 401-446.

Pysklywec, D., K. Davar, and D. Bray: 1968, "Snowmelt at an index plot," *Wat. Res. Research* 4, pp. 937-945.

Rahn, J. and D. M. Brown: 1971, "Corn canopy temperatures during freezing or near freezing conditions," *Can. J. Pl. Sci.* 51, pp. 173-175.

Rakipova, L. R.: 1966, "The influence of the Arctic ice cover on the zonal distribution of atmospheric temperature," in Rand Corporation 1966, pp. 411-441.

Reed, R. J. and B. A. Kunkel: 1960, "The Arctic circulation in summer," *Jl. Met.* 17, pp. 489-506.

Richards, T. L. and V. S. Derco: 1963, "The role of lake effect storms in the distribution of snowfall in Southern Ontario." *Proc. E. Snow Conf.* Quebec City, pp. 61-85.

Robertson, G. W.: 1955, "Low temperature fog at the Edmonton airport as influenced by moisture from the combustion of natural gas." *Q. Jl. Roy. Met. Soc.* 81, pp. 190-197.

Robertson, G. W.: 1968, "A biometeorological time scale for a cereal crop involving day and night temperatures photo-periodism," *Int. Jl. Biomet.* 12, pp. 191-223.

Robertson, G. W. and R. Holmes: 1958, "A new concept of the measurement of evaporation for climatic purposes," Gen. Assembly I.A.S.H. Toronto 1957, No. 45, pp. 399-406.

Rouse, W. and J. McCutcheon: 1972, "The diurnal behaviour of incoming solar and infrared radiation in Hamilton, Canada," *International Geography.* (Adams and Helleiner Eds.), 1, pp. 191-196.

Rouse, W. and R. B. Stewart: 1972, "A simple model for determining evaporation from high-latitude upland sites," *Jl. Appl. Met.* 11, pp. 1063-70.

Sagar, R. B.: 1960, "Glacial-meteorological observations in Northern Ellesmere Island during phase III Operation Hazen, May-Aug. 1958," *Arctic Met. Res. Group* 29, McGill Univ.

Sagar, R. B.: 1966, "Glaciological and climatological studies on the Barnes Ice Cap, 1962-64," *Geog. Bull.* 8, pp. 3-47.

Sanderson, M.: 1948a, "The climate of Canada according to the new Thornthwaite classification," *Scient. Agric.* 28, pp. 501-517.

Sanderson, M.: 1948b, "Drought in the Canadian Northwest," *Geographical Review (Geog. Rev.)* 38, pp. 289-99.

Sater, J. E. and collaborators: 1963, "The Arctic Basin," *Arctic Institute of North America,* Washington, 319 pp.

Schaerer, P. A.: 1962, "Planning avalanche defence works for the Trans-Canada Highway at Rogers Pass," *Engng. J.* 45, pp. 31-38.

Scherhag, R.: 1958, "The role of tropospheric cold-air poles and of stratospheric high-pressure centres in the Arctic weather," *Polar Atmosphere Symposium.* Part 1. Meteorology Section, pp. 101-117.

Schwerdtfeger, P. and E. R. Pounder: 1963, "Energy exchange through an annual sea ice cover," I.A.S.C., I.U.G.G., Pub. No. 61, 109 pp.

Selirio, I. S. and D. M. Brown: 1971, "Moisture budgeting technique for a fallow soil in spring," *Canadian Journal of Soil Science (Can. Jl. Soil Sci.)* 51, pp. 516-518.

Sellers, W. D.: 1965, *Physical Climatology,* University of Chicago Press, 272 pp.

Shaw, R. H. and G. Thurtell: 1971, "Turbulent transport within crop canopy layers," *Annual Progress Report 1970.* Can. Comm. I.B.P. University of Guelph.

Shpolyanskaya, N. A.: 1962, "The influence of the heat exchange between the soil and the atmosphere on permafrost in Transbaikalia," *Vestnik Moskovskogo Universiteta,* No. 2, pp. 37-42, (Text in Russian).

Stanley, A. D.: 1971, "Combined balance studies of selected glacier basins in Canada," *Glaciers: Proc. of Workshop Seminar 1970.* Can. Nat. Com. IHD. pp. 5-10.

Stearns, S. R.: 1966, "Cold regions science and engineering," Part 1 Section A2. *Permafrost (Perennially Frozen Ground),* pp. 6-11, U.S. Army Cold Regions Research and Engineering Lab., Hanover, New Hampshire.

Stern, A. C.: 1970, "Utilization of air pollution models," in *Proceedings of Symposium on Multiple Source Urban Diffusion Models* (Stern, A.C., Ed.), U.S. Environmental Protection Agency, AP-86, pp. 13.1-13.8.

Stupart, R. F.: 1905, "The Canadian climate," *Rep. of 8th Int. Geog. Cong. 1904,* pp. 294-307.

Sumbin, M. I., S. P. Kachurin, N. I. Tolstikhin, and V. F. Tumel: 1940, "General permafrost studies," Acad. Sci. USSR, Moscow, pp. 160-165 (Text in Russian).

Summers, P. W.: 1962, "Smoke concentration in Montreal related to local meteorological factors," Symposium: *Air Over Cities,* U.S. Public Health Service. SEC Technical Report, Cincinnati, A.62-5, pp. 89-113.

Summers, P. W.: 1963, "Urban ventilation in Montreal," *Smokeless Air* 34, pp. 118-123.

Summers, P. W.: 1964, "An urban ventilation model applied to Montreal," Unpubl. Ph.D. thesis, McGill University.

Summers, P. W.: 1966, "The seasonal, weekly and daily cycles of atmospheric smoke content in central Montreal," *J. Air Pollut. Contr. Assoc.* 16, pp. 432-438.

Sverdrup, H. U.: 1936, "The eddy conductivity of the air over a smooth snow field," *Geofys. Publr.* 11, 67 p.

Swinbank, W. C.: 1963, "Longwave radiation from clear skies," *Q. Jl. Roy. Met. Soc.* 89, pp. 339-48.

Swithinbank, C. W. M.: 1960, *Ice atlas of Arctic Canada,* Defence Res. Board Canada Pub. DR3-1060.

Sychev, K. A.: 1959, "Thermal balance of the active layer of permafrost in summer," *Problemy Arktiki I Antarktiki No. 1,* pp. 87-92, (Text in Russian).

Terjung, W. H.: 1970, "Urban energy balance climatology: a preliminary investigation of the city-man system in downtown Los Angeles," *Geog. Rev.* 60, pp, 31-53.

Tikhomirov, B. A.: 1952, "The importance of moss cover in the vegetation of the far North," *Botanicheskiy Zhur.,* pp. 629-938. (Text in Russian.)

Tikhomirov, B. A.: 1963, "Principal stages of vegetation development in northern U.S.S.R. as related to climatic fluctuations and the activity of man," *Can. Geog.* 7, pp. 55-71.

Tissot, J.: 1963, *80W mean cross sections 1959, 1960,* Pub. Met. McGill University, 55, 37 pp.

Thomas, M. K.: 1953, *Winter temperatures at Toronto,* Royal Meteorological Society (R. Met. Soc.) Canadian Branch 4, 10 pp.

Thomas, M. K.: 1961, *A bibliography of Canadian climate 1763-1957,* Dept. of Transport, Met. Br., 114 pp.

Thomas, M. K.: 1964, *Snowfall in Canada,* Dept. of Transport, Met. Br.

Thompson, F. D.: 1969. *Lake-effect snowstorms in Southern Ontario during November and December 1968,* Dept. of Transport, Met. Br. Memo. 726.

Tyrtikov, A. P.: 1959, "Permafrost and vegetation," *Fundamentals of Geocryology,* Vol. 1, Chap. 12, Acad. Sci. U.S.S.R., Moscow, 1959, pp. 399-421, (Text in Russian).

Tyrtikov, A. P.: 1956, "The effect of vegetation on perennially frozen soil," *Materialy k osno-*

vam ucheniya o merzlykh zonakh zemnoy kory (Materials for the basis of study of the frozen zones of the earth's crust). Issue III, 1956, pp. 85-108.

Untersteiner, N.: 1962, "Mass and heat budget of Arctic sea ice," *Arch. Met. Geophys. Bioklim.* Series A 12, pp. 153-180.

Untersteiner, N.: 1964, "Calculation of temperature regime and heat budget of sea ice in the central Arctic," *Jl. Geophys. Res.* 69, pp. 4755-4766.

Untersteiner, N.: 1966, "Calculating thermal regime and mass budget of sea ice," in Rand Corporation 1966, pp. 203-213.

Unstead, J.: 1912, "The climatic limits of wheat cultivation, with special reference to N.A.," *Geographical Journal (Geog. Jl.)* 39, pp. 347-366, 421-46.

U.S. Army Corps of Engineers, 1956: *Snow hydrology, summary report of the snow investigations, North Pacific Division.*

Van Mieghem, J.: 1961, *Zonal harmonic analysis of the Northern Hemisphere geostrophic wind field*, Monogrs. U.G.G.I., 8, 57 p.

Villimow, J.: 1956, "The nature and origin of the Canadian dry belt," *Ann. Assoc. Am. Geogr.* 46, pp. 211-32.

Vowinckel, E.: 1962, *Cloud amount and type over the Arctic*, Pub. Met. McGill University 51, 27 pp.

Vowinckel, E.: 1964, *Atmosphere energy advection in the Arctic*, Pub. Met., McGill University 71, 17 pp.

Vowinckel, E.: 1966, *The surface heat budgets at Ottawa and Resolute, N.W.T.*, Pub. Met. McGill University 84 pp.

Vowinckel, E. and S. Orvig: 1962, "Water balance and heat flow of the Arctic Ocean," *Arctic* 15, pp. 205-223.

Vowinckel, E. and S. Orvig: 1964, "Energy balance of the Arctic. I, incoming and absorbed solar radiation at the ground in the Arctic," *Arch. Met. Geophys. Bioklim.* 13, pp. 352-377.

Vowinckel, E. and S. Orvig: 1965a, "Energy balance of the Arctic. II, long wave radiation and total radiation balance at the surface in the Arctic," ibid 13, pp. 451-479.

Vowinckel, E. and S. Orvig: 1965b, "Energy balance of the Arctic. III, radiation balance of the troposphere and of the earth-atmosphere system in the Arctic," ibid 13, pp. 480-502.

Vowinckel, E. and S. Orvig: 1966, "Energy balance of the Arctic. V, the heat budget over the Arctic Ocean," ibid 14, pp. 303-325.

Vowinckel, E. and S. Orvig: 1967, "Climate change over the Polar Ocean. I, the radiation budget," ibid 15, pp. 1-23.

Vowinckel, E. and B. Taylor: 1965, "Energy balance of the Arctic. IV, evaporation and sensible heat flux over the Arctic Ocean," ibid 14, pp. 36-52.

Ward, W. and S. Orvig: 1953, "The glaciological studies of the Baffin Island Expedition, 1950. Part IV: the heat exchange at the surface of the Barnes Ice Cap during the ablation period," *Jl. Glac.* 2, pp. 158-168.

Williams, G. D.: 1969a, "Weather and Prairie wheat production," *Canadian Journal of Agricultural Economics* 17, pp. 99-109.

Williams, G. D.: 1969b, "Applying estimated temperature normals to the zonation of the Canadian Great Plains for wheat," *Can. Jl. Soil Sci.* 49, pp. 263-276.

Williams, G. D.: 1971, "Wheat phenology in relation to latitude, longitude and elevation on the Canadian Great Plains," *Can. J. Pl. Sci.* 51, pp. 1-12.

Williams, G. D.: 1972, "Agrometeorology and geography," *International Geography* (Adams and Helleiner Eds.), 1, pp. 204-206.

Williams, G. P.: 1959, *Two notes relating to frazil ice formation—1. an empirical method of estimating total heat losses from open-water surfaces; 2. some observations on super-cooling and frazil ice production. Seminar on Ice Problems in Hydraulic Structures, August 1959.* Proc. 8th Cong. Int. Assn. Hydraul. Res., DBR Res. Paper 162.

Williams, G. P.: 1959, *Evaporation from snow covers in Eastern Canada*, Proc. E. Snow Conf. 1958, NRC. 5003 (DBR Res. Paper No. 73).

Williams, G. P.: 1961, *Evaporation from water, snow and ice,* Proc. of Hydrol. Symp. No. 2, Assoc. Com. on Geodesy and Geophysics. pp. 31-51.

Williams, G. P.: 1963, "Probability charts for predicting ice thickness," *Engng. J.* 46, pp. 31-35.

Williams, G. P.: 1963, *Ice growth rates and heat exchange.* Symposium on Heat Exchange at Snow and Ice Surfaces, Assoc. on Soil and Snow Mechanics, Natl. Res. Coun., Tech. Memo. 78.

Williams, G. P.: 1963, *Heat transfer coefficients for natural water surfaces,* Int. Assn. of Sci. Hydrol., I.G.G.U., Pub. No. 62.

Williams, G. P.: 1965, "Correlating freeze up and break up with weather conditions," *Canadian Geotechnical Journal (Can. Geotech. Jl.)* 2, pp. 313-326.

Williams, G. P. and L. Gold: 1963, "The use of dust to advance the break-up of ice in lakes and rivers," *Proc. E. Snow Conf. 1963,* pp. 31-56.

Wilson, C. V.: 1958, *Synoptic regimes in the lower Arctic troposphere during 1955,* Pub. Met. McGill University 8.

Wilson, C. V.: 1967, *Cold regions science and engineering Part 1, Section A3: climatology introduction Northern Hemisphere 1,* U.S. Army, Cold Regions Research and Engineering Laboratory, Hanover, New Hampshire, pp. 27-69.

Wilson, C. V. and M. A. MacFarlane: 1971, *A preliminary study of radiation balance at Poste-de-la-Baleine (Great Whale) Quebec,* Pub. No. 32, Laval University, Quebec.

Wilson, H. J.: 1964, *Atmospheric pollution at Saint John, N.B.,* Dept. of Transport, Met. Br., CIR-3962, TEC-497, 7 pp.

Wittman, W. I. and J. J. Schule, Jr.: 1966, "Comment on the mass budget of Arctic pack ice," in Rand Corporation 1966, pp. 215-246.

Wright, J. B.: 1966, *Precipitation patterns over Vancouver City and the Lower Fraser Valley,* Dept. of Transport, Met. Br., CIR-4474, TEC-623, 4 pp.

Yamamoto, G.: 1957, "Estimation of additional downward radiation from aerosols over large cities," *J. Met. Soc. Japan,* 75th Anniversity Vol. 1-4.

Yap, D., K. L. Gunn, and C. East: 1969, "Vertical temperature distribution over Montreal," *Naturaliste Can.* 96, pp. 561-80.

Zacks, M. B.: 1945, "Oats and climate in Southern Ontario," *Canadian Journal of Research* 23, pp. 45-75.

Zdunkowski, W., D. Henderson, and J. V. Hales: 1966, "The effect of atmospheric haze on infra-red radiative cooling rates," *Journal of Atmospheric Science* 23, pp. 297-304.

NOTE

Information Canada: This agency now publishes the reports of government departments and is the official publisher for the following items:

Can. Nat. Comm. IHD. 1970, *Snow Hydrology.* Proc. Workshop Seminar 1968, 82 pp.

Can. Nat. Comm. IHD. 1971, *Canadian Progress Report 1970.* 199 pp.

Can. Nat. Comm. IHD. 1972, *Glaciers.* Proc. of Workshop Seminar 1970, 61 pp.

Dept. of Transport, Meteorological Branch (now A.E.S.)

— Climatological Studies Series of Canadian Provinces and Cities.

— Snow Cover Data Reports 1954-1972.

— Selected references to papers concerning Canadian Urban Climates (D.S. #2-69).

Dept. of Energy, Mines and Resources (now Dept. of Environment) Inland Waters Branch: 1969, Ice Studies in the Dept. of E.M.R. 1969, (Report Series No. 7).

Dept. of Energy, Mines and Resources (now Dept. of Environment) 1971, Research Projects in Glaciology (Report Series No. 15).

Dept. of Energy, Mines and Resources (now Dept. of Environment): 1972, Research Projects in Glaciology 1972 (Report Series No. 23).

Section 4
Soils and Vegetation

Chapter 12

Soils in the Canadian Landscape

Despite the fact that soils constitute a vitally important component of the landscape, the paucity of published material in the professional journals of geography underscores the lack of attention afforded them by physical geographers. This is demonstrated by the necessity on occasion to resort to papers written by non-geographers, a situation which in no way detracts from the original intentions of this book, but rather serves to illustrate an area of research where physical geographers could play a more important role. The stage for the discussion and presentation of the chosen papers is set in the following paragraphs which furnish a brief historical background concerning the growth in our understanding of the basic concepts in soil science in Canada and the United States.

Historical Background

Geographers have traditionally been interested in soil classification, soil genesis, and the areal distribution of discrete soil units. Most authors of conventional texts in Canadian regional geography make reference either directly or indirectly to these units by employing the terms soil zone or great soil group (Putman et al 1963; Warkentin 1968). These two categories were based on work initiated by Russian scientists during the latter half of the nineteenth century, in particular Dokuchaiev, Sibirtzev, and Glinka. Dokuchaiev, who is frequently heralded as the founder of modern scientific pedology, was trained as a geologist and began his career as a soil scientist in 1877 by investigating soils characteristic of parts of the Russian steppe. Through his discovery that in many cases a close correspondence existed between soil units and other factors of the natural landscape, notably climate and vegetation, he came to recognize soil as an independent natural body whose formation was governed by what he called "soil formers," identified by him as parent rock, topography, climate, organisms, and the length of time the soil had to develop. Of importance to the physical geographer was the recognition by these early Russian investigators that soils tended to conform to a broad zonal pattern with climate and vegetation playing a dominant role in the geographic distribution of each zone. In mountainous areas the Russians believed that climatic control was demonstrated by the existence of altitudinally differentiated soil zones. In accepting a causal relationship between climate and soil, Dokuchaiev was able to produce the first soil classification scheme in which he distinguished three major categories: normal soils reflecting a strong climatic influence by virtue of their relative maturity and widespread distribution; transitional soils or those influenced by local conditions; abnormal soils constituting immature varieties lacking distinctive profile characteristics. Later he modified his classification so that the normal category comprised seven soil types or great soil groups corresponding with major vegetation zones, with the remaining six groups being equally divided between the transitional and normal categories. Sibirtzev, a close associate of Dokuchaiev, suggested that the major soil categories be redesignated zonal, intrazonal, and azonal respectively, a change that was later adopted by western investigators. Dokuchaiev's immediate successors often overemphasized the role of climate in accounting for the type of soil likely to be found at a given locality, when in fact Dokuchaiev himself advocated that each soil-former could be equally important in local situations; only when large areas were being considered did the climatic factor tend to overshadow the effect of the others. It should also be stressed that Dokuchaiev and his followers maintained a considerable advantage over their contemporaries in other areas such as Europe; in their quest to study possible cause and effect relationships between soil, vegetation, and climate, they were able to view elements of the natural landscape over vast distances unobstructed by major political boundaries.

Unfortunately, soil scientists in the Western world had to wait until 1914, following the translation into German of a text written by Glinka, before they could take advantage of pedological concepts stemming from the Russian school. Prior to C. F. Marbut's appointment to the United States soil survey in 1914, soil investigations in that country laboured under a strong geological bias. Quickly realizing the significance of the Russian work, Marbut translated the German edition of Glinka's book into English and, through his efforts, there was a major shift in emphasis in pedological research away from the traditional geological approach to one in which climate assumed pre-eminence. This in turn led to the formulation of a classification scheme in which similar soils were grouped on the basis of prevailing climate. Marbut introduced two terms for soils occupying the top level of the newly established hierarchy: pedalfers, characterized by accumulations of iron and aluminum compounds in the lower horizons; and pedocals, distinguished by accumulations of calcium and magnesium in the profile. Below this level there were five further categories, two of which corresponded approximately with the zonal and intrazonal soil groups recognized by the Russians. After further modifications and revision (Baldwin et al 1938) the system finally proposed comprised six categories, the first three of which were the Order (zonal, intrazonal, and azonal), the Suborder, and the Great Soil Group respectively (Thorp and Smith 1949). Because this hierarchy resembled that employed by biologists to classify living organisms, it became known as the taxonomic classification.

Those soils in the great group category were defined on the basis of their pedological properties and grouped together by virtue of their assumed common origin. This constituted an important milestone in the quest for a universally acceptable soil classification as it marked the first serious attempt to view soils at this level in terms of their morphological characteristics. Immediately below the Great Soil Group was the Soil Family, each of which possessed similar profile characteristics and comprised one or more distinct Soil Series. The series constituted soils developing on a common parent material, while soils of the same series but with distinctively different textural properties in the surface horizons were designated as Soil Types. The scheme proposed by Thorp and Smith lacked any mention of pedalfers and pedocals.

Despite the general acceptance of the classification scheme outlined above, it suffered from a number of important disadvantages. In particular it still tended to rely rather heavily upon genetic factors such as vegetation and climate rather than on actual soil characteristics. Moreover, it was frequently debated whether a soil whose properties had been modified or altered by cultivation practices could be incorporated into a classification derived from the investigation of virgin soils. Further problems with the old scheme were encountered when it was discovered that new soil series often did not conform to existing great soil groups. Thus it was advocated that a new classification was necessary in order to resolve these problems satisfactorily. The outcome was the appearance in 1960 of the Seventh Approximation, a system which was heralded as a natural classification based on actual soil properties. While a cursory examination may suggest that this work is simply an essay in word building, it is a very detailed and carefully contrived scheme that is essentially devoid of the traditional bias towards genetic factors evident in previous systems. As clearly demonstrated in the title, the system does not claim to be the final work in soil classification, but simply an approximation that must undergo vigorous analysis and testing in the field with the view towards potential modification before being completely accepted.

In Canada early attempts at developing an acceptable soil classification scheme together with analyzing and interpreting the geographical pattern of soils across the country were strongly influenced by developments in the United States (Joel 1926; Ruhnke 1926; Morwick 1933). According to Putnam (1951), the first integrated and organized soil survey work was initiated in 1914 through the efforts of the Ontario Agricultural College at Guelph. Provincially controlled surveys began in Saskatchewan and Alberta in 1929 and, by 1940, most provinces had commenced soil survey investigations on one scale or another. As in the United States, soil texture and parent material served as the basis for early soil survey work in Canada, an approach which was well suited to local studies in small areas but was not readily applicable to comparative studies involving broad groupings of soils. The result was an unfortunate lack in uniformity between soil survey investigations in the different provinces. While soil zones ostensibly reflecting variations in climate and vegetation were established for some areas in the Prairie provinces (Joel 1928), the situation was not conducive to establishing a nation-wide soil distribution picture. In 1945 the National Soil Survey Committee, which had been formed in 1940 with the express purpose of improving the taxonomic classification system of Canadian soils, proposed that the soils be grouped into seven major categories. At the top of the hierarchy were the Soil Regions, comprising the Tundra Soil Region, the Grassland Soil Region, and the Woodland Soil Region, followed in descending order by the Soil Zone and Subzone. The former was defined as "a broad belt of soils in which the dominant normal soils correspond to the great soil group" (National Soil Survey Committee 1945). However, because soil zones often embraced more than one great soil group, the two terms could not be used interchangeably. This situation brought about a major switch in emphasis in the quest for a scientifically acceptable but readily adaptable classification, as the term zone, despite assertions to the contrary, implied the existence of a causal relationship among climate, vegetation, and soils. By 1948, the concept of the great soil group was proposed in lieu of the soil zone and subzone which opened the way towards the establishment of a truly taxonomic classification. However, the soil zone was subsequently retained to facilitate the broad scale representation of the soil pattern.

A more rigorous taxonomic classification scheme designed to incorporate soil morphology as well as genesis was proposed by the National Soil Survey Committee in 1955 for nation-wide use and in 1960 a hierarchy constituting six categories was adopted; the Soil Type is the lowest category in the system but the most fundamental since it is the one about which the most detailed statements can be made. Soil Types are grouped together into progressively higher categories comprising the Series, the Family, the Subgroup, the Great Group, and the Soil Order respectively. In 1970, after a few minor changes, the committee published a handbook which listed 8 Soil Orders, 22 Great Groups, and 138 Subgroups (Canada Department of Agriculture 1970). It was anticipated that no further alterations would be necessary for the next five years so that the system could be thoroughly tested. According to Leahey (1963), the family and series terms roughly correspond with the same two terms in the Seventh Approximation, but here the similarity ends. While appreciating the efforts of their American counterparts to develop a universally adaptable classification system based on actual soil properties, as well as the potential applicability of the Seventh Approximation to Canadian soils, Canadian pedologists have expressed little desire to adopt the scheme. As Stobbe (1962) pointed out, it must be remembered that the Canadian land surface is unique in being relatively youthful from the pedologic point of view, as much of the country was not subjected to pedogenic processes until deglaciation some 10,000 years ago. Moreover, while initial approaches to soil classification and related geographic distributions relied heavily upon genetically related environmental factors, almost fifty years have elapsed since soil investigations began in earnest. Over this period, a large amount of information has been gathered on the profile characteristics of Canadian soils. Thus there is a general belief that the Canadian taxonomic classification system established in 1960, and incorporating the subsequent modifications made in 1963, 1968, and 1970, has served as a suitable and adequate framework for the categorization of the multitude of soil types throughout the country. While the taxonomic scheme has facilitated soil mapping programs, physical geographers still rely on the traditional "geographical" approach for the documentation and interpretation of soil patterns at the continental scale. However they frequently resort to taxonomic nomenclature when dealing with detailed investigations at the regional or local level.

Soils at the Sub-Continental Scale

Information on the soils of Canada at the country-wide level can be found in papers by Putnam (1951), Stobbe (1960), and Leahey (1961). While recognizing that a large number of soils can be identified in the field, Stobbe (1960) states that general similarities in profile characteristics enable the myriad of soil types to be reduced to a comparatively few great soil groups. However, on the map accompanying his paper he subdivides the country into soil zones which correspond approximately with major climatic and vegetation patterns. Likewise, Leahey (1961) employs the term zone in his discussion of Canadian soil geography, defining it in essentially the same manner as did the National Soil Survey Committee. In this context the term retains certain genetic implications insofar as each soil zone is often considered to reflect regional climate and vegetation. We have already noted that contemporary views on soil classification have shifted toward morphological description rather than documenting the relationships between soils and other environmental factors. Nonetheless, we still need a term that enables taxonomic units to be grouped in a manner that facilitates the depiction of the soil pattern on small-scale maps.

The following paper summarizes current thinking on some of the issues raised in the preceding paragraphs. After defining certain terms used in describing basic soil morphology, the author goes on to outline the major environmental factors in the Canadian landscape which affect soil formation. He briefly describes the eight soil orders that have been identified at the sub-continental scale, but draws the fundamental distinction between geographical and taxonomic classifications. It is noteworthy that in his discussion of the soil pattern the author avoids any genetic bias by assigning the term region to the major subdivisions.

THE CANADIAN SOILS SYSTEM*

Douglas W. Hoffman

Centre for Resources Development
University of Guelph

With the development of society and the proliferation of the human race, man's role on earth has evolved from one in which he was subordinate to his environment, accepting the fortunes bestowed on him by nature, to one in which he is dominant, modifying his environment to suit his needs.

Man is an important instrument of change. He is a prime ecological, geological and biological agent having an effect on, among other things, the land. The way in which human populations utilize the land is dictated by their stage of cultural development and the technology available to them as a result of that culture. The identification, use and care of the land is of concern to all, and in the end, it is a problem of human values and human behaviour.

The world's resources of arable soil are strictly limited. At present about 4,000 million acres of land, or 3 percent of the earth's surface, are in crop production. Even with ample time and reasonable resources it is doubtful whether this latter figure could be raised above 5 percent. Canada's supply of arable soil is little better than the world average. Less than 10 percent of Canada's 2,272 million acres is in farmland and about half that is in cultivation. At least two-thirds of Canada has little or no capability for commercial agriculture because of the cold climate and, of the remining one-third which is climatically favoured, large areas are unusable agriculturally because of unsuitable soils.

Soils, then, have considerable influence on productivity and land use. The characteristics of soil profiles are used to classify soils, and these characteristics along with those outside the soil profile, such as slope and stoniness, have a bearing on soil productivity and land use.

The profile is basic in any classification of soils, and some idea of its components is needed to understand soil classification systems. A soil profile can be identified by digging a pit into the surface of the earth and carrying out visual observations of the pit face. The depth of the pit is determined by the nature of the soil itself, but commonly this varies from one to three metres, below which is the relatively unaltered material often termed parent material. Once the pit face is exposed a characteristic pattern of layers is revealed. The layers are most easily recognized by colour. Each individual layer is known as an horizon and the set of layers is called a soil profile.

In some cases the contrast between horizons is dramatic while in others it is very subtle. Some of the symbols used to designate horizons and a brief definition of each are listed as follows.

L — uppermost layer; consists of fresh plant litter

F — dark brown fermented layer; partially decomposed plant material

H — black, well decomposed, amorphous organic matter

Ah — a dark coloured mineral horizon in which humus has accumulated

Ae — a light coloured eluviated (leached) horizon

Bf — an illuvial horizon in which iron and aluminum are the main accumulating materials

Bt — an illuvial horizon in which clay is the major accumulating substance

Bg — a gleyed (mottled) horizon

Bn — an illuvial horizon containing one or more salts. Often said to be saline

C — varies in colour; only slightly altered; often called parent material

Soils are natural products of the environment. They form in the uppermost part of the earth's crust from rocks of many kinds or from transported mixed rock debris. Transportation of debris may be by ice, water, wind, or gravity. Deposits such as till, alluvium, loess, or colluvium as well as rock are the parent materials of soils. Physical, chemical and biological processes combine to alter the original rock or rock debris and the resulting weathering and formation of soil horizons constitutes soil formation. The intensity of these processes largely determines the kind of soil that will develop from a particular parent material. This intensity is controlled by environmental factors such as climate, vegetation and topography working over time. It is therefore important to know something about climate, vegetation, geology and geomorphic development of the landscape in order to understand the basic reasons why soils differ from one another and why specific soils are found at specific places.

Accordingly a very general account of some of the natural features of Canada is presented as a prelude to the discussion of its soils.

Physiography

North America has at times been described as resembling a saucer with the Appalachian and Pacific Coast mountain ranges forming the rim and the central plains occupying the depression in the middle. Canada occupies part of that "saucer" and thus has a morphology that is something of that nature.

Canada can be divided into seven physiographic regions. The largest of these is the Canadian Shield, which is characterized by knobs and ridges of siliceous rocks with an intermingling of more basic rocks of metavolcanic origin. Once mountainous, this region has been planed by erosion and now has a skyline that is relatively even. Much of the Shield stands between 500 and 2,000 feet above sea level.

The Interior Plains are marked by three steppes rising westward. For the most part drainage is eastward, the rivers falling from 4,000 feet at the foothills to about 1,000 feet above sea level in Manitoba. The Cordilleran Region comprises the whole series of mountain ranges running in a northerly direction through British Columbia and the Yukon. Mountain peaks and slopes, plateaux and valleys provide a diversified landscape of great scenic beauty.

The Great Lakes and St. Lawrence Lowland is the smallest of Canada's physiographic regions and is probably best known as a region of deep glacial drift except for some tracts of shallow soils over limestone. One of the best known topographic breaks in the region is the Niagara Escarpment. The highest part of the upland occurs near the village of Dundalk, where the relief stands at 1,800 feet above sea level and from whence plains slope towards Lakes Huron and Erie.

The Appalachian Region occupies Canada's Atlantic provinces and consists of old mountains so worn that their skylines are quite regular. Most of the region is composed of rolling plateaux, but there are some large areas with more level topography. The best soils in the region are developed from the limestone, shale and calcareous slates common in the Eastern Townships and Gaspé Peninsula of Quebec and in northern New Brunswick. The quartzite and granite that make up much of the rest of the region tend to be very thinly covered with soil. The lowlands of Nova Scotia and Prince Edward Island are underlain by predominantly red sandstone which gives a characteristic colour to the land.

The two remaining physiographic regions are located in the northerly parts of Canada. The Hudson Bay Lowland is characterized by flat boggy plains and gravel beaches. On the other hand the Arctic Archipelago presents quite a different picture. Here are lowlands underlain by limestone and shale and rubble-strewn uplands with mountains in some places.

A general idea of the distribution of the physiographic regions can be gained from Figure 1.

Geology

Both the bedrock and surface geology influence the properties, classification, use and management of soils. The composition of the soil parent materials is related to the bedrock and to the various surface deposits. In addition, topography is controlled by both surface and bedrock deposits.

In Canada, the rocks of the Precambrian Era form the foundation of the country. Granites and gneisses having a very low base content are most common although there are extensive areas of metavolcanic rocks such as diabase and basalt. In addition metamorphic

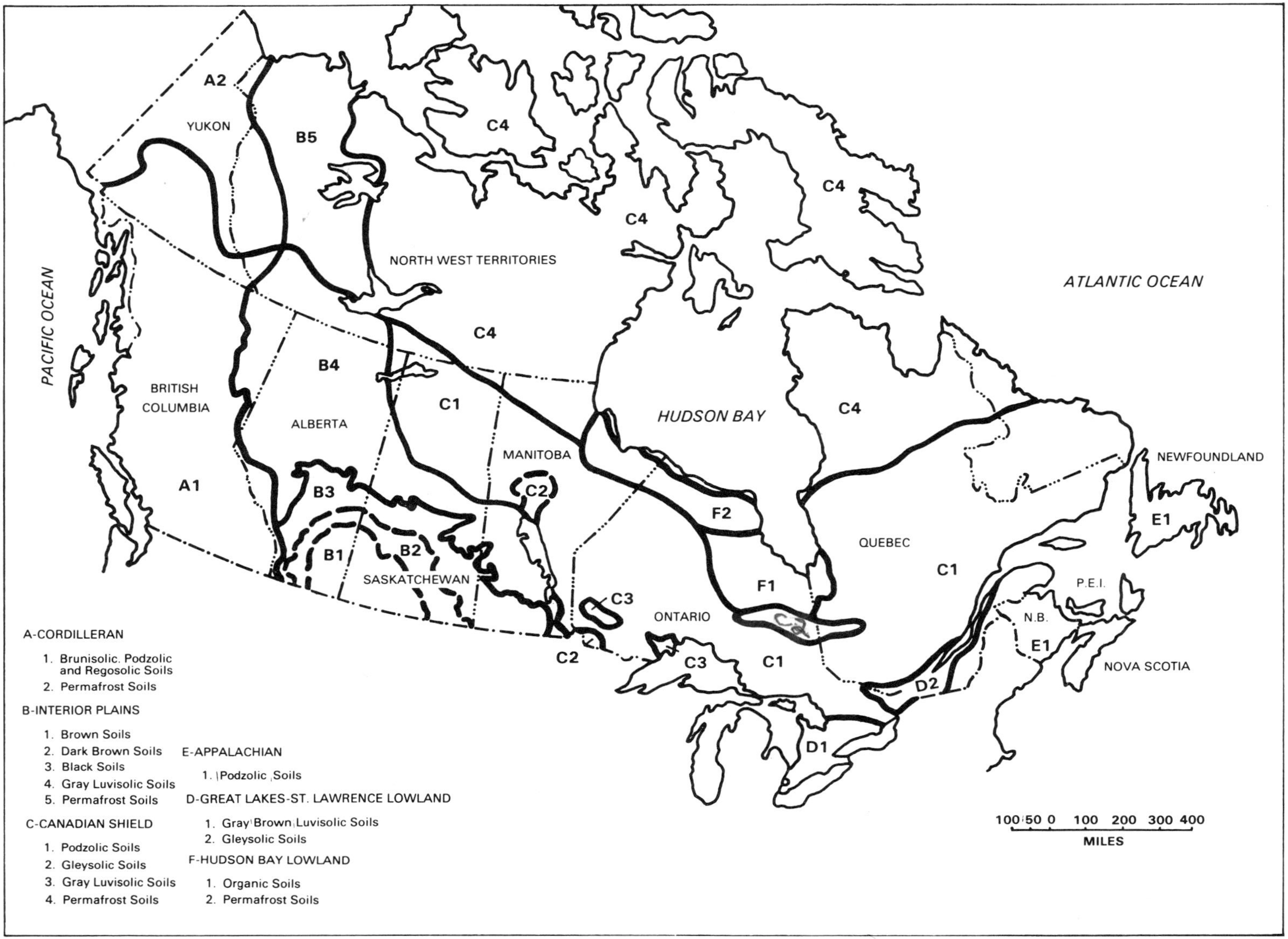

PACIFIC OCEAN
ATLANTIC OCEAN
HUDSON BAY
YUKON
NORTH WEST TERRITORIES
BRITISH COLUMBIA
ALBERTA
SASKATCHEWAN
MANITOBA
ONTARIO
QUEBEC
NEWFOUNDLAND
N.B.
P.E.I.
NOVA SCOTIA
A1
A2
B1
B2
B3
B4
B5
C1
C2
C3
C4
D1
D2
E1
F1
F2
100 50 0 100 200 300 400
MILES
A-CORDILLERAN
1. Brunisolic, Podzolic and Regosolic Soils
2. Permafrost Soils
B-INTERIOR PLAINS
1. Brown Soils
2. Dark Brown Soils
3. Black Soils
4. Gray Luvisolic Soils
5. Permafrost Soils
C-CANADIAN SHIELD
1. Podzolic Soils
2. Gleysolic Soils
3. Gray Luvisolic Soils
4. Permafrost Soils
E-APPALACHIAN
1. Podzolic Soils
D-GREAT LAKES-ST. LAWRENCE LOWLAND
1. Gray Brown Luvisolic Soils
2. Gleysolic Soils
F-HUDSON BAY LOWLAND
1. Organic Soils
2. Permafrost Soils

Fig. 1. Canada: Major physiographic and soil regions.

limestones, shales and sandstones occur but not to the same degree as those mentioned previously. Precambrian rocks occur at the surface of the Canadian Shield but are often buried under other formations in some physiographic regions.

The Interior Plains are underlain by Proterozoic, Paleozoic, Mesozoic and Tertiary strata and these are largely limestones, shales and sandstones. Similar sedimentary rocks form the Arctic Lowlands, the Hudson Bay Lowlands and the St. Lawrence Lowlands. The Cordillera in Canada is divided longitudinally into three great belts each having its own characteristic geology.

The eastern belt is composed almost entirely of folded sedimentary strata, the interior section is made of folded sedimentary and volcanic strata, massive metamorphic rocks all intruded by bodies of igneous rocks, and the western belt is composed mainly of plutonic rocks.

In general, the best soils for biological production are those developed from sedimentary rocks, especially the limestones. Conversely, the granites and gneisses of Precambrian formations contribute comparatively low amounts of nutrients for plant growth. Soils from volcanic and metamorphic rocks are intermediate in productivity.

Climate

Soils and vegetation of any region show a direct response to topography, parent materials and climate. While topography and parent materials are major influences on local variations in soil and vegetation, climate is the dominant factor affecting soil and vegetative variability among large regions.

The climate of Canada is very complex and the general comments that follow do not adequately describe this complexity. However, it may be of interest to point out a few of the major characteristics of the climate in each of the physiographic regions.

In the far north occurs the Arctic climate. It is dry and cold with long winters and short summers. South of this is the subarctic climate characterized by very cold winters and short but surprisingly warm summers. The northern prairies, the southern part of the Canadian Shield, the St. Lawrence Lowlands and the Appalachian region all have cool humid climates. Winters are cold but summers are warm.

The more southerly part of the Interior Plains region is characterized by dry, cold winters, warm summers, and chinooks bringing rapid changes in temperature. Low rainfall and high evaporation combined with periodic droughts are handicaps to agricultural production. In the Great Lakes Lowland around peninsular Ontario and the Montreal plain, warmer climate prevails and precipitation varies from 30 to about 38 inches per year.

By far the greatest amount of rain falls on the west coast where the annual amount varies from 30 to 100 inches. Here, summers are cool, however. The mountain climate is humid, subarctic to polar but the intermontane areas are dry and somewhat warmer.

Vegetation

Natural vegetation regions occur in bands across the country more or less occupying the same areas as the major soil groups shown in Figure 1. Tundra vegetation of grasses, sedges and low-growing shrubs occupies the northern part of Canada. To the south is the Forest-Tundra transition zone extending from one side of the country to the other. The trees are the same as in the Boreal Forest but they are stunted and stands are sparse.

The Boreal Forest is an association of conifers, aspen and white birch. Of the conifers, black spruce, white spruce, balsam fir, larch and jack pine are most common. South of the Boreal Forest in the east is the Eastern Hardwood zone consisting mainly of sugar maple and yellow birch in the north and sugar maple and beech in the south. In Manitoba, Saskatchewan and Alberta, the Boreal Forest gives way to the grasslands.

Throughout the Cordilleran region the vegetation zones are related to altitude in a complex manner. On the Pacific coast, however, are some of the finest forests in Canada. Western red cedars and western hemlock are abundant together with Douglas fir in the south and Sitka spruce in the north.

Soil Classification

The soil mantle is three-dimensional; it has length, breadth and depth. It is impossible, in practice, to see all parts of this mantle, and the best that can be done is to sample it at intervals by studying vertical sections called soil profiles. The classification of soils is based on the features of the soil profile including the soil parent material and on certain concepts of soil genesis.

The biological sciences have influenced the construction of methods of soil classification. They employ a hierarchical system comprising a series of categories at different levels each having fewer components than the category below, achieved by grouping the members of the lower category on the basis of mutually exclusive properties. Attempts to produce a similar "taxonomic" hierarchical system of soil classification have not been so successful. Consequently, it has always been necessary to make compromises with the properties selected.

Most recent systems of soil classification are definitive and based upon the intrinsic characteristics of the soil, but they also use genesis where important. Published systems of this sort include those of the Canada Soil Survey and the Seventh Approximation in the U.S.A. The system used in Canada incorporates six categories: the Order, Great Group, Subgroup, Family, Series and Type. The number of classes in each category increases from eight in the Order to thousands in the Soil Type.

The eight classes[1] in the order are named as follows:

Regosolic	—	A, C
Brunisolic	—	A, B, C
Chernozemic	—	A, B, C
Luvisolic	—	L-H, or A, Ae, Bt, C
Podzolic	—	L-H, Ae, Bf, C
Solonetzic	—	A, Bn, C
Gleysolic	—	A, Bg, C
Organic	—	Of, Om, Oh

The major horizons identified with each soil order are indicated beside the name. The horizons serve to define the orders but a brief description may help to show a few of the basic similarities and differences among them.

The Regosolic Order contains well and imperfectly drained soils that lack noticeable horizon development except, in some cases, for a mineral-organic layer at the surface. Soils of the Brunisolic and Chernozemic Orders are often very much alike but the former is found in forested regions, whereas the latter occurs where grass forms the main vegetative cover. Different vegetation is presumed to make the soils of these two orders genetically dissimilar and, therefore, soils of like morphology are, in this case, classified in two separate orders. Luvisolic and Podzolic soils bear some similarities and some differences. The most striking difference is between the B horizons; in Luvisolic soils the major accumulating agent in the B is clay, while in the Podzolic soils iron, aluminum and organic matter, singly or in combination, are the dominant accumulation products. The Solonetzic Order contains those soils which have developed from saline parent material or under the influence of saline waters. All of the poorly drained soils appear in the one Order (Gleysolic), and soils developed from materials containing 30 percent or more of organic matter are placed in the Organic Order.

An illustration of two of the subdivisions within the Canadian System is shown in Table 1.

One of the interests of the soil scientist is to determine the spatial distribution of the various kinds of soil. To this purpose he makes soil maps of many different kinds ranging from those which give a generalized small-scale picture of soil conditions of a whole country to detailed maps showing the distribution of soils on a single building site. Soil survey entails recognizing and delineating on a map soil bodies that are meaningful in terms of the objectives of the survey. The map legend relates these bodies to the classes of a taxonomic system which are not mapping or geographic units. Areas on a map are seldom identical with the classes of the taxonomic classification because of the nature of the soil continuum. Changes in soil morphology often occur over very short distances and the soil series cannot be used as a mapping unit except on very large scale maps. Even then some mapped soil areas con-

[1]The number of orders may be increased to nine. The groupings of permafrost soils into a Cryosolic Order is being considered by the Canada Soil Survey Committee.

TABLE 1
The Orders and Great Groups of the Soil Classification for Canada

Order	Great Group	Order	Great Group
Chernozemic	Brown	Brunisolic	Melanic Brunisol
	Dark Brown		Eutric Brunisol
	Black		Sombric Brunisol
	Dark Gray		Dystric Brunisol
Solonetzic	Solonetz	Regosolic	Regosol
	Solod	Gleysolic	Humic Gleysol
Luvisolic	Gray Brown Luvisol		Gleysol
	Gray Luvisol		Eluviated Gleysol
Podzolic	Humic Podzol	Organic	Fibrisol
	Ferro-Humic Podzol		Mesisol
	Humo-Ferric Podzol		Humisol

tain soil series not mentioned in the mapping unit name. The mapping (geographic) units may contain inclusions or may range from a subdivision of series to groups of series depending on the scale of the map and complexity of the soil pattern. The classification serves the soil survey by making possible identification, correlation and naming of soil bodies.

The Distribution of Soils in Canada

A very general view of the distribution of the soils of Canada is given in Figure 1. More detailed information can be found in publications issued by the different soil survey organizations in Canada but soil maps are not available for all sections of the country. Little is known about much of the unsettled parts of Canada.

In the Arctic and Sub-Arctic regions profile development is strongly influenced by the presence of permafrost. Although only a few areas have been mapped, the soils investigated seem to have morphological features like those of the Brunisolic, Gleysolic, Regosolic and Organic orders. However, the soils with permafrost appear to differ greatly in their overall characteristics from the modal concepts of Regosolic, Brunisolic and Gleysolic soils, and new groupings in the classification system may become necessary as more information becomes available.

The complex nature of geology, vegetation and climate characterizing the Cordillera is reflected in the soils and their arrangement. Both horizontal and vertical zoning occurs. In humid, cool areas Podzols are dominant whereas Brunisolic and Luvisolic soils tend to favour less humid areas. Regosolic soils are found in the many areas where comparatively young parent materials occur due to rock slides, avalanches and so forth.

Chernozemic soils cover the grassland areas in the Interior Plains region. Soils of the Brown group lie in the most arid part of the region in the south. North of the Brown soils distributed in semi-circular fashion are the dark Brown soils followed by the Black. Although these are the dominant soils in the grasslands there are also large sections covered by Solonetzic, Gleysolic and Organic soils. The Solonetzic soils occupy the sites where salts are a common component of the parent materials while the Gleysolic and Organic soils are found in the wetter locations.

The section of the Interior Plains region between the grasslands and the permafrost line is covered by Luvisolic soils. These are dominantly Gray Luvisolic. Gleysolic soils, mainly of the Gleysol and Eluviated Gleysol great groups, occur on the wetter sites but the major soils of the wet sites are organic.

In that part of the Canadian Shield south of the permafrost line the dominant soils are Podzolic. The most common are the Ferro-Humic and Humo-Ferric Podzols. Gleysolic soils occur in the level clay plains scattered throughout the region. Some of these clay deposits are rolling and Gray Luvisolic soils are dominant. A large part of the Shield is covered by organic soils commonly referred to as muskeg. Fibrisols and Mesisols are the main great groups found in Organic soil areas. Other soils of significance in the Canadian Shield are the Brunisolics. These occur mainly along the southern edge of the Shield where Dystric Brunisols are most common.

The Great Lakes-St. Lawrence Lowland contains representatives of Luvisolic, Gleysolic, Brunisolic, Regosolic and Organic soils. Gray-Brown Luvisolic soils are dominant in peninsular Ontario occupying the imperfectly and well drained sites. This part of Ontario also contains Melanic Brunisols in the upland areas while Humic Gleysols occupy much of the level wetland. Ontario and Quebec portions of the Ottawa and St. Lawrence river valleys display mainly Gleysolic soils. However, Melanic Brunisols are dominant on the better drained sites. Humisols are the dominant Organic soils found and these are scattered throughout the whole of the region. Some idea of the great variety of soils in Southern Ontario can be gained from Figure 2.

Podzolic soils dominate the Appalachian region. Gleysols and Eluviated Gleysols occupy the poorly drained locations and, of course, Organic soils, especially Fibrisols, are found where water accumulates and very wet conditions exist. Humic Podzols are widespread in Newfoundland but are of lesser importance in Nova Scotia and New Brunswick.

The Hudson Bay Lowland is characterized by large expanses of Organic soils broken by gravel or till uplands on which Podzolic soils have developed. In the more northerly part of the Lowland, permafrost is a feature of soil development.

Limitations To Use

From the foregoing it can be seen that soils develop in widely varying climatic regions under different kinds of plant cover and from various parent materials. Consequently there

are many distinct kinds of soils, each as part of its own system, each with its own capabilities and each presenting peculiar problems to man.

Without question the soil is of great importance to mankind. Food crops are obtained directly from the soil or from livestock fattened from feed taken from the land. Roads, highways, buildings and parking lots are all built upon the soil and the soil is utilized for various kinds of recreation. Each use presents a different set of soil problems.

In Canada the major limitation to use is not one directly associated with the soil. It is climate. Long, cold winters affect a large proportion of the country and make the growing of a variety of crops difficult. In addition, frost movement creates problems with above and below ground facilities. Frost boils in the spring make quagmires of the roads, and pipelines sink in unstable permafrost. However, not all the climatic problems are due to cold alone; for example, in the areas of chernozemic soils in the Interior Plains region, lack of moisture limits crop production.

There are other limitations to land use. Steep slopes limit the use of mountainous areas, shallow soils over bedrock cover much of the country making crop production impossible, and in other regions wind and water erosion are common. Erosion, once the scourge of the arid lands in the west, is considered to be under control. However, new erosion problems, even more severe than those of the 1930s, are occurring in some parts of the country. These are associated with the building of houses, factories and transportation corridors when the vegetation is swept away and the soil surface is left without cover.

Salty parent materials create problems for plant growth on the Solonetzic soils while on many Podzolics stoniness and a small supply of plant nutrients reduce productivity. Many of the Regosolic soils lie in the bottomlands along the rivers, and these are flooded each year preventing their use for anything but recreation or grazing purposes.

Of course, not all the soils of Canada have limitations to use. Indeed, there are soils which have a high capability for use. These areas are delineated, as well as those with a low cap-

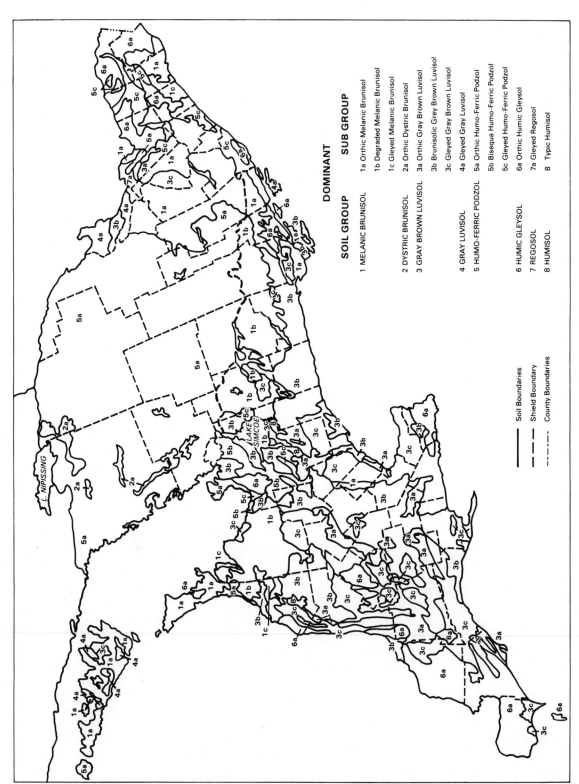

SOIL GROUP

1 MELANIC BRUNISOL

2 DYSTRIC BRUNISOL

3 GRAY BROWN LUVISOL

4 GRAY LUVISOL

5 HUMO-FERRIC PODZOL

6 HUMIC GLEYSOL

7 REGOSOL

8 HUMISOL

DOMINANT

SUB GROUP

1a Orthic Melanic Brunisol
1b Degraded Melanic Brunisol
1c Gleyed Melanic Brunisol
2a Orthic Dystric Brunisol
3a Orthic Gray Brown Luvisol
3b Brunisolic Grey Brown Luvisol
3c Gleyed Gray Brown Luvisol
4a Gleyed Gray Luvisol
5a Orthic Humo-Ferric Podzol
5b Bisequa Humo-Ferric Podzol
5c Gleyed Humo-Ferric Podzol
6a Orthic Humic Gleysol
7a Gleyed Humic Gleysol
8 Typic Humisol

———— Soil Boundaries
– – – Shield Boundary
–·–·– County Boundaries

Fig. 2. Southern Ontario.

ability, on maps produced under the Agricultural and Rural Development Act and are available from Information Canada. Capability maps have been prepared for such uses as agriculture, forestry, recreation and wildlife and they provide a measure of the land resources of Canada.

Soils at the Regional and Local Scale

Prior to 1960, but with the exceptions of work by Feustel et al (1939) and Leahey (1947, 1954), Canadian pedologists had paid little attention to soils in the Arctic and Subarctic. However, the recent discovery of economically recoverable oil and gas reserves in the Canadian Arctic, and the very distinct possibility that their development may lead to severe and perhaps irreversible damage to the fragile tundra environment, points to the obvious need for more information on all aspects of Arctic soils. American soil scientists have been conducting investigations into Alaskan soils for over a decade. A team of pedologists headed by J.C.F. Tedrow of Rutgers University, New Jersey, deserves much of the credit for expanding our knowledge of Arctic soils (Tedrow et al 1958). Although they are concerned with Alaskan soils, their work demonstrates how soils in the Arctic can be viewed within the zonal framework established for soils elsewhere on the continent.

Tedrow and his co-authors have two major objectives: to offer a comprehensive description of soils characteristic of the Arctic slope of Alaska, and to provide information on the processes active in soil genesis in an Arctic tundra environment. They found that on the Arctic slope of Alaska, soil formation is not characterized by a unique and readily identifiable process as is the case with zonal soils in tropical, semi-arid, and cool temperate environments. Instead podzolization appears to weaken northward away from the boreal forest. The authors distinguish a number of great soil groups across the Arctic slope with the most widespread being the tundra soil, a wet mineral variety found in imperfectly-to-poorly drained locations generally underlain by permafrost. Mature soils developing in stable well-drained sites were labelled as Arctic Brown. This was the only group the authors felt could be safely assigned to a zonal status, on the grounds that zonality is generally equated with maturity in soil profile development, and this in turn implies adequate drainage. In many instances, the widespread occurrence of an essentially impermeable permafrost layer in the Arctic impedes the vertical migration of soil water; this gives rise to saturated conditions in the profile above the permafrost table during the thaw season. Bog soils are typical of these environments. The Arctic Brown, the Tundra, and the Bog soils were regarded by the authors as the northern counterparts of the Podzol, Glei, and Bog soils respectively (see also Tedrow and Cantlon 1958). The Tundra and Bog varieties are both intrazonal. Lithosols occur on relatively steep slopes where drainage is adequate but where profile development is very weak; by definition they are azonal. Thus drainage, which reflects the degree of permafrost development together with the topographic situation, is a primary controlling factor in the genesis of Arctic soils.

Not all researchers have been in total agreement with the proposals forwarded by Tedrow and his colleagues. For instance, on the basis of four soil profile descriptions from four localities in the Queen Elizabeth Island group of the high Canadian Arctic, McMillan (1960) maintained that existing classification schemes for Arctic soils might need revising. He concurred with the view that Tundra soils are encountered north of the Podzols, but at the same time, none of his descriptions seemed to comply with the requirements of this group. Beyond the tundra girdle he called for the recognition of a "raw" soil which he characterized through the adoption of terms established by European pedologists. Later, Tedrow and Douglas (1964) and Tedrow (1966) conducted soil investigations on Banks Island and Prince Patrick Island respectively. They borrowed the term Polar Desert from the Russian literature and applied it to those well-drained soils developing in low precipitation areas which exhibited a desert-like appearance with a sparse vegetation cover and occasional expanses of desert pavement. By 1968 Tedrow proposed that polar regions could be divided into three major pedogenic zones: an Arctic Brown zone comprising zonal soils developing on well-drained sites north of the Podzols; a Polar Desert zone with soils typical of far northern ice-free regions; and a Cold Desert zone found only in the Antarctic.

The progress made by Tedrow and his associates in their quest to rationalize polar soils has provided the stimulus for a number of recent pedological studies elsewhere in Arctic Canada (e.g., James 1970; Cruickshank 1971). James, whose paper is reprinted below, describes the soils of the Rankin Inlet area, Northwest Territories; from observations made of representative profiles, he distinguishes soils whose pedological characteristics correspond closely with Tedrow's Half Bog, Tundra, and Arctic Brown varieties. He also finds evidence to indicate that some of the better drained soils tend to exhibit weakly podzolized profiles. Towards the end of the paper, he discusses the consequences of frost action during seasonal freezing and thawing above the permafrost table, another process which can impart a distinctive character to Arctic soils.

THE SOILS OF THE RANKIN INLET AREA, KEEWATIN, N.W.T., CANADA*

P. A. James

Abstract

Profiles representative of the soils of an area in the Canadian Arctic Barren Lands bordering Hudson Bay are described. The soils are classified according to the degree of disturbance by frost action and according to the state of soil drainage. Relatively undisturbed soils include those similar to Half Bog, Meadow Tundra, Upland Tundra, and Arctic Brown found elsewhere in the Arctic. Disruption of soil horizons are associated with a number of periglacial (frost) processes. Evidence of podzolization in the well-drained soil comprises a slightly-bleached horizon and, in one profile examined, an iron pan. The degree of podzolization is believed to be representative of the "low arctic" position of the area. No significant difference is apparent between soils which are 4,000 years old and soils which are appreciably younger.

Introduction

This paper is offered as a contribution to the present meager knowledge of the arctic soils of Keewatin. It is based on data collected from an area of 2,500 km² centered on the settlement of Rankin Inlet, which lies 480 km north of the arctic tree line and the southern boundary of continuous permafrost (Figure 1). No previous study of soils near Rankin Inlet has been undertaken. Feustal et al. (1939) analyzed soil samples collected from limited coastal areas between Churchill, Manitoba, and Craig Harbour, Ellesmere Island, including one sample from Chesterfield Inlet. A lack of profile development and certain chemical and physical

* Reprinted from *Arctic and Alpine Research*, Vol. 2, No. 4, 1970, pp. 293-302 with the permission of the author and publishers, Institute of Arctic and Alpine Research, University of Colorado.

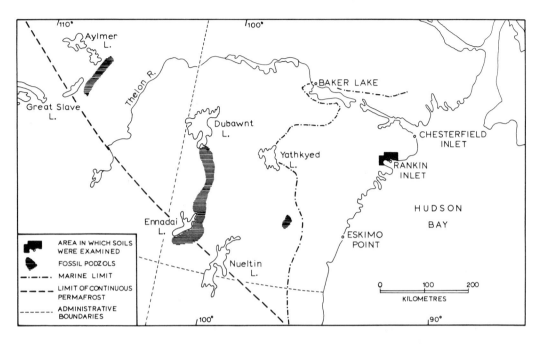

Fig. 1. Location map showing distribution of fossil podzils as mapped by Larsen (1964), the approximate position of the marine limit, and the southern boundary of continuous permafrost.

characteristics of the samples indicated the immaturity of the soils examined.

The Physical Setting

The physical landscape of the Rankin Inlet area has as its base Precambrian intrusive, volcanic, and metamorphosed rocks; upon these have been deposited and moulded glacial and glacio-fluvial forms, which in turn have undergone modification during late-glacial and postglacial marine submergence. Subsequent to postglacial uplift of the land, recent geomorphic processes and the growth of vegetation have produced the present details of the surface, together with the soils. The present landscape is monotonous and of low relief (Figure 2); more than half the area lies below 60 m, and nowhere does altitude reach the limit of late-glacial and marine submergence.

The glacial geomorphology has been sum-marized and the periglacial geomorphology described in detail elsewhere (James, 1972). The glacial, glaciofluvial, and marine deposits comprising the soil parent materials may be classified as follows:

Glacial till

Lodgement till: forming sheets, and drum-linoid and morainic ridges. Generally, this comprises a very bouldery deposit with a coarse sandy matrix; pockets of fine sand and silty fine sand are common and im-portant in determining the nature of certain soil profiles.

Ablation till: boulders with coarse sandy and gravelly matrix; also includes exten-sive boulder fields.

Glaciofluvial

Eskers: the constituent material varies from coarse sand and fine gravel to large erratic boulders.

Fig. 2. Aerial view of the landscape northwest of Rankin Inlet. Freely drained, brown soils occur on the sandy ridges which are cut by fissures of ice-wedge polygons, and colonized by lichen-heath tundra. A frost-churned soil is associated with the vegetation nets and sedges and mosses occupying the intervening dead-ice hollows. (Photo by P. A. James, June 1967).

Kame-like plateaus: similar to eskers.

Crevasse-fillings: comprising ridges of smaller dimensions than eskers, but of similar materials.

Marine

Gravel, coarse sand, and boulders: of raised beaches.

Fine and silty fine sand: these are widespread and susceptible to vigorous frost processes.

The climate of Keewatin is extremely continental: at Chesterfield Inlet (Figure 1) the absolute temperature range recorded between 1921 and 1955 was 63°C (−33° to 30°C); the mean daily temperature for January and February was −14°C, and for July and August, below 10°C. Total annual precipitation (rain plus water equivalent of snow) is approximately 280 mm (Hare, 1963). Permafrost, which is approximately 300 m deep at Rankin Inlet (Geological Survey of Canada, 1967), is a significant influence on soil drainage. The depth of seasonal thaw varies from 25 cm in marshy depressions to over 1.8 m on esker crests. Soil drainage is least impeded on ridges of gravel and coarse sand which comprise glaciofluvial landforms and raised beaches; it is impeded on the ridges and sheets of lodgement till and is very poor in the marine deposits of depressions such as dead-ice hollows.

The vegetation of the Rankin Inlet area is continuous except for small areas of bare sand on some glaciofluvial ridges and the boulders and cobbles of certain eskers and raised beaches on which only crustose lichens grow. A sparse cover of lichens occurs on the driest sandy and gravelly sites; where conditions are moister, moss and vascular plants, particularly ericaceous species, grow with the lichens to form a lichen-heath vegetation. In poorly drained habitats, sedges and mosses are mixed with lichens and heath plants; microrelief is commonly hummocky. A dense growth of sedges and mosses grows on marshy depression floors which remain wet for much of the summer.

The Soils

The following classification of the soils of the Rankin Inlet area is based on field observations and may have to be modified after further fieldwork and laboratory analyses. As Tedrow (1962) has shown, any consideration of arctic soils must inevitably take into account the effects of frost disturbance which disrupts normal soil-forming processes and destroys the horizons of arctic soils. A two-fold division of the soils is therefore thought to be convenient: those which are relatively undisturbed by frost and those which are frost-disrupted. The undisturbed soils are subdivided below into A, B, C, and D. A comparison (which at present is only tentative) may be drawn between soils A, B, and C and those described as major genetic

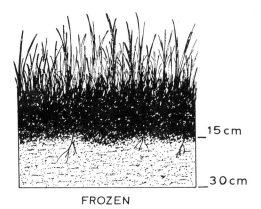

Fig. 3. Profile 1: very poorly drained soil.

soils by Tedrow (1962): soil A is similar to the Half-Bog, soil B to Meadow and Upland Tundra, and soil C to the Arctic Brown soil.

Soils Relatively Undisturbed by Frost   In very poorly drained soils (A) which occupy marshy depressions, anaerobic conditions beneath a mat of organic matter, mosses, and sedges have induced gleying throughout the whole of the mineral soil above the permafrost table. Where soil drainage improves, an oxidized horizon appears and deepens in an upslope direction (soil B). With further improvement in drainage and a grading to a moss-lichen vegetation, the top horizon becomes thinner; gleying is restricted to the base of the active layer. In the most thoroughly drained sands beneath a lichen-heath vegetation, gleying is absent: a profile with brownish colors has developed (soil C). Here a bleached horizon indicating podzolization is not uncommon beneath the surface organic matter. Translocation of iron was very evident in one profile which contained a distinct iron pan. Soil consisting of organic matter with only wind-incorporated sand grains has developed in some crevices and depressions on rock outcrops (soil D).

A. Very poorly drained soils (including Half-Bog)

These soils occur in extensive marshy depressions occupied by sedges and mosses. Such depressions are common and are generally water-logged following the spring thaw but be-

come dry during the summer. The permafrost table lies within 45 cm of the surface.

PROFILE 1 (Figure 3)
Site: floor of large depression 3.2 km west-northwest of Rankin Inlet.
Slope: $< 1°$
Vegetation: moss and sedges with other vascular plants on small hummocks.

Horizons:
1. 0-15 cm   Wet, dark brown organic matter consisting of slightly decomposed plant remains; merging into
2. 15 cm+   Wet, bluish-gray fine sand; platy structure; several unweathered stones up to 6 cm long. Frozen at 30 cm, August 22.

B. Imperfectly drained soils (Meadow Tundra and Upland Tundra)

These soils form on gentle slopes with moderate drainage. Upslope from large depressions occupied by soil A, soil drainage improves and a brownish-yellow or olive-brown oxidized horizon appears above gleyed, bluish-gray sand and is deeper where the depth of seasonal thaw increases; the surface layer of organic matter becomes thinner, more sharply defined, darker, and more thoroughly decomposed. This soil is typical of poorly drained surfaces of the side-slopes of marshy depressions and lower segments of slopes of many glacial and glacio-fluvial landforms; it also occurs on extensive plateau areas of till and marine sand. The permafrost table lies between 50 and 80 cm. Mosses and sedges are the dominant plants. Profiles 2 and 3 represent a gradation in the direction of improved drainage and deeper active layer.

PROFILE 2 (Figure 4)
Site: Slope to depression from which profile 1 was described.
Slope: ½° southwest.
Vegetation: Mosses and sedges are dominant with _Dryas intergrifolia, Rhododendron lapponicum,_ and _Cetraria_ sp. being the dominant vascular plants on hummocks.

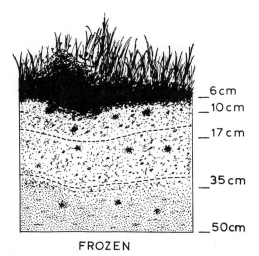

FROZEN

Fig. 4. Profile 2: imperfectly drained soil.

Horizons: 1, 2, and 3 vary in depth as shown in Figure 4. Depths quoted here were measured down center of profile.

1. 0-6 cm   Reddish-brown organic matter; living and dead moss and other plant remains; merging into

2. 6-10 cm   Dark brown to near black organic matter; fibrous with many plant remains; merging into

3. 10-17 cm Moist, yellowish-brown fine sand containing gravel; patches of near black fibrous to fluffy organic matter; merging into

4. 17-35 cm Moist, grayish-brown fine sand containing gravel; rusty and grey mottles and patches; rusty lining along root channels; merging imperceptibly into

5. 35 cm+   Very moist, bluish-gray, slightly silty fine sand; rusty patches and linings along root channels; dark gray mottles; becoming wet and thixotropic toward the base. Frozen at 50 cm, August 22.

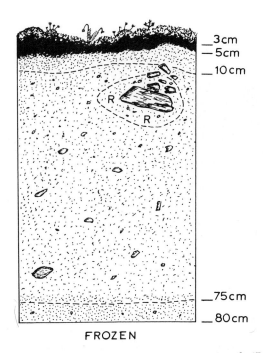

FROZEN

Fig. 5. Profile 3: imperfectly drained soil. (R: zone of rusty staining).

PROFILE 3 (Figure 5)

Site: boulder-strewn slope extending to sea level, 1.6 km north of Rankin Inlet.

Slope: 2° east.

Vegetation: mosses, *Cetraria* sp., *Alectoria* sp,. *Dryas integrifolia, Cassiope tetragona, Rhododendron lapponicum, Arctostaphylos alpina, Salix* sp., and a little grass.

Horizons beneath litter:

1. 0-3 cm   Dark brown organic matter; undecomposed and slightly decomposed plant remains; merging quickly into

2. 3-5 cm   Near black, moist, fluffy organic matter; sharp boundary

3. 5-10 cm   Olive-gray medium-fine sand; merging gradually into

4. 10-75 cm Olive-gray medium-fine sand, containing gravel and shells; weathering boulder surrounded by iron-stained rusty sand;

numerous stones, beneath some of which are deposited brown carbonate crusts; merging gradually into

5. 75-80 cm   Very moist, bluish-gray, slightly silty fine sand, containing shells; thin silty and sandy "cutans" capping some stones.
Frozen at 80 cm, August 21. Ice veins up to 2 mm wide in the frozen sand.

C. Freely drained soils

Soils of brown or brownish-yellow colors with no evidence of gleying occur on the most thoroughly drained, sandy, glaciofluvial ridges and raised beaches where the depth of seasonal thaw exceeds 1 m. Vegetation varies from lichen-heath overlying 7 cm of decomposed organic matter to a discontinuous cover of heath and lichen species where bare sand comprises much of the surface. A distinct bleached horizon underlying the surface organic matter of a number of profiles examined suggests a podzolic process:

PROFILE 4 (Figure 6)
Site: raised beach on Thomson Island to the east of Rankin Inlet.
Slope: 1½° west.
Vegetation: *Alectoria* sp., *Cetraria* sp., *Empetrum nigrum, Arctostaphylos alpina, Ledum palustre,* a little grass.
Horizons beneath litter:

1. 0-5 cm   Reddish - brown - undecomposed plant remains; merging into
2. 5-10 cm   Near black fluffy organic matter; sharp boundary with
3. 10-15 cm   Brownish - gray coarse sand; bleached appearance, with white quartz grains clearly visible; includes streaks and patches of dark organic matter.
4. 15-21 cm   Dull yellowish - brown coarse sand; merging quickly into

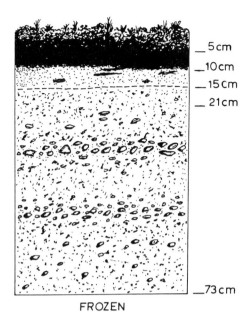

FROZEN

Fig. 6. Profile 4: freely drained brown soil.

5. 21 cm+   Dull yellowish - brown gravel and very coarse sand; abundant stones; roots occurring to 45 cm. Frozen at 73 cm, July 7.

In one place, a deeper layer of bleached sand was underlain by a distinct iron pan:

PROFILE 5 (Figure 7)
Site: near edge of lake, 32 km west of Rankin Inlet.
Slope: 0°
Vegetation: *Vaccinium uliginosum, Empetrum nigrum, Cassiope tetragona, Arctostaphylos alpina, Alectoria* sp.
Horizons:

1. 0-2.5 cm   Dark brown undecomposed organic matter.
2. 2.5-18 cm   Grayish-white, very well-sorted fine sand; stoneless; distinctly bleached.
3. 18-18.5 cm Distinct, thin, soft, rusty iron pan; undulating.
4. 18.5 cm+   Yellow fine sand; rusty mottles; soft, dark rusty iron concretions between

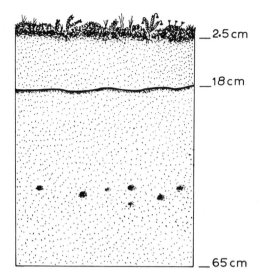

_2.5 cm

_18 cm

_65 cm

Fig. 7. Profile 5: freely drained soil with iron pan.

43 and 48 cm. Pit to 65 cm: not frozen, July 19.

Elsewhere in the tundra of the Districts of Keewatin and Mackenzie, evidence of podzolization has been attributed to processes active under former forests. Areas of fossil podzols mapped by Larsen (1964) are shown in Figure 1; these occur nearer to the present tree line than does Rankin Inlet. Bryson and others (1965) report fossil podzols near Dubawnt Lake. Buried charcoal found in association with the podzols enabled the latter to be dated and correlated with the Climatic Optimum (ca. 3,500 BP). However, profiles 4 and 5 were dug at altitudes of less than 30 m above present sea level, which places them below the position of sea level during the Climatic Optimum (James, 1972). Therefore, the podzolic profiles of the coastal Rankin Inlet area are probably due to present-day pedogenesis.

Evidence of present-day podzolization in well-drained arctic soils has been reported from Alaska and from various parts of the Soviet Arctic. Tedrow and his colleagues describe a weakening of the podzolic process poleward from the North American tree line and state: "If the summer precipitation were greater in the areas of Alaskan Arctic Brown soil and if the soils had sufficient permeability

and drainage, podzols would probably dominate the well-drained sites. The degree of this podzolic development would be in proportion to the increase in rainfall" (Tedrow *et al.*, 1958, p. 42). Annual precipitation on the arctic slope of Alaska amounts to 10 to 20 cm, as compared with approximately 28 cm in eastern Keewatin. In the Rankin Inlet area, conditions of climate and, in places, of parent material and soil drainage are such that podzolization may occur sporadically.

D. Organic soil on rock

Accumulations of decomposed organic matter are common in crevices and depressions in rock outcrops. Mixed with the organic matter are fragments of weathered bedrock at the base, and wind-deposited sand grains throughout the profile.

PROFILE 6 (Figure 8)

Site: summit of rocky hills 1.5 km southeast of Rankin Inlet.

Slope: <1°

Vegetation: mainly mosses and *Ledum palustre* with *Alectoria* sp., *Arctostaphylos alpina, Cetraria* sp., *Empetrum nigrum.*

Horizons (depths in deepest part of profile shown in Figure 8)

1. 0-6 cm   Reddish - brown organic matter; undecomposed moss and other plant remains; merging into

2. 6-13 cm   Dark brown organic matter consisting of partly decomposed plant remains; merging into

3. 13-23 cm   Near black fibrous organic matter, becoming moister and darker with depth; contains wind - deposited sand grains and a few

_6 cm

_13 cm

_23 cm

Fig. 8. Profile 6: organic soil on rock.

fragments of angular, fine gravel weathered from underlying rock.

4. 23 cm+   Fine grained crystalline rock.

Effects of Frost in Relatively Undisturbed Soils
Most of the soils are inevitably affected by frost. Drainage is particularly impeded where the active layer is shallow beneath an insulating cover of peaty organic matter (as in profile 1, Figure 3): the mineral soil remains wet throughout the summer and is gleyed. Seasonal freezing of the active layer influences soil structure and results in the disintegration of stones of certain lithological types. The growth of abundant "sirloin" ice veins (*see* Hamelin and Cook, 1967, p. 21) during the winter freezing of horizons of silty fine sand results in a fine platy structure which is commonly well defined when the active layer has thawed. On the other hand, most medium to coarse sands, which are cemented with separate ice grains when frozen, have single-grain structure when thawed. Frost weathering has produced shattered, angular fragments in soils containing stones derived from fissile metamorphic rock. However, the degree of pedogenic weathering is very slight: soils throughout the area contain very small proportions of clay. No conclusive evidence has yet been found in the Rankin Inlet area of particle size sorting by freeze-thaw processes which has been demonstrated by the experiments of Corte (1966). In a number of profiles examined, grain-size increased toward the surface, but this sorting may be an effect of former marine wave-action and deposition.

Frost-Disrupted Soils Most landforms and most types of surface in the Rankin Inlet area exhibit some evidence of frost action in the form of patterned ground or other periglacial feature. With each periglacial form is associated a particular profile in which mineral and organic material has been intermixed or in some other way disturbed. The most common frost-disrupted profiles are described below. Some represent a disturbance which is restricted to the immediate vicinity of the periglacial feature: for example, the distortion of

soil horizons caused by the formation of frost-fissure polygons ("tundra polygons") is restricted to the trenches and rims which form the borders of the polygons. Other forms of disruption, such as the involutions of peat and fine sand of the "frost-churned soil" (profile 7, Figure 9) cover extensive areas. The "churned soil" is associated with the formation of nonsorted circles ("mud boils") and is common on sheets and ridges of glacial till. The fine, silty sand which forms the nonsorted circles and related forms occurs abundantly in pockets in till and as a marine deposit.

PROFILE 7 (Figure 9): "frost-churned soil"
Site: crest of drumlinoid ridge, 18 km northwest of Rankin Inlet.
Slope: <1°
Vegetation: mosses, *Alectoria* sp., *Dryas integrifolia, Cassiope tetragona, Empetrum nigrum,* and grass.
Horizons (as shown in Figure 8):
1. Brown, little decomposed moss and other plant remains.
2. Fibrous to fluffy organic matter, becoming black with depth; surrounded by gray-stained sand.
3. Light yellowish-gray, silty fine sand with patches of gray and inclusions of

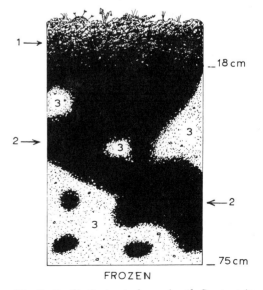

Fig. 9. Profile 7: frost-churned soil. See text for description of 1, 2, and 3.

Fig. 10. (Right) Sections through a solifluction lobe. 1—Nonsorted circles and wet muck; 2—Sand; 3—Involuted organic matter; 4—Organic matter which has been overrun by sand; 5—Moss in process of being overrun; 6—Frozen sand and gravel.

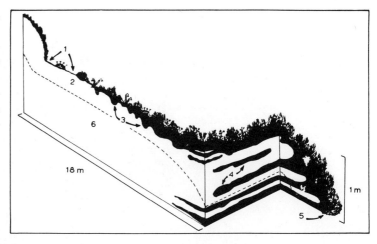

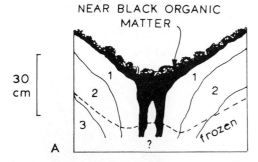

Fig. 11. Section through a hummock. 1—Moss and disoriented sedges; 2—Bare muck; 3—Coarse sand and gravel.

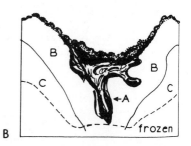

Fig. 12. Sections across border trenches of tundra polygons. A: 1—Brown, coarse sand with high organic matter content; 2—Yellowish-brown, coarse sand and gravel; 3—Yellowish-brown, coarse sand.
B: A—Involuted, near-black, wet organic matter; B—Yellowish-brown, coarse sand; C—Brownish-yellow, coarse sand and gravel.

organic matter; angular gravel fairly abundant. Frozen at 75 cm, August 15.

Solifluction, both on a large and small scale, is an important process in the area and causes folding and burial of soil horizons. On a large scale, the displacement is associated with solifluction lobes (Figure 10), and on a small scale, with nonsorted steps.

Hummocks represent another form of disturbance. A hill of sand is overlain by a peaty cover; the latter may show signs of downslope movement, as illustrated in Figure 11.

Horizons of relatively well-drained soils on glaciofluvial ridges and raised beaches are deformed in the rims which border frost-fissure polygons. Beneath the damp trench-floors, involutions and columns of organic matter extend into the underlying sand (Figure 12). The border trenches of large, tetragonal polygons developed in marshy depressions are wider and lack rims. Organic matter is commonly absent from the waterlogged trench-floors, whereas on the flat polygon-centers it is as thick as is shown in profile 1, Figure 3.

Distribution of Soils and Extent of Soil Development Since Emergence from the Postglacial Sea

The local patterns of parent materials, slope, drainage, vegetation, and periglacial surface features follow closely the distribution of glacial and glaciofluvial landforms. Each type of landform is distributed widely throughout the area; the patterns of soils are repeated accordingly. Soils commonly change rapidly and repeatedly across small areas. Very detailed soil maps would therefore need to be of large scale, produced entirely by field mapping. Less detailed maps could be produced without difficulty mainly by air photograph interpretation. A system of soil mapping complexes based upon relationships between soils, type of landform, vegetation, depth of active layer, and periglacial form would be of value both to pedologists and engineers.

The fact that the Rankin Inlet area has emerged from the sea in late-glacial and postglacial times provided an opportunity to examine the extent of soil development in relation to the length of time during which the deposits at various heights above present sea level have been exposed. Soil profiles at many points throughout the area between sea level and the highest parts (approximately 150 m) were compared, and profiles were dug along transects up long slopes from present sea level. No essential differences were apparent between soils at low elevations and those at high elevations (i.e., between soils which have been developing for more than 4,000 years and soils which have had appreciably less time to develop.) In every case, *local* conditions of parent material, slope, drainage, vegetation, and periglacial process appeared to be the determining factors in soil development.

Conclusion

The soils of the Rankin Inlet area are typical of the arctic: they are immature and reflect the presence of permafrost and the effects of frost processes in the active layer. They may be classified provisionally according to quality of soil drainage and the degree of frost-disturbance. Extensive areas of soil have been disrupted by the displacement by frost of mineral and organic material. Elsewhere, disruption of soil horizons has been confined to the vicinity of certain periglacial patterned ground forms.

Because of the presence of permafrost and the nature of many of the deposits, imperfectly and poorly drained soils, similar to the Meadow Tundra and Half-Bog soils, cover a much greater area than does the freely-drained soil. The former are associated with a vegetation of mosses and sedges mixed with other vascular plants, and the latter with marsh vegetation of mosses and sedges. The freely drained soil, developed in the most stable deposits of glaciofluvial ridges and sandy raised beaches, is associated with a lichen-heath vegetation and approaches the nearest to what might be called the climax soil of the area. It bears similarities to the Arctic Brown soil of Alaska. The degree of podzolization encountered in certain of the profiles examined reflects the "low arctic" position of the Rankin Inlet area. No significant difference is apparent between soils which are some 4,000 years old and soils which are appreciably younger.

ACKNOWLEDGMENTS

The observations presented in this paper were made as part of a field study of the Rankin Inlet area carried out in the summer of 1967 when the author was a graduate student of the Geography Department of the University of Saskatchewan. The study was generously financed by the Institute for Northern Studies, University of Saskatchewan; the Institute's Arctic Research and Training Centre served as a field base. The help of the Geological Survey of Canada in providing helicopter transport and accommodation at their base camp in the area is also gratefully acknowledged. Dr. J. C. F. Tedrow read the manuscript and made several constructive suggestions. Mrs. Hilary McCullagh kindly typed the text.

Examples illustrating changing viewpoints on the characterization of soil distribution patterns at the regional level are available in the literature for Ontario and the Prairie provinces. Published works on the soils of Ontario, for example, date back to 1926 when Ruhnke, who was interested in the soil from the standpoint of agricultural productivity, recognized two soil provinces in southern Ontario: The Glacial Drift Province and the Glacial Lake Province. He attributed the heterogeneity of the soil mosaic at the series level within each province to the wide range of rock types serving as parent material. Morwick (1933), on the other hand, was not in favour of stressing the geological factor to the same degree, maintaining that many factors are at work in determining the soil pattern of southern Ontario. Utilizing available soil data on both sides of the border, he extrapolated into Ontario soil boundaries established by the Americans. This enabled him to distinguish two major soil groups, notably the Podzols and the Gray Brown Podzolic soils. Podzols were also recognized at the regional level by Chapman and Putnam (1937). Locally, however, they felt they could list thirty-five to forty distinct soil types, although these could be conveniently grouped into thirteen "land types" based on the physical and chemical properties of the profile. In a special publication for the Royal Society of Canada, Richards (1961) identified, mapped, and described seven soils at the great group level and thirteen at the subgroup level for southern Ontario. The major orders represented are the Brunisols, the Podzols, the Gleysols, the Regosols, and the Organic soils. A similar trend in the evolution of classification schemes in the case of the Prairie provinces can be traced in the works of Shutt (1910), Hubbard (1950), Moss (1954), Bowser (1960), Pawluk and Lindsay (1964), and Dudas and Pawluk (1969).

In those branches of physical geography dealt with so far in this book, we have witnessed a tendency for studies to become increasingly more specific and detailed in their objectives, and at the same time frequently more limited in their areal scope. Pedological research is no exception. On the basis of rather fragmented and descriptive profile data, early researchers sought to unravel the complexity of the soil pattern by establishing convenient categories into which similar soils could be grouped. Detailed investigations became more fashionable once the broad regional patterns had been identified. They not only enabled previously established soil groupings to be tested and, if necessary, redefined, but they also added to our knowledge of the soil as a resource base. This restriction in geographic scope of soil studies was accompanied by the development of more refined and sophisticated field and laboratory techniques for describing and analyzing soil properties. The growing specialization within soil science is clearly portrayed in many of the papers which have appeared in the *Canadian Journal of Soil Science* since its inception in 1957. Most of the topics are of only marginal or academic interest to the physical geographer. However, the physical geographer must be aware of certain aspects of methodology and some of the standard procedures used in contemporary research in soil science. The following paper by McKeague et al (1972) is included with this in mind. It also illustrates the status of knowledge in soil science from the standpoint of soil classification as it employs the nomenclature established in the 1970 Canadian soil classification system. Moreover, it clearly demonstrates the shift from a philosophy concerned primarily with the establishment of genetic relationships to one embodying the actual chemical, mineralogical and mechanical properties gleaned from careful field observation and precise laboratory analyses.

The authors' objective is to compare Luvisolic soils in three widely separated regions of Canada. In the 1970 classification two great groups of Luvisols are recognized: the Gray Brown Luvisols and the Gray Luvisols, formerly known as the Gray Brown Podzolic and the Gray Wooded soils respectively under the old Podzolic order. It is interesting to note that in the case of the former group, despite the name change and the additional data now available, the authors feel that the basic concepts of this group have remained unchanged. They also point out that many pedologists have questioned the validity of distinguishing Gray Brown Luvisols as a separate entity on the grounds that greater variations seem to exist among the Gray Luvisols developing on dissimilar parent materials and in different climates. Thus the primary objective of the authors is to determine whether in fact the current status of the Luvisolic order in the existing classification is truly valid. They work towards this objective by comparing the physical, chemical and mineralogical properties of representative profiles in central Alberta, southern Ontario, and Nova Scotia. The paper has been reprinted in full in order to illustrate the various techniques of the soil scientists which are available to the physical geographer if he wishes to tackle specific or applied pedological problems. Those interested in becoming more familiar with these techniques can do so by reading the appropriate literature referenced throughout the paper. Despite the large amount of soil data presented, the authors did not feel that the subsequent comparisons had successfully resolved the issue of whether or not these soils were correctly categorized, except to advocate that their distinction at the great group level seemed reasonable.

A COMPARISON OF LUVISOLIC SOILS FROM THREE REGIONS IN CANADA*

J. A. McKeague, N. M. Miles, T. W. Peters and D. W. Hoffman

Abstract

Luvisolic soils (Alfisols) developed in calcareous glacial till in cold continental subhumid, moderately warm continental humid and moderately cool perhumid regions of Canada were compared. The properties and degrees of development of the soils could be related mainly to climatic and parent material factors, but also to organisms. The soils from perhumid Nova Scotia had upper sola of low base saturation and appreciable weathering of clay minerals had occurred in one of the soils. Base saturation was high in the Luvisolic soils from subhumid Alberta and weathering of clays was slight. A high dolomite content in the parent materials of one of the soils from humid Ontario resulted apparently in weak development as indicated by high base saturation, weak weathering and minimal clay translocation. The development of clay skins was most evident in the strongly-structured B horizons of the montmorillonitic Alberta soils, although areal percentages of oriented clay were higher in the weakly-structured Nova Scotia soils and in one of the Ontario soils. The mineral—organic surface horizons of the Gray Brown Luvisols from Ontario differed from comparable horizons of the Gray Luvisols from Alberta in exhibiting a more intimate association of organic and inorganic constituents and in containing extractable organic matter of lower humic/fulvic acid ratio.

It is suggested that Gray Luvisols having low base saturation in the solum should be separated at the subgroup level in the Canadian system of classification.

Introduction

The Luvisolic Order in the Canadian system of soil classification (Canada Department of Agriculture, 1970a) includes well to imperfectly drained soils having eluvial horizons (Ae) and illuvial horizons (Bt) in which clay is the main accumulation product. They correspond to Boralfs and Udalfs in the United States system

---

* Reprinted from *Geoderma*, 7, pp. 49-69, 1972, by permission of the Elsevier Publishing Company, Geo Sciences Section.

(Soil Survey Staff, 1967) and to Albic Luvisols in the F.A.O. classification units (World Soil Resource Report 33, 1968). Luvisolic soils occur extensively in Canada in forested areas where parent materials are neutral to alkaline in reaction and finer in texture than loamy sand. They also occur, however, in some acidic materials (Clark and Green, 1964). Two great groups of Luvisolic soils have been separated: Gray Brown Luvisols, formerly called Gray Brown Podzolic (mainly Udalfs) which have a forest-mull Ah horizon and occur where the mean annual temperature is above 5.5°C; and Gray Luvisols, formerly Gray Wooded, (mainly Boralfs) which lack a forest-mull Ah horizon, although they may have an Ah horizon,

and occur where the mean annual temperature is less than 5.5°C (Canada Department of Agriculture, 1970a). According to data summarized for the Canadian portion of the world soil map being prepared by FAO (World Resource Report 33, 1968), Gray Brown Luvisols are the dominant soils in about 44,000 km² in southern Ontario and Quebec and they occur locally in British Columbia. Gray Luvisols are the dominant soils in about 720,000 km², approximately 8% of the land area of Canada. They occur in all of the provinces, as well as in the Yukon and the Northwest Territories where knowledge of their extent is incomplete.

Gray Luvisols were first described and char-

Fig. 1. Landscape showing an area of Gray Luvisolic soils in Alberta. The native vegetation is aspen, Populus tremuloides.

Fig. 2. Landscape showing cultivated Guelph soils on the slopes and the associated Gleysolic soils in the depressions.

acterized in Alberta and Saskatchewan and the early concepts of these soils were summarized by Moss and St.Arnaud (1955), and by Williams and Bowser (1952) who compared Gray Wooded (Gray Luvisols) soils in Canada and the United States. More recently Ehrlich et al. (1955), Pawluk (1961), St.Arnaud and Whiteside (1964) and St.Arnaud and Mortland (1963) provided detailed characterization data for typical Gray Luvisols. Most of the concepts associated with the Gray Luvisol great group are based upon characteristics of soils in central Alberta, Saskatchewan and Manitoba. These concepts have been modified somewhat by information on Gray Luvisols from other areas (Clark and Green, 1964; McKeague et al., 1967).

Stobbe (1952) characterized the Gray Brown Podzolic soils in Canada and related them to associated soils and to similar soils in the United States. Concepts of the great group have not changed appreciably although more detailed information is available on some of these soils (Willis, 1949; Gillespie and Elrick, 1968).

Some of the major subgroups of Luvisolic soils are: Brunisolic subgroups (have chromas of three or more in part of the Ae horizon); Bisequa subgroups (have a Podzolic upper sequum); Gleyed subgroups (have dull colors accompanied by rusty mottles in the solum); Dark Gray Luvisols (have surface mineral—organic horizons more than 5 cm thick).

The desirability of separating Gray Brown

Luvisols at the great group level has been questioned by many pedologists some of whom feel that greater differences exist among Gray Luvisols developed in widely different parent materials in various climatic regions. Also, opinions differ on the adequacy of the bases of some of the subgroup separations, particularly those for Brunisolic subgroups. Thus an assessment of the characteristics of Luvisolic soils from different regions of Canada seemed desirable. This assessment could be done largely by studying published material but methods and kinds of analyses as well as criteria of classification have changed through the years and many published data on Luvisolic soils are not strictly comparable. Typical Luvisolic soils from Alberta and southern Ontario were analyzed and compared with Luvisolic soils from the Atlantic provinces studied recently (McKeague et al., 1969; McKeague and Cann, 1969; McKeague and Brydon, 1970) by the same methods. The purpose of the investigation was to compare typical Luvisolic soils from different regions of Canada, to consider their genesis and to assess the logic of their present classification.

Soils The soils studied included two Gray Brown Luvisols from southern Ontario, an Orthic Gray Luvisol and two Dark Gray Luvisols from central Alberta, and two Luvisolic soils from Nova Scotia (Table I). Typical landscapes in each of the three regions are shown in Fig. 1, 2 and 3. Brief descriptions are

Fig. 3. Landscape showing cultivated soils of the Queens catena in Nova Scotia and a spruce forest in the background. (Photograph courtesy of J. L. Nowland, Nova Scotia Soil Survey).

tabulated (Table II) but more details on macromorphology have been reported in the references indicated. Monoliths were taken in the field where the soils were described. They were studied in the laboratory under a stereomicroscope at magnifications of 10 to 30 times and samples were taken for preparation of thin sections, bulk density determinations and other analyses.

Methods of analysis Bulk density: saran clod method of Brasher et al. (1966). Particle-size distribution pipette method after pre treatments with HCl to remove carbonates, H_2O_2 to remove organic matter and dithionite-citrate-bicarbonate (Mehra and Jackson, 1958) to remove iron oxides.

- Organic matter: dichromate oxidation (Atkinson et al., 1958).
- Humic and fulvic acids, ultimate and functional group analysis of organic matter extracted by 0.5 N NaOH: by methods cited by Schnitzer (1969).
- Nitrogen: Kjeldahl method.
- pH: measured in 0.01 M $CaCl_2$.
- Exchangeable Ca + Mg and Al: displacement with NaCl (Clark, 1965).
- Extractable Fe and Al: dithionite-citrate bicarbonate (Mehra and Jackson, 1960) and by oxalate (McKeague and Day, 1966).
- Carbonates: by the method of Skinner et al. (1959).
- Micromorphology: thin sections were prepared as outlined by Guertin and Bourbeau (1971), areal estimates of oriented clay were made as described previously (McKeague et al., 1968). Brewer's (1964) terminology was used.
- Sand mineralogy: as described previously (McKeague and Brydon, 1970).
- Clay mineralogy: by X-ray diffraction supported by infrared spectrometry to aid in the identification of kaolinite (Brydon et al., 1968).

Results

Stereomicroscopic observations Some distinctive morphological features were revealed by examination of soil monoliths under a stereomicroscope. Features not indicated in Table II are mentioned.

Cooking Lake: Ped surfaces coated with

white sand grains were common in the Ae and AB horizons. In the upper Bt, argillans were mainly in voids within peds; in the lower Bt, dark argillans coated vertical ped faces nearly continuously. Ped and clod interiors within the BC and Ck horizons were mottled gray and yellow.

Uncas 1: Surfaces of some peds in the AB horizons were silt coated. Ped surfaces in the upper Bt horizon were mottled gray and brown; lower cutans were darker. Cutans occurred along some major voids in the CB horizons.

Uncas 2: Bases of plates in the upper Ae horizon were coarser in texture than the tops. Sand-surfaced peds occurred in the upper Ae, silt coated peds in the AB, and clay coated peds in the Bt horizon. Ped surfaces in the Bt were more uniform in color than those of Uncas 1.

Oneida: Color of the Ah horizon was much more uniform than that of the Ah and Ahe horizons of the Gray Luvisols. Silt coatings occurred in the Ae and AB horizons. Argillans were common in the Bt, and they occurred rarely in the Ck horizons.

Guelph: Peds in the Bt horizon were not as continuously clay coated as those of Cooking Lake and Uncas soils.

Queens 1: Silt coatings were common in the lower Ae. Some argillans in the Bt consisted of a yellowish brown layer superimposed on a dark brown layer.

Queens 2: Below about 70 cm, the till was very dense and the only large voids were a few vertical cracks.

Micromorphology Most of the micromorphological data are based upon observations of two vertically- and two horizontally-oriented thin sections from each subhorizon.

Cooking Lake: The Ah horizon was dominated by partly decomposed organic fragments apparently unassociated with the mineral material. Plasma was concentrated in the upper part of some of the 1 to 1.5 mm thick plates in the upper Ae horizon, but isoband fabric (Dumanski and St.Arnaud, 1966) dominated in the lower Ae. Argillans accounted for about 2% of the area of sections of the Bt horizons; values

TABLE I
Location, classification and some site characteristics of the Luvisolic soils

Soil[1]	Classification	Location		Topography	Dominant vegetation	Annual[2] precipitation (cm)	Summer[3] precipitation (cm)	P.E.[4] (cm)	Temp. (°C)		climatic[5] class	References
									annual air	summer soil		
Cooking Lake	Orthic Gray Luvisol	53°22'N	113°10'W	undulating	aspen	47	31	43	2	13	cold continental subhumid	Pawluk, 1961; Bowser et al., 1962
Uncas 1	Dark Gray Luvisol	53°00'N	113°51'W	gently undulating	aspen	47	31	43	2	13		
Uncas 2	Dark Gray Luvisol	53°23'N	113°18'W	gently undulating	aspen	47	31	43	2	13		
Guelph	Orthic Gray Brown Luvisol	43°41'N	80°15'W	gently rolling	maple	83	38	47	7	16	moderately warm continental humid	Hoffman et al., 1963
Oneida	Orthic Gray Brown Luvisol	43°26'N	79°50'W	undulating	maple, oak, ash	83	28	47	7	16		Gillespie and Elrick, 1968
Queens 1	Gleyed Brunisolic Gray Luvisol	45°56'N	61°00'W	gently rolling	spruce	125	46	42	6	—	moderately cool perhumid	McKeague et al., 1967
Queens 2	Brunisolic Gray Luvisol	45°14'N	63°16'W	gently sloping	spruce and fir	106	39	43	6	—		McKeague and Cann, 1969; McKeague et al., 1969

1 Soils series except Queens (which is a catena name).
2 Precipitation and temperature data are values for the closest Meteorological Station (Boughner, 1964).
3 Precipitation during June, July and August.
4 Potential evapotranspiration.
5 Canada Soil Survey Committee Report (1970).

TABLE II
Description of the Luvisolic soils*

Horizon	Depth (cm)	Color (moist)	Color (dry)	Structure	Consistence	Other features
Cooking Lake (Orthic Gray Luvisol):						
LF	4–0	10YR2/1	10YR3.5/2	partly decomposed leaves and roots	very friable, soft	many roots; white sand grains
Ah	0–3	10YR2/1	10YR3.5/1	mod. fine granular	very friable, soft	color variable; light to dark gray
Ahe	3–8	10YR3.5/2	10YR5.5/15	comp. weak, med. platy to granular	very friable, soft	plentiful roots; few gravels
Ae1	8–13	10YR4.2	10YR6/1.5	mod., med. platy	friable, slightly hard	few roots
Ae2	13–18	10YR5/2.5	10YR7/2	comp. weak subangular blocky to platy	firm, hard	compact, low porosity
AB	18–23	10YR5/2.5	10YR6/2.5	weak subangular blocky	firm, hard	tongues of Ae; some argillans
BA	23–33	10YR4.5/3	10YR6/2.5	strong, med. subangular blocky	firm, very hard	
Bt	33–73	10YR4/3	10YR6/3	comp. coarse prismatic to strong blocky	firm, very hard	nearly continuous thin, dark argillans
BC	73–92	1YR4.5/3	1YR6/2.5	comp. coarse prismatic to blocky	firm, very hard	argillans not continuous
BCk	92–105	2.5Y4.5/3	2.5Y5.5/2	comp. coarse prismatic to pseudo blocky	firm, very hard	few argillans, carbonate flecks
Ck	105–118	2.5Y4.5/2	2.5Y5.5/2	comp. pseudo blocky to coarse platy	firm, very hard	calcareous c.l. till
Uncas 1 (Dark Gray Luvisol):						
LH	4–0			partly decomposed leaf litter mixed with mineral soil	very friable, soft	many roots
Ah	0–4	10YR2/1	10YR3/1	mod. fine granular	very friable	color value decreases with depth
Ahe1	4–9	10YR2/2	10YR4/1.5	weak, fine granular	friable, slightly hard	plentiful roots
Ahe2	9–19	10YR3/2	10YR5/2	mod. fine platy	friable, hard	
Ae	19–24	10YR4.5/2	10YR6/1.5	comp. weak blocky to platy	friable, hard	very compact, low porosity
AB1	24–29	10YR5/2.5	10YR6/1.5	comp. weak prismatic to mod. med. blocky	firm, hard	ped surfaces light gray, mottles
ABgj	29–39	10YR4.5/2	10YR6/2.5			
Btgj	39–79	10YR5/2	10YR5.5/2	comp. prismatic to strong med. blocky	firm, very hard	argillans, mottles
CBkgj	79–92	2.5Y5/2	2.5Y5.5/2.5	coarse pseudo platy	firm, very hard	weakly calcareous c.l. till
Uncas 2 (Dark Gray Luvisol):						
Ah	0–5	10YR2/1	10YR3.5/1	mod. fine granular	very friable, soft	many roots
Ahe	5–10	10YR2.5/1	10YR4/1	mod. fine platy	very friable, soft	variable, light to dark gray
Ae1	10–20	10YR4.5/2	10YR6/1	strong fine to med. platy	friable, slightly hard	fewer roots than above
Ae2	20–25	2.5Y5/2	2.5Y1/2	strong, medium platy	friable, hard	plates darker at base
AB	25–35	10YR4.5/2	10YR6/2	strong, fine to medium blocky	firm, very hard	some argillans
Bt1	35–60	10YR5/3	10YR6/2.5	comp. coarse prismatic to strong blocky	firm, very hard	continuous, thin, dark cutans

Horizon	Depth (cm)	Color (moist)	Color (dry)	Structure	Consistence	Features
Bt2	60–85	10YR5/2.5	10YR6/2	comp. coarse prismatic to strong med. blocky	firm, hard	continuous vertical channels
BCgj	85–110	10YR5/2	10YR6/2	comp. coarse prismatic to blocky	firm, hard	distinct mottles
CBkgj	110–114	10YR4.5/2	10YR6/2	comp. coarse prismatic to pseudo blocky	firm, hard	weakly calcareous, c.l. till
Oneida (Orthic Gray Brown Luvisol):						
Ah	0–10	10YR4/1.5	10YR6/1.5	strong, coarse granular	friable, hard	many roots, wormholes
Ae	10–20	10YR5.5/2.5	10YR7/2	Weak, medium blocky	friable, hard	variable color
AB	20–30	10YR6/3	10YR7/3	weak to moderate blocky	friable, hard	some cutans
Bt	30–50	10YR4/3	10YR6/3	moderate to strong blocky	firm, hard	gray and dark brown argillans
BCk	50–75	10YR4/3	10YR6/3	moderate medium blocky	firm, very hard	fewer argillans, carbonate
Ck	75–95	10YR5/2.5	10YR6/2.5	amorphous, pseudo blocky	firm, very hard	high calcite c.l. till
Guelph (Orthic Gray Brown Luvisol):						
Ah	0–15	10YR3.5/1.5	10YR5/1.5	moderate, fine to med. granular	very friable, soft	many roots
Aeh	15–20	10YR4/2	10YR6/2.5	weak, fine blocky	very friable, slightly hard	mixture of Ah and Ae
AB	20–25	10YR4.5/3	10YR6/3	weak, medium blocky	friable, hard	tongues down
Bt	25–50	10YR4/3	10YR5.5/4	weak, medium blocky	friable, hard	some argillans
BCk	50–65	10YR4.5/3	10YR6/3.5	amorphous	friable, hard	very few argillans
Ck	65–100	10YR4.5/3	10YR6.5/3	amorphous	firm, hard	extremely calcareous l. till
Queens catena 1 (Gleyed Brunisolic Gray Luvisol):						
LF	5–0			undecomposed and partly decomposed leaves and twigs		massive at base
Ae	0–9	10YR7/2	7.5YR8/1	weak, medium platy	friable, slightly hard	common distinct mottles
Bmgj	9–22	7.5YR5/3	7.5YR7/3	weak to moderate blocky	friable, hard	gray, silt lined voids
Ae	22–30	7.5YR5/3	7.5YR7/2	weak, fine blocky	friable, slightly hard	argillans, MnO_2 flecks
Bt	30–39	5YR4/3	7.5YR5.5/4	weak, coarse blocky	firm, extremely hard	argillans, roots in main voids
BC	39–57	5YR4/2.5	7.5YR5/4	very weak, coarse blocky	firm, extremely hard	moderately calcareous c.l. till
Ck	57–76	5YR4/3	7.5YR5.5/4	amorphous	very firm, extremely hard	
Queens catena 2 (Brunisolic Gray Luvisol):						
LF	5–0	10YR3/1		peaty leaf litter	very friable, soft	many roots
Ae	0–4	7.5YR4.5/2	7.5YR6/2	weak coarse platy	very friable, slightly hard	
Bm	4–20	5YR5/4	5YR6.5/4	weak granular		
Aegj	20–25	5YR5/3	5YR6.5/3	weak, medium blocky	very friable, soft	common, distinct mottles
AB	25–42	5YR4/4	7.5YR6/4	weak to moderate, fine blocky	friable, slightly hard	some argillans
Bt	42–100	5YR4/3.5	5YR5/4	moderate medium blocky	firm, hard	common argillans
CB	100–130	2.5YR4/3	2.5YR5/4	amorphous	firm, very hard	MnO_2 mottles
Ck	130–180	2.5YR4/3	2.5YR5/4	amorphous	firm, very hard	weakly calcareous c.l. till

* More detailed descriptions of some of the soils can be found in the reference indicated in Table I.

for 11 sections ranged from 1.1 to 2.8%. Most of the argillans were moderately oriented and thin; usually less than 50 microns, but occasionally as much as 200 microns thick. Thin black streaks within the argillans were common. Calcitans occurred rarely in the Ck horizon. The fabric of this soil and of the two Uncas soils was dominantly silasepic in the Ae horizon, vo-masepic porphyroskelic in the Bt horizons and in-masepic porphyroskelic in the C horizons.

Uncas 1: The Ahe horizon contained many sand-sized organic fragments. Isoband fabric was weakly developed in the Ae horizon. Moderately oriented argillans occupied about 3% of the area of the Bt horizon sections (2.3–3.9% in nine sections) and some contained black streaks. Diffuse sesquioxide nodules occurred from the AB to the Ck horizons.

Uncas 2: The micromorphology was generally similar to that of Uncas 1. Isoband to weakly banded fabric was evident in the Ae; the bands were 0.5–1.5 mm thick. Argillans occupied about 2.8% of the area of the Bt horizon sections (2.1–4.3% in eight sections). Moderately thick (up to 0.4 mm) argillans extended along major voids in the CB horizon. Sesquioxidic nodules were most common in the lower Ae and the AB horizons.

Oneida: The Ah horizon consisted largely of silasepic granules 2–10 mm in diameter that contained relatively few coarse organic fragments. A few silan–argillans partially lined some voids in the dense Ae. Deposits of oriented clay increased with depth occupying about 2% of the area of the lower AB horizon, 4% of the upper Bt and 8% of the middle of the Bt where some were thicker than 0.5 mm. They decreased with depth through the lower Bt and BC horizons and were absent in the Ck. Bodies of oriented clay occurred within the matrix as well as associated with voids in the vo-masepic porphyroskelic Bt. A few sesquioxidic nodules within and below the Ae horizon and small manganiferous accumulations occurred in and below the Bt. A few calcitans occurred in the silasepic to crystic Ck horizon but most of the carbonate was primary.

Guelph: The upper Ah horizon consisted of distinct, dark granules about 1–5 mm in diam-

eter. They became less distinct with depth. No clear pattern of voids was evident in the silasepic Ae horizon where argillans lining vughs and vesicles accounted for about 0.6% of the area. Thin, moderately to strongly oriented argillans lining vughs and planar voids in the insepic Bt accounted for about 1.8% of the area (0.8–2.8% in eight sections). Sesquioxidic nodules were few in the Bt and more common in the BCk horizon. Very little secondary carbonate was evident in the BCk and the largely crystic Ck horizons.

Queens catena 1: The micromorphology was summarized previously (McKeague et al., 1968). The principal features were the numerous sesquioxidic nodules in the upper Bmgj horizon (designated as Bfgj previously), the silans in the upper solum and the moderately to strongly oriented argillans lining voids in the Bt horizon and occupying nearly 8% of the area of the sections.

Queens catena 2: The main micromorphological features were summarized in a previous publication (McKeague et al., 1969, Soil 1). They were broadly similar to those of Queens catena 1 but the fabric of the Bt horizon was masepic to omnisepic porphyroskelic. About 6% of the area of the upper Bt and less than 3% of the lower Bt were occupied by argillans.

Bulk density, particle-size and chemical data: Bulk densities of all horizons except for a few near surface horizons were between 1.5 and 2.1 and there were no marked differences among soils (Table III). Bulk densities of dry peds or clods are usually somewhat higher than average values for the moist soil both because coarse voids are not usually included in the clods, and because some shrinkage commonly occurs on drying. These factors would affect appreciably the values for the B horizons of the Cooking Lake and Uncas soils as they had strongly developed structure with clearly defined cleavage separating peds, and they shrank on drying. Bulk densities based on moist samples of the Oneida, Guelph and Queens C horizons would probably be only slightly lower than those reported.

Plots of the distribution of five sand fractions with depth such as that shown previously for Queens catena 2 (McKeague and Cann,

TABLE III
Bulk density, particle-size and some chemical data for seven Luvisolic soils

Horizon	Depth (cm)	Bulk density (%)	Sand 2000–50 μ (%)	Silt 50–2 μ (%)	Clay 2–0 μ (%)	Fine clay <0.2 μ (%)	Fine clay / Total clay (%)	Organic matter (%)	N (%)	pH	Exchangeable Ca+Mg (mequiv./100 g)	Al	Oxalate Fe (%)	Oxalate Al (%)	Dithionite Fe (%)
Cooking Lake:															
LF	4–0	–	–	–	–	–	–	30	1.0	6.9	42	0	0.19	0.10	0.68
Ah	0–3	–	–	–	–	–	–	24	0.86	6.6	37	0	0.14	0.08	0.38
Ahe	3–8	–	54	34	12	6.1	51	1.6	0.09	5.9	6.3	0	0.14	0.06	0.42
Ae1	8–13	–	54	36	10	2.3	23	0.6	0.04	5.8	4.1	0	0.09	0.05	0.36
Ae2	13–18	1.9	50	28	22	9.7	44	0.7	–	5.5	10	0.1	0.09	0.08	0.50
AB	18–23	1.9	43	27	30	16	53	0.8	–	5.1	13	0.1	0.09	0.10	0.59
BA	23–33	1.8	38	26	36	22	61	0.8	–	4.6	17	0.4	0.16	0.18	1.1
Bt	33–53	1.9	40	25	35	22	63	1.0	0.07	4.5	17	0.4	0.20	0.15	1.2
Bt	53–73	1.9	41	25	34	21	62	0.9	–	5.0	19	0	0.20	0.10	1.2
BC	73–92	1.9	41	26	33	19	57	0.8	–	5.3	19	0	0.18	0.09	1.1
BCk	92–105	1.9	–	–	–	–	–	0.9	–	>7	–	–	0.15	0.06	1.1
Ck	105–118	–	46	25	29	16	55	0.9	0.06	>7	–	–	0.15	0.06	1.0

The CaCO₃ equivalent of the Ck horizon was 6.0% (3.7% was calcite)

Horizon	Depth (cm)	Bulk density (%)	Sand 2000–50 μ (%)	Silt 50–2 μ (%)	Clay 2–0 μ (%)	Fine clay <0.2 μ (%)	Fine clay / Total clay (%)	Organic matter (%)	N (%)	pH	Exchangeable Ca+Mg (mequiv./100 g)	Al	Oxalate Fe (%)	Oxalate Al (%)	Dithionite Fe (%)
Uncas 1:															
LH	4–0	–	–	–	–	–	–	28	1.0	6.2	50	0	0.19	0.09	0.46
Ah	0–4	–	–	–	–	–	–	21	0.82	5.8	30	0	0.36	0.19	0.61
Ahe1	4–9	1.2	39	44	17	8.2	48	5.4	0.27	4.9	12	0.1	0.39	0.23	0.67
Ahe2	9–19	1.5	41	45	14	5.3	38	1.8	–	4.7	7.1	0.1	0.30	0.16	0.66
Ae	19–24	1.7	42	41	17	6.1	36	0.7	0.07	4.5	7.5	0.3	0.19	0.11	0.59
AB	24–29	1.7	42	33	25	12	48	0.8	–	4.4	12	0.7	0.24	0.16	0.73
ABgj	29–39	1.7	42	28	30	17	57	0.8	–	4.3	14	1.0	0.28	0.16	0.80
Btgj	39–59	1.8	44	28	28	17	60	0.7	0.05	4.9	18	0.2	0.34	0.14	0.91
Btgj	59–79	1.8	43	28	29	16	55	0.6	–	6.8	24	0	0.30	0.14	1.0
CBkgj	79–92	1.8	42	31	27	15	55	0.6	0.04	>7	–	–	0.27	0.12	1.0

The CaCO₃ equivalent of the deepest 5 cm of the CBkgj horizon was 0.5% (all was calcite)

Uncas 2:

Horizon	Depth														
Ah	0–5	1.0	—	—	—	—	—	10	0.53	6.3	26	0	0.36	0.18	0.66
Ahe	5–10	1.3	32	45	23	9.1	40	4.8	0.28	6.3	17	0	0.36	0.18	0.66
Ae1	10–20	1.5	23	60	17	5.1	30	1.3	0.09	6.1	9	0	0.24	0.10	0.48
Ae2	20–25	1.7	16	60	24	9.6	40	0.5	—	6.0	11	0	0.11	0.09	0.35
AB	25–35	1.8	12	52	36	22	61	0.7	—	5.9	22	0	0.19	0.14	0.70
Bt1	35–60	1.9	26	29	45	27	60	0.7	0.06	5.0	26	0.1	0.25	0.16	1.1
Bt2	60–85	1.9	33	26	41	22	54	0.6	—	5.2	26	0	0.23	0.13	1.0
BCgj	85–110	1.8	40	27	33	17	52	0.5	—	6.8	22	0	0.19	0.10	0.93
CBkgj	110–114	—	40	30	30	16	53	—	0.04	>7	—	—	0.17	0.10	1.0

The CaCO₃ equivalent of the CBkgj horizon was 2.8 (1.1% was calcite)

Oneida:

Horizon	Depth														
Ah	0–10	—	24	55	21	6.3	30	6.7	0.21	5.6	12	0.1	0.39	0.18	0.92
Ae	10–20	1.6	22	57	21	5.1	24	1.4	—	4.1	3.4	2.5	0.40	0.22	1.0
AB	20–30	1.9	20	53	27	8.0	30	0.9	—	4.0	4.6	2.6	0.48	0.24	1.3
Bt	30–50	1.8	15	55	40	17	42	0.9	0.08	4.8	13	0.6	0.43	0.22	1.7
BCk	50–75	1.9	21	43	36	15	42	0.7	—	>7	—	—	0.27	0.14	1.5
Ck	75–95	2.0	26	44	30	12	40	0.6	0.04	>7	—	—	0.18	0.08	1.2

The CaCO₃ equivalent of the deepest 5 cm of Ck horizon was 25% (23% was calcite)

Guelph:

Horizon	Depth														
Ah	0–15	—	36	48	16	9.4	58	7.0	0.30	6.8	18	0	0.38	0.27	0.94
Aeh	15–20	—	35	50	15	6.1	41	2.7	0.13	6.6	11	0	0.40	0.23	0.92
AB	20–25	—	38	39	23	11	48	3.2	—	6.6	11	0	0.40	0.27	1.1
Bt	25–50	1.7	35	40	25	14	56	1.3	0.11	6.9	13	0	0.39	0.30	1.3
BCk	50–65	1.8	44	40	16	8.7	54	0.8	—	>7	—	0	0.25	0.18	1.0
Ck	65–100	2.0	47	39	14	7.2	52	0.3	0.03	>7	—	—	0.10	0.08	0.60

The CaCO₃ equivalent of the deepest 10 cm of Ck horizon was 45% (20% was calcite)

Queens catena 1:

Horizon	Depth													
Ae	0–15	—	—	—	—	—	—	5.0	—	3.5	2.2	2.2	0.20	0.11	0.50
Ae	1.5–9	1.8	45	50	5.2	1.7	33	1.2	—	3.7	1.6	2.2	0.07	0.07	0.19
Bmgj	9–22	1.7	40	44	16	6.5	41	1.2	—	3.8	2.1	4.2	0.69	0.22	1.3
Ae	22–30	—	41	43	16	7.0	44	0.6	—	4.1	1.5	1.8	0.38	0.33	1.0
Bt	30–39	1.9	32	36	32	18	56	0.3	—	4.1	3.6	2.3	0.31	0.32	1.5
BC	39–57	1.9	30	38	31	18	58	0.3	—	6.2	11	0	0.17	0.16	1.3
Ck	57–76	2.0	29	40	31	16	52	0.3	—	>7	—	—	0.12	0.11	1.1

The $CaCO_3$ equivalent of the Ck horizon was 9.5% (all was calcite)

Queens catena 2:

Horizon	Depth													
LF	5–0	—	—	—	—	—	—	67	—	—	—	—	0.24	0.06	0.32
Ae	0–4	1.4	43	41	16	4.6	29	5.0	—	3.5	2.8	4.2	0.56	0.26	1.6
Bm	4–20	1.6	44	41	15	3.8	25	2.0	—	4.0	1.6	2.7	0.40	0.28	1.6
Aegj	20–25	1.7	53	36	11	2.5	23	0.7	—	4.0	1.5	2.7	0.25	0.16	1.0
AB	25–42	1.7	45	37	18	6.2	34	0.6	—	4.1	0.8	2.2	0.60	0.29	1.6
Bt	42–65	1.9	44	32	24	11	46	0.3	—	4.0	3.2	3.3	0.50	0.26	1.7
Bt	65–100	2.0	38	40	22	10	46	0.2	—	4.4	6.0	1.6	0.26	0.13	1.6
CB	100–130	2.0	39	32	29	12	41	0.2	—	4.8	13	0	0.22	0.07	1.6
Ck	130–180	2.1	40	31	29	12	41	0.2	—	>7	—	—	0.13	0.06	1.5

The $CaCO_3$ equivalent of the deepest 20 cm of Ck horizon was 3.1% (all was calcite)

1969, Soil 1) indicated only minor or no inherited discontinuities with depth in Cooking Lake, Uncas 1, Oneida, Queens catena 2. Discontinuities were indicated in the AB horizon of Uncas 2 and in the upper part of the Bt horizons of Guelph. The dominant sand fractions in all of the soils were fine and very fine sand. The particle-size distribution data (Table III) show that all of the soils contained appreciably more clay and more fine clay in the B horizons than in the A horizons. Except for the Queens catena soils, the Bt horizons contained more clay than the corresponding C horizons. Ratios of fine to total clay were slightly higher in the Bt horizons than in the corresponding C horizons, and for most of the soils the ratio was the lowest in an Ae horizon.

Organic matter accumulation was restricted essentially to the surface and to the upper 3 cm of the Cooking Lake mineral soil, and to the upper 10 cm of the Uncas soils (Table III). The organic matter contents of the Bt horizons of these three soils were similar to or only slightly greater than those of the C horizons which may have contained some coal. In the Guelph soil but not appreciably so in the Oneida the accumulation of organic matter extended to a greater depth than in the Gray Luvisols discussed above. In the Queens catena soils, accumulation of organic matter was slight below the upper few centimeters of mineral soil. If it is assumed that about 58% of the organic matter is C, the C/N ratios of the LFH and Ah horizons are well above 10 and those of the B and C horizons are below 10.

Most of the horizons of the three Gray Luvi-

sols from Alberta were 100% base saturated and all of them were over 90% base saturated based upon neutral salt extraction (Table III). The lowest pH values and the highest exchangeable A1 values occurred in the AB, BA or Bt horizons of these soils. The pH values reported were those obtained after five days equilibration with 0.01 M CaC1$_2$ and they were usually from 0.2 to 0.7 units higher than the values obtained 15 min after mixing the soils with CaC1$_2$. Oneida, a Gray Brown Luvisol was more acid and lower in base saturation than the three Gray Luvisols. Guelph on the other hand was nearly neutral and 100% base saturated in all horizons. The two Queens catena soils were acid and less than 40% base saturated in most horizons of the upper sola.

Accumulation of oxalate-extractable Fe and Al in the sola of the three Gray Luvisols from Alberta was minor relative to the values for the C horizons (Table III). Dithionite-extractable Fe was depleted from the upper sola of these soils, and in some but not all horizons the depletion was roughly proportional to the depletion of clay. Ratios of oxalate to dithionite-extractable Fe were considerably higher in the upper sola than in the Bt and C horizons of these soils. Oxalate-extractable Fe and A1 values were somewhat higher for the sola of Oneida and Guelph than for the sola of the Gray Luvisols from Alberta but there were no horizons of distinct maxima of these elements. Maxima of dithionite-extractable Fe values occurred in the Bt horizons of both Guelph and Oneida. As in the Gray Luvisols, the ratios of oxalate to dithionite Fe were higher in the sola than in

TABLE IV
Carbon extractable by 0.5 N NaOH, C in humic and fulvic acids as percent of total C in some Ah horizons of five Luvisolic soils

Sample		C extracted (%)	C humic (%)	C fulvic (%)	C humic / C fulvic
Cooking Lake	Ah	21	13	8	1.6
Uncas 1	Ahe	44	29	15	1.9
Uncas 2	Ah	31	20	11	1.8
Oneida	Ah	36	20	16	1.2
Guelph	Ah	32	15	17	0.9

TABLE V

Ultimate and functional group analysis on dry, ash-free basis of humic and fulvic acids extracted by 0.5 N NaOH from the mineral-organic horizons of five Luvisolic soils

Soil Sample		Fraction	C (%)	H (%)	N (%)	COOH (mequiv./g)	Total acidity (mequiv./g)	Phenolic OH (mequiv./g)	C=O
Cooking Lake	Ah	ha	57.7	5.5	3.9	4.3	5.6	1.3	1.8
		fa	49.3	6.0	2.3	7.1	10.0	2.9	5.6
Uncas 1	Ahe	ha	55.4	5.2	4.5	3.8	7.4	3.6	3.5
		fa	47.9	6.4	0.9	8.7	12.6	3.9	9.0
Uncas 2	Ah	ha	56.8	5.1	4.6	3.9	7.3	3.4	2.6
		fa	47.9	6.4	2.4	7.0	11.3	4.3	6.2
Uncas 2	Ahe	ha	55.4	5.8	5.2	4.4	8.7	4.3	1.8
		fa	44.8	5.8	2.6	6.9	12.0	5.1	3.4
Oneida	Ah	ha	57.1	7.0	3.5	3.5	7.7	4.2	3.7
		fa	52.3	6.7	1.6	7.0	10.5	3.5	4.0
Guelph	Ah	ha	53.2	6.2	4.6	3.9	7.5	3.6	3.6
		fa	47.8	6.1	2.4	8.0	10.0	2.0	8.3

TABLE VI

Ratios of quartz to feldspars in the 0.25–0.50 mm fractions of the A and C horizons of the Luvisolic soils

Soil	Horizon	Quartz / Feldspar
Cooking Lake	Ahe	6.1
	Ck	8.0
Uncas 1	Ahe	11
	CBkgj	9.0
Uncas 2	Ahe	13
	CBkgj	15
Oneida	Ah	2.6
	Ck	1.9
Guelph	Ah	2.2
	Ck	4.0
Queens catena 2	Ae	6.2
	Ck	3.7

the C horizons. The highest contents of oxalate-extractable Fe, 0.6 to 0.7%, occurred in subhorizons of the sola of the Queens catena soils. Accumulations of oxalate-extractable Al were less marked than those of Fe and they were similar to those in the Guelph and Oneida soils. The upper Ae horizon of Queens catena 1 was strongly depleted of dithionite-extractable Fe but this was not the case in Queens catena 2.

Less than 40% of the organic carbon was extracted by NaOH from all except one of the surface mineral-organic horizons tested (Table IV). The ratios of humic to fulvic acid extracted were higher for the Ah horizons of the three Gray Luvisols than for those of the two Gray Brown Luvisols.

Ultimate and functional group analysis of the humic and fulvic acids extracted by NaOH from the mineral–organic horizons revealed no major differences in the nature of the extractable organic matter in the soils (Table V). The fulvic acids were lower than the humic acids in C and N contents and higher in carboxyl and carbonyl groups.

The proportions of sand-sized quartz and feldspar differed considerably among the soils, but the ratios of quartz to feldspar in the A

and C horizons revealed little weathering of the 250 to 500 micron feldspar in any of the Ae horizons except that of the Queens catena 2 soil (Table VI). Observations of heavy minerals in thin sections indicated pronounced weathering only in the upper horizons of the Queens catena soils.

The Ahe horizons of the Gray Luvisols from Alberta, Cooking Lake and Uncas, contained higher proportions of kaolinite and mica and lower proportions of montmorillonite than the corresponding C horizons (Table VII). Similarly in Oneida, mica was enriched in the A horizons at the expense of expansible minerals. Chlorite was depleted in the upper horizons of Queens catena 1 and expansible minerals had developed in these horizons. However, little weathering was evident from the clay mineralogy of Queens catena 2.

Discussions and Conclusions

Some obvious differences are apparent in the morphological, physical, chemical and mineralogical properties of the three groups of soils: the Gray Luvisols, Cooking Lake and Uncas, from Alberta; the Gray Brown Luvisols, Oneida and Guelph, from Ontario; and the Brunisolic Gray Luvisols, Queens soils, from Nova Scotia. The Cooking Lake and Uncas soils had

TABLE VII

Minerals in the clay fraction ($< 2\ \mu$) of seven Luvisolic soils

Soil	Horizon	Phyllosilicates[1]			exp.[2]	Remarks
		kaol.	mica	chlorite		
Cooking Lake	Ahe	3	3	0	3	Montmorillonite is the expansible mineral. It is poorly crystalline in the Ae and highly crystalline in the Ck. Collapse on heating is impeded.
	Ae2	3	3	0	4	
	Bt	2–3	3	0	4	
	Ck	2	2	0	4	
Uncas 1	Ahe	3	3	0	4	Montmorillonite is the expansible mineral throughout. Quartz accounts for more than 15% of the clay of all horizons. Feldspar is absent.
	AB	2	3	0	4–5	
	Btgj	2	3	0	4–5	
	CBkgj	2	3	1	4–5	
Uncas 2	Ahe	3	3	0	3	Montmorillonite is the expansible mineral throughout. Quartz accounts for more than 15% of the clay at the surface and decreases with depth.
	Ae	4	3	1	3	
	Bt	2–3	2–3	0	4	
	Ck	2–3	2–3	0	4–5	
Oneida	Ah	0	4–5	2	3	Vermiculite is the major component of the expansible minerals. It is interstratified with mica and chlorite. Quartz accounts for 20% or more of the clay fraction in all horizons. Feldspar is absent.
	Ae	0	3	2–3	4	
	Bt	0	2–3	1–2	5	
	Ck	0	2–3	3	4	
Guelph	Ah	1	3	2	4	The expansibles include 18 Å material interstratified with 14 Å material in the Ahe, and probably vermiculite. Chlorite is interstratified with mica in the Bt and largely discrete in the Ck. About 10 to 15% quartz and a trace of feldspar are present throughout.
	AB	1	2–3	2	4	
	Bt	0	2	2	5	
	Ck	0	3–4	3	3	
Queens catena 1	Ae	0	3	0	5	Montmorillonite is interstratified with mica in the upper Ae, and with vermiculite in the Bmgj. Vermiculite accounts for most of the expansible material below the Ae. Quartz, present in all horizons, was least abundant in the upper Ae.
	Bmgj	0	3	2	4	
	Ae	0	4	4	2	
	Bt	0	4	4	2	
	Ck	0	4	4	2	
Queens catena 2	Ae	0	4	3	3	The expansible material lacked discrete montmorillonite and was probably interstratified.
	Aegj	0	3	3	3	
	AB	0	4	3	3	
	Ck	0	4	2	3	

[1] The proportions of the phyllosilicate minerals are indicated as follows: 0 = absent or not detected; 1 = trace to 5%; 2 = 5–20%; 3 = 20–40%; 4 = 40–60%; 5 = 60–80%.

[2] Expansible minerals: those that expand on treatment with glycerol and collapse to about 10 Å on heating.

more strongly developed structure, especially in the B horizons, and more obvious cutans than the other soils. The strong structure of the B horizon of Uncas 1 might be related to its 5% exchangeable Na (by NH_4OAc extraction) but the other soils from Alberta had less than 1% exchangeable Na. Their generally gray colors were of only slightly lower chroma, and in the lower horizons of yellower hue, than those of Oneida and Guelph, but they contrasted sharply with the inherited reddish hues of the Queens catena soils. The Bt horizons of all of the soils except Guelph contained about two times as much clay as the corresponding A horizons, but the Bt horizons of Cooking Lake, Uncas 1 and 2, and Queens catena 2 were considerably thicker than those of the other soils and clay skins were most continuous on ped surfaces of Cooking Lake and Uncas. The thinner Bt horizons were associated with C horizons having relatively high carbonate contents. The carbonate, especially the dolomite, in the Guelph material probably accounted for the relatively weak Bt development, the high pH and the 100% base saturation. Ehrlich et al. (1955) pointed out the influence of carbonate content on soil development in Manitoba. Except for the Guelph soil, the pH values and degrees of base saturation of the soils were consistent with expectations based upon climatic data (Table I). The Gray Luvisols from the cold, subhumid region of Alberta were nearly fully base saturated at all depths, the upper horizons of Oneida (mild, humid) were about 60% base saturated, and some horizons of the Queens catena soils (moderately cool, perhumid) were less than 40% base saturated. The same trend of increasing degree of development from the Gray Luvisols through the Gray Brown Luvisols to the Brunisolic Gray Luvisols is indicated by the oxalate-extractable Fe and Al data.

Gray Brown Luvisols such as Oneida and Guelph are distinguished from Dark Gray Luvisols such as the Uncas soils partly on the basis of the forest mull Ah of the Gray Brown Luvisols (Canada Department of Agriculture, 1970a). The main morphological difference between the Uncas Ah or Ahe horizons and the Guelph or Oneida Ah horizons was the

more intimate association of organic and inorganic materials in the Gray Brown Luvisols. Much of the organic matter in the Dark Gray Luvisol Ahe horizons as viewed in thin section occurred as relatively coarse, partly humified fragments unassociated with mineral material. In the Gray Brown Luvisols relatively few discrete organic fragments were visible in thin section and the homogeneous appearance of the granules suggested a close association of mineral and organic materials presumably resulting partly from passage through earthworms.

The attempt at chemical characterization of the organic matter in Ah and Ahe horizons of these Luvisolic soils (Tables IV and V) showed few differences except that the humic/fulvic acid ratios of the Gray Luvisol horizons were higher than those of the Gray Brown Luvisols. Subsequent testing showed that the relationship held although the ratios were lower when sodium hydroxide-pyrophosphate was used as an extractant. Most of the organic matter in these horizons was not extracted by NaOH and it is possible that the material that was not extracted was characteristic of specific kinds of soils. Ultimate and functional group analysis did not reveal major differences in the organic matter extractable from Gray Luvisols and Gray Brown Luvisols.

The mineralogical data indicated significant differences in the parent materials. The Uncas and Cooking Lake soils had appreciable contents of montmorillonite, mica and kaolinite but little or no chlorite. Pawluk (1961) reported similar data for clays from a number of Gray Luvisols in Alberta except that he found appreciable chlorite. St.Arnaud and Mortland (1963) reported generally similar clay mineralogy for a Chernozemic to Luvisolic sequence of soils from Saskatchewan. The 0.2–0.04 μ clay from seven widely separated soils, including Cooking Lake, in western Canada was found by Kodama and Brydon (1965) to be composed mainly (93%) of randomly interstratified montmorillonite-mica. The Guelph and Oneida clays were broadly similar to the Queens catena clays which contained mica, chlorite and interstratified material including vermiculite. Similar results were reported by

Willis (1949) for the clays from several soils from southern Ontario. The high proportion of mica in the A relative to the C horizons of Oneida, Cooking Lake and Uncas could be attributed to preferential translocation of montmorillonite from the A to the B, or to illitization of expansible minerals. Others have reported illitization in surface horizons (Rice et al., 1959; Pawluk, 1961; St. Arnaud and Mortland, (1963). Some weathering of chlorite may have occurred in the sola of Guelph and Oneida. The most intense weathering was evident in Queens catena 1. Chlorite was absent and montmorillonite was dominant in the upper Ae whereas chlorite was a major component and montmorillonite was absent in the lower horizons. Thus the upper Ae had the clay mineralogy characteristic of Podzolic soils (Brydon et al., 1968) although the underlying B horizons lacked the organic matter accumulation, extractable Al and amorphous clay characteristic of Podzol B horizons (McKeague and Cann, 1969; Brydon et al., 1968).

The differences in properties and degrees of development of these Luvisolic soils can be related, in part, to differences in the factors of soil formation, particularly climate and parent material.

Topography is not considered to be a major differentiating factor as all of the soils were well- to moderately-well drained. All of the soils were derived from glacial tills and, although the ice may have receded from the Alberta sites somewhat earlier than from the Ontario and Nova Scotia sites, the time involved was probably within the range of 10,000 to 15,000 years. The kinds of forests in the three regions differed (Table I) and thus vegetation presumably had some influence on soil development. Undoubtedly, earthworms affected the characteristics of the Ah horizons of Oneida and Guelph, as indicated by Thorp (1949) for similar soils. However, differences in climates (Table I) and in parent materials (Table II) are considered to be the major factors to which the differences in properties and degrees of development of the soils can be attributed.

The strongly-developed structure of the B horizons of the Cooking Lake and Uncas soils

may be related to their montmorillonite composition and repeated shrinkage and cracking along the same planes of weakness on drying, as well as on freezing. The durability of the resulting peds was probably enhanced by the relatively continuous clay skins that were deposited on their surfaces. Although no quantitative measurements were made, it was evident that the shrink-swell capacities of the Oneida, Guelph and Queens catena soil materials were far less than those of the Cooking Lake and Uncas soils. Clay skins were less evident in the Oneida, Guelph and Queens soils perhaps because of their weaker structure. The areal proportions of oriented clay were, however, greater in thin sections of the B horizons of the Oneida and the Queens catena soils than in those of Cooking Lake and Uncas although clay skins were distributed throughout greater depths in the latter soils.

The relationship between increasingly humid climate and base saturation of the sola has been mentioned. Comparison of the Guelph and Oneida soils shows that a high carbonate content obscures this relationship. The low base saturation of the upper horizons of the Queens catena soils may be associated with vegetation as well as climate. The coniferous forest would be expected to cycle bases less than the deciduous forests at the other sites. It is noteworthy that the pH values of the Queens catena soils were lowest at or near the surface whereas those of the other soils were lowest in the AB or Bt horizons.

Although this comparison of properties does not resolve the question of the logic of the present classification of these soils, it does show that the Queens catena soils differ in many respects from the typical Gray Luvisols such as Cooking Lake. They differ less from Luvisolic soils such as those characterized by Clark and Green (1964) that inherited their acidity and low degree of base saturation. Because depletion of bases marks a stage in the development of soils that is relevant to soil–plant relationships as well as to soil genesis, there would be some merit in separating base-saturated from unsaturated Luvisolic soils at the subgroup level. Brunisolic subgroups could be deleted from the Luvisolic Order.

Although the present study revealed little new information on the essential differences between Gray Luvisols and Gray Brown Luvisols their differentiation at the great group level seems reasonable. Micromorphological examination of the Ah horizons supported the conclusion based on macromorphology that the degree of incorporation of mineral and organic fractions was much greater in the Gray Brown Luvisols. This difference might be obscured in cultivated soils but the climatic difference, which is similar to the one used in separating Boralfs and Udalfs (Soil Survey Staff, 1967) would remain.

ACKNOWLEDGEMENTS

We wish to acknowledge the technical assistance of R. K. Guertin, B. Sheldrick and D. Graham. We thank M. Schnitzer for the functional group analysis and J. E. Brydon for constructive criticism of the interpretation of the mineralogy.

An example of the application of appropriate procedures by physical geographers to contemporary field problems is provided in the following paper by Rutherford and Sullivan (1970). They employed the 1968 National Soil Survey Committee of Canada nomenclature to classify soils developing on a variety of Quaternary surficial deposits across a quartzite ridge near Kingston, Ontario, and discovered a close correspondence between soil type and surficial material. Their survey illustrates that the soil classification system can be successfully adapted to differentiate soils within a very small area. It also shows that as the areal scale of operation is reduced, identifiable soil patterns override climatic controls and more closely reflect the effect of other soil forming factors, in this case parent material.

PROPERTIES AND GEOMORPHIC RELATIONSHIPS OF SOILS DEVELOPED ON A QUARTZITE RIDGE NEAR KINGSTON, ONTARIO*

G. K. Rutherford and D. K. Sullivan

Abstract

The work of Ruhe, Butler, Churchward and others in temperate and tropical environments has shown that soil profiles are intimately associated in age with their soil surfaces. A sequence of soils forming a soil surface across a Precambrian ridge near Kingston, Ontario was investigated to consider soil characteristics, soil genesis and stratigraphic relationships. Lateral variations in the parent materials along the cross-section indicate the presence of depositional layers of different relative ages and constitution. The stratigraphic as well as the pedological relationships of these deposits were investigated to determine a depositional stratigraphic history of the ridge.

Introduction

Butler (2) proposed a study of soils based upon the periodicity of land surface development. Identification of the ground surface reveals the chronological development of the area. Other workers have applied his principles to soil studies in Australia (3, 11, 18, 19). Similar work has been attempted in North America in the American Middle West, where such workers as Ruhe (15) and Foss and Rust (4) have studied geomorphic surfaces and soils related to loess deposits.

The basic approach in studying the chronological development of this cross-section is thus unique, although the principles of Butler apply whether new surfaces are the result of glacial deposition, lacustrine deposition, or common hillslope erosion.

Although the general characteristics of the soils of southeastern Ontario are known, little consideration has been given to the pedogeo-

* Reprinted from *Canadian Journal of Soil Science*, Vol. 50, 1970, pp. 419-429 with the permission of the senior author and the publisher, Agricultural Institute of Canada.

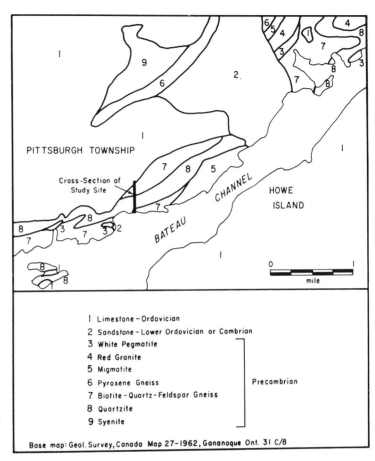

PITTSBURGH TOWNSHIP

Cross-Section of
Study Site

CHANNEL

HOWE
ISLAND

BATEAU

0   1
mile

1 Limestone – Ordovician
2 Sandstone – Lower Ordovician or Cambrian
3 White Pegmatite
4 Red Granite
5 Migmatite
6 Pyroxene Gneiss   Precambrian
7 Biotite – Quartz – Feldspar Gneiss
8 Quartzite
9 Syenite

Base map: Geol. Survey, Canada Map 27-1962, Gananoque Ont. 31 C/8

Fig. 1. Bedrock geology of an area surrounding Kingston, Ont. (1 mile = 1.609 cm).

morphic relationships of the major soil types. The objectives of this study of a cross-section on a Precambrian quartzite ridge were to determine the main soil properties, and indicate the geomorphic relationships obtaining between the soil profiles.

Methods

The study site is located approximately 11 km east of Kingston. A cross-section was taken transverse to a ridge-like outcrop of Precambrian rock with slopes of between 5° to 10° aligned in a northeast-southwest direction (Fig. 1). The stratigraphy and nature of the unconsolidated deposits were studied and soil profiles representative of significantly observable differences in soil parent material, drainage and profile development across the section were described, sampled and analyzed.

Particle-size analyses were undertaken following the pipette method of Kilmer and Alexander (9), and bulk density was determined using paraffin wax. Loss on ignition to 550 C, soil moisture using standard pressure plates and membrane, conductance, pH (liquid : soil = 25:1), exchange cations in ammonium acetate, dithionite Al and Fe, carbon and carbonates using carbon induction apparatus, and total B and Zr were all determined using methods suggested by Jackson (7, 8). Oxalate Al and Fe were determined using a method by McKeague and Day (10). Total nitrogen was determined by a colorimetric method employing Nessler's reagent. Clay mineral analyses were carried out on the 2–0.5 μ and < 0.5 μ fractions of 16 soil samples using CuK alpharadiation (8). The clay fractions were saturated with magnesium acetate and diffractograms

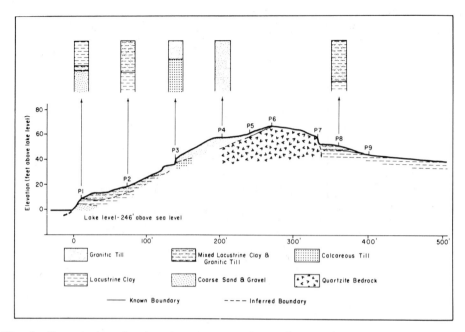

Fig. 2. Cross-section showing deposits on a Precambrian ridge near Kingston, Ont. (1 foot = 30.48 cm)

were made with (*i*) untreated, (*ii*) glycol-treated, (*iii*) heated to 550 C, and (*iv*) 5 *N* HCl-treated samples.

Heavy and light minerals were separated using bromoform with specific gravity of 2.80, and the heavy minerals in the 125–250 μ fraction were differentiated using a Franz isodynamic separator (14). Thin sections were prepared and described using methods recommended by Brewer (1). DTA curves were prepared from the < 74 μ fraction of selected air-dry samples.

The general morphological characteristics of the soils are presented in Table 1. Profiles were classified using the nomenclature of the National Soil Survey Committee of Canada (12) and the 7th Approximation (17), while descriptive terms and abbreviations are those outlined in the U.S.D.A. Soil Survey Manual (16) with soil consistency terms referring to consistency when moist. Moist colors are described using a Munsell color chart. The profile sites and parent materials on the cross-section are shown in Fig. 2.

Results

Although quartzite is the only rock outcropping on the ridge (20), acid igneous and metamorphic rocks comprise the flanks of this ridge for its whole length. The till material which is dominated by such fragments is referred to as the granitic till. The individual deposits overlying the Precambrian bedrock are described (Fig. 2). Inman's (6) six coefficients were calculated from cumulative particle size curves to indicate the degree of sorting and nature of the particle size distributions of the deposits. Pettersson (13), however, has shown that soil stones alone are an unreliable guide to the parent material of the soil as a whole.

Surficial deposits In the granitic till, below the Ah horizon, the till is structureless and loose with a bulk density averaging from 1.16 g/cc in the upper layers to 1.60 g/cc or greater, lower in the deposit. The texture varies from sandy loam to loam with clay contents less than 15%, except where illuviated clays

Table 1. Summary of soil profile morphology of soils formed on a quartzite ridge near Kingston, Ontario (abbreviations according to Soil Survey Manual, 1951 and National Soil Survey of Canada, 1968)

Horizon	Depth – cm	Color	Texture	Structure consistency
Profile 1: Gleyed Regosol (Humaquept)				
Ah	0–15	IOYR3/2	c	f2sbk;mfr
AC	15–30	IOYR3/2	c	mlsbk;mfr
Ckg	30–91	IOYR4/1	c	— ;mvfi
		IOYR5/2	sic	
IICkgj	91–104	IOYR4/2	l	— ;mvfi
IICk	104–163	IOYR3/2	sl	— ;dl
Profile 2: Gleyed Sombric Brunisol (Dystric Eutrochrept)				
Ah	0–7.6	IOYR2/2	sl	flsbk;mfr
AB	7.6–30	IOYR3/3	l	flsbk;mfr
Bmgj	30–56	IOYR4/4	cl	— ;dl
Btg	56–97	IOYR4/1	c	f2sbk;mfr
Cg	97–112	IOYR4/3	cl	— ;dl
IICcag	112–163	IOYR4/1	c	— ;mvfi
Profile 3: Orthic Sombric Brunisol (Typic Eutrochrept)				
Ah	0–5.1	IOYR3/3	l	flsbk;mfr
AB	5.1–20	IOYR4/2	sl	flsbk;mfr
Bm2	20–48	IOYR4/4	sl	— ;dl
C	48–84	IOYR4/2	sl	— ;dl
IICk	84–130	IOYR4/2	scl	mlsbk;dl
IICkg	130–198	IOYR4/1	l	m ;mvfi
Profile 4: Brunisolic Gray Brown Luvisol (Dystrochrept)				
Ah	0–7.6	IOYR3/2	sil	flcr;mvfi
Aej	7.6–30	IOYR5/4	sl	— ;dl
Bm	30–66	IOYR4/4	sl	— ;dl
Bt	66–109	IOYR4/2	l	— ;dl
C	109–137	IOYR4/2	sl	— ;dl
Profile 5: Orthic Sombric Brunisol (Dystrochrept)				
Ah	0–7.6	IOYR3/3	l	flcr;dl
AB	7.6–30	IOYR4/3	sil	— ;dl
Bm	30–46	IOYR4/3	l	m ;mfi
C	46–61	IOYR4/3	sl	m ;mfi
Profile 6: Lithic Sombric Brunisol				
Ah	0–5.1	IOYR3/2	sl	flsbk;mvfr
Ah2	5.1–10	IOYR3/2	sl	flsbk;dl
Bm	10–15	7.5YR4/4	sl	— ;dl

Profile 7

| Ah | 0–3.8 | IOYR4/3 | sic | — | ;dl |
| Bm | 3.8–7.6 | IOYR4/4 | sic | — | ;dl |

Profile 8: Humic Eluviated Gleysol (Aquic Dystrochrept)

Ah	0–18	IOYR3/3	sicl	f2cr;mfr
Bmg	18–53	IOYR4/1	c	f2sbk;mfr
Btg	53–58	IOYR4/1	c	f2sbk;mfr
Cg	84–122	IOYR4/1	c	f3sbk;mfr

Profile 9: Humic Gleysol (Humaquept)

Ap	0–15	IOYR2/2	c	m3sbk;mfi
Aej	15–23	IOYR3/2	c	m3sbk;mfi
Btjg	23–66	IOYR4/1	c	mlsbk;mfi
Cgk1	66–109	IOYR5/1	sic	mlsbk;mfi
Cgk2	109–122	IOYR5/1	c	;mfi

have accumulated. Although this till is generally sandy, the heterogeneous nature is shown by high sorting coefficients (Table 3) which range from 2.40 to 4.45 and the mean particle size diameter which ranges from 78.04 to 17.35 μ. Subangular to angular granitic and quartzite rock fragments greater than 2 mm in diameter comprise up to 21.3% (by weight) of the till. Rotting granite rocks as large as 60 cm in diameter are found in the till and indicate rapid weathering of the larger Precambrian rock fragments. Heavy minerals account for approximately 13.0% of the 125–250 μ sand fraction. Mica and vermiculite are the dominant clay-size minerals, with some chlorite and mixed mica minerals also present in small amounts.

Mottles increase in size and prominence with depth in the lacustrine clays. Below the surface horizon, a medium, weak, subangular blocky structure is present. Sorting coefficients indicate that the degree of sorting increases with depth and that the sorting of material has been considerable especially below 30 cm in depth. Below 66 cm some secondary calcium carbonate nodules are forming. Light minerals comprise 93.6% of the 125–250 μ sand fraction and the content of minerals recoverable with a hand magnet is low in contrast to the tills having a granitic component. Greater than 50% of the

heavy minerals are recoverable only at amperages greater than 0.5 on a Frantz separator. Vermiculite is the dominant clay-size mineral, with much less mica than in the granitic till.

At the lakeshore, lacustrine clays 30 cm in thickness overlie a series of clay bands 2.5 to 7.6 cm in thickness which alternate with silty clay bands 2.5 cm in thickness between 30 and 91 cm. The composition of the heavy mineral fraction which comprises less than 10% of the 125–250 μ sand fraction, is similar to that of the granitic till on top of the ridge. Mica and vermiculite are the dominant clay-size minerals.

On both flanks of the ridge, lacustrine clays and granitic till have been mixed together. Although the sand content is considerably greater at profile 2 than at profile 8, both profiles are typical of mixed deposits where granitic till has been eroded and deposited on top of, and to some degree mixed with, lacustrine clays. Sorting coefficients are high and values of kurtosis are low, indicating poor sorting of the material.

Buried deposits Below the 69-cm level in profile 3, a grayish brown (10 YR 5/2) calcareous till is present marked by high pH and calcium carbonate contents. Medium, very weak, subangular blocky structure in the upper layers grades to structureless massive at lower levels. High bulk density (greater than 2.0 g/cc) reflects the highly compacted nature of the till. Sandy clay loam texture in the upper levels grades to loam below 130 cm. High contents of subangular rock fragments greater than 2 mm in diameter comprise up to 19.4% of the weight of the till. Limestone fragments make up 65% of this fraction, while quartzite and granitic rock fragments account for the remainder. Above 69 cm in profile 3, only quartzite and granite rock fragments were observed. Sorting coefficients indicate that only modest sorting of the material has taken place. The residual heavy minerals after separation at 0.5 amps on a Franz separator are more prominent than in the overlying material. Mica and vermiculite are the predominant clay-size minerals. This material cannot without doubt qualify as a separate till, different from the material above 69 cm in profile 3, but has in the text and in Fig. 2 been called "calcareous till".

Below the banded layers in profile 1 (Fig. 2) there is a deposit of poorly sorted gravel and sandy loam. Between 91 and 94 cm in this layer, rotting granitic rocks and sand form a loose, structureless layer. Underlying this layer, between 94 and 104 cm, laminated calcareous clays are found with some rotting granitic rock fragments dispersed throughout. Rock fragments at these levels are subangular to subrounded with limestone comprising 10% and quartzite and granitic fragments 90%. Below 104 cm, a coarse layer of structureless, loose gravel and sandy loam is present with up to 35.5% of the material comprised of rock fragments greater than 2 mm in diameter. These subrounded to subangular fragments are 26% of Precambrian origin. Although the heavy mineral content of the 125–250 μ sand fraction is greater than that of the granitic till covering the ridge, the magnetic division of the heavy minerals is similar.

Comments on properties of the soils The general morphological characteristics and classification of the soils are presented in Table 1 and the results of some relevant physical and chemical analysis are rendered in Tables 2, 3 and 4.

CHEMICAL AND PHYSICAL PROPERTIES. The results of particle size analyses show that a sharp distinction exists between the soils derived from granitic till on the top of the ridge and those formed from lacustrine clays on the flanks. In general, the soils exhibit considerable textural fluctuations throughout their profiles, especially in the sandy granitic till. Much of this variation is probably due to textural differences within the original parent material. However, in profiles 2, 4 and 9, horizons of clay concentration are accompanied by an increase in the clay less than 0.5 μ in diameter.

Acidic pH values characterize those soils with a granitic compound while soils high in lacustrine clays are alkaline, although pH differences may be related to position as much as parent material. Base saturation is 100% in the lacustrine clays except in the surface horizons, while in the granitic soils it increases with depth. Ca is the dominant exchange cation although the high Mg may reflect a dolomitic element within the calcareous materials. K is of greater relative importance in the granitic soils.

Minor concentrations of soluble Fe and Al are registered in the upper horizons of most soil profiles. Dithionite-iron is not usually concentrated markedly with depth, except in those horizons where mottling is best developed.

MINERALOGICAL PROPERTIES. There is little difference between the 2-0.5 μ and the < 0.5 μ fractions; quartz and feldspar are expectedly less in the finer than the coarse fraction. Generally, biotite and vermiculite are the dominant clay minerals with minor amounts of muscovite, chlorite and illite. Vermiculite is higher in the fine than in the coarse fractions in surface compared with sub-surface horizons and in the better drained sites, suggesting its origin from biotite. Muscovite and illite are relatively rare, which is not surprising as biotite gneiss dominates the rocks of this ridge for 16 km to the east (20). Mixed-layer lattice minerals were not indisputably confirmed.

Differential thermal analysis curves support other evidence for the changing nature of parent materials, both horizontally and vertically.

Table 2. Particle-size and bulk density for some soils on a quartzite ridge near Kingston, Ont.

Horizon	Particle size analysis					Bulk density g/cc
	2000–250μ	250–50μ	50–2μ	2–0.5μ	>0.5μ	
Profile 1						
Ah	1.8	2.9	56.4	20.0	20.9	1.1
Ckg	0.9	0.6	32.7	28.1	37.7	1.7
IICkgj	12.6	11.2	50.3	16.9	9.9	1.6
IICk	35.7	24.8	24.8	8.8	5.9	1.9
Profile 3						
AB	25.5	28.5	34.4	5.6	6.1	1.1
Bm2	30.6	25.3	35.1	7.3	1.7	1.5
C	30.1	26.5	31.0	4.5	7.9	1.7
IICK	26.2	29.2	29.5	7.6	7.5	2.0
IICkg	25.8	19.6	37.1	8.9	8.6	2.0
Profile 4						
Aej	12.7	19.4	58.6	6.1	3.2	1.2
Bm	10.3	19.9	58.9	8.5	2.4	1.3
Bm2	16.3	25.9	45.3	5.9	6.6	1.6
Bt	20.4	20.9	34.1	8.0	16.6	1.4
C	28.6	27.3	29.8	5.3	9.0	1.5
Profile 8						
Ah	7.5	4.5	51.4	30.4	6.2	1.0
Btg	2.1	1.8	3.9	35.0	61.1	1.6
Cg	1.2	0.9	24.8	42.7	30.4	1.3
Profile 9						
Aej	0.8	2.0	31.8	25.3	40.1	1.6
Btjg	0.2	0.3	0.5	33.3	66.2	1.6
Cgkl	4.4	3.8	8.2	41.1	50.7	1.4

Characteristic curves were obtained for each of the parent materials so that the presence of more than one of these in one soil profile could be readily confirmed.

Soils derived from granitic parent material (Table 3) exhibit higher heavy mineral contents than those formed on lacustrine clays. The strongly magnetic minerals are highest in those samples with a significant granitic component.

MICROMORPHOLOGICAL PROPERTIES. Twenty-three thin sections from various horizons in profiles 1, 3, 4 and 9 were examined. Some relevant micrographs are shown in Fig. 3.

Thin sections of the laminated clays between 94 and 104 cm in profile 1 showed the presence of oriented clays, as well as horizontal layering of both clay and subangular to subrounded skeleton grains ranging in size from 0.02 to 0.15 mm (Fig. 3f).

Unorganized asepic fabric characterizes the 13–20 and 28–36 cm depths in profile 3, while at 48–58 and 69–84 cm well developed ferri-argillans and iron nodules occur (Fig. 3b, c, d). These pedological features occur only in the noncalcareous soil material and at and near the boundary with the underlying calcareous material. Those occurring at 69–84 cm may be

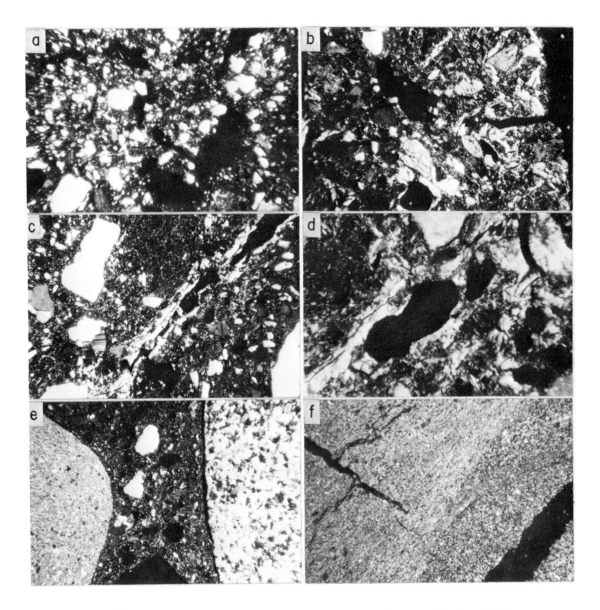

Fig. 3. Micromorphology of some horizons near Kingston, Ont., Profile 3: <u>a</u>, 13-29 cm, x 100; <u>b</u>, 67-84 cm, x 100; <u>c</u>, 64-84 cm, x 100; <u>d</u>, 64-84 cm, x 300; <u>e</u>, 99-114 cm, x 300. Profile 1: <u>f</u>, 91-104 cm, x 100.

Table 3. Heavy mineral analyses and sorting coefficients for some soils on a quartzite ridge near Kingston, Ontario

| Horizon | Sorting co-efficient (Inman) | Total Sp. Gr. 2.80 | Heavy minerals as % of 125–250μ fraction | | | |
| | | | Magnet | Recovery range, amps | | |
				0.4	0.4–0.5	0.5
Profile 1						
Ah	2.5	8.5	0.8	0.8	3.0	3.9
Ckg	2.5	7.1	0.4	0.8	1.5	4.4
IICkgj	3.8	14.7	1.4	2.5	4.9	5.9
IICk	4.1	12.8	1.4	1.5	3.0	6.9
Profile 3						
AB	3.4	13.0	1.3	1.9	5.0	4.8
Bm2	3.4	12.8	1.1	2.1	5.2	4.4
C	3.8	13.3	1.5	2.2	3.2	6.4
IICk	3.8	12.3	1.0	1.9	3.0	6.4
IICkg	4.2	11.3	1.1	1.9	3.2	5.1
Profile 4						
Aej	2.9	14.1	1.4	2.3	5.0	5.4
Bm	2.5	13.9	1.8	1.9	3.8	6.4
Bm2	3.0	14.4	1.5	1.5	4.7	6.7
Bt	4.4	13.0	1.6	2.3	4.6	4.5
C	3.8	17.6	1.4	2.2	4.3	9.7
Profile 8						
Ah	2.5	5.1	0.4	0.7	2.0	2.0
Btg	2.8	5.0	0.2	0.6	1.5	2.7
Cg	2.3	4.3				
Profile 9						
Aej	2.9	5.1	0.4	0.7	2.0	2.0
Btjg	2.3	6.5				
Ckgl	2.2	6.1	0.4	1.3	1.3	3.1

formed by illuviation waters being checked at the contact with the denser calcareous till. Those at 45–58 cm could either be formed in pedologic B horizons or could have been formed when the calcareous/noncalcareous weathering front stood at that level. Cutannic material was also observed between crystals in a fragment of rotting granitic rock at 69–94 cm. Pedo- and granotubules are quite common in the calcareous till.

Profile 4 has mainly asepic fabric with simple packing voids and skeleton grains. There are, however, some minor amounts of skelsepic fabric between 41–51 cm.

Geomorphic relationships of the soils   An association exists between the nature of the surficial deposit, the drainage and the kind of soil forming at a site: the drainage may be a function of the nature of the deposit and not merely that of the topographic situation. The granitic surficial till covering the major area of the ridge is the parent material for Sombric

Table 4.   Some relevant chemical data for soils on a quartzite ridge near Kingston, Ontario

Horizon	pH 1:2.5 H_2O	Exchangeable cations meq/100 g soil				Base saturation %	Fe Dithionite ./.	Fe Oxalate ./.	Organic C %	CaCO$_3$ %
		Ca	Mg	Na	K					
Profile 1										
Ah	7.8	17.3	5.0	0.1	0.5	100.0	0.8	0.5	0.5	13.9
Ckgj	7.9	19.7	5.2	0.0	0.3	100.0			0.2	15.9
IICkgj	7.4	9.0	2.9	0.0	0.2	100.0			0.1	12.5
IICk	7.8	8.8	2.4	0.0	0.2	100.0			0.1	16.1
Profile 3										
AB	5.0	2.5	0.6	0.1	0.3	61.2	0.7	0.5	2.9	—
Bm2	6.0	1.8	0.3	0.1	0.2	79.1	0.8	0.5	0.9	—
C	6.1	3.8	2.4	0.1	0.2	82.7	1.0	0.4	0.6	0.1
IICk	7.2	4.7	3.6	0.1	0.2	100.0	0.9	0.4	0.4	4.0
IICkg	7.3	16.2	2.6	0.1	0.1	100.0	0.9	0.4	0.3	14.9
Profile 4										
Aej	4.2	0.5	0.2	0.1	0.1	43.1	1.1	0.6	1.2	—
Bm	4.3	0.5	0.2	0.1	0.1	38.9	1.0	0.6	0.6	—
Bm2	4.7	1.2	1.3	0.1	0.2	79.6	1.0	0.5	0.3	—
Bt	5.3	3.5	2.4	0.1	0.2	93.7	1.5	0.5	0.4	—
C	7.0	5.4	3.1	0.1	0.1	100.0	0.9	0.3	0.3	—
Profile 8										
Ah	4.8	3.4	1.4	0.1	0.5	56.1	1.8	1.5	2.7	—
Btg	4.0	6.8	4.8	0.1	0.6	89.7	2.0	1.7	0.4	—
Cg	5.6	8.8	7.5	0.2	0.5	94.6	1.2	0.6	0.2	—
Profile 9										
Aej	7.0	12.6	11.8	0.1	0.4	97.1	1.9	1.1	2.0	—
Btjg	7.6	18.0	15.0	0.1	0.3	100.0	0.8	0.2	0.3	—
Ckgl	7.9	18.2	6.4	0.1	0.3	100.0	1.0	0.3	0.2	7.2

Brunisols and Brunisolic Luvisols, and Orthic Regosols form where the till cover is thin. On the lacustrine clays north of the ridge, Humic Gleysols have formed while on those at the lakeshore, a Gleyed Regosol is found. On the south flank of the ridge where sandy materials have been mixed with the lacustrine clays, Brunisolic Luvisols and Sombric Brunisols, respectively, have formed. No evidence of buried soils was found in this cross-section, although soil forming processes may have acted upon those deposits immediately after deposition. Although such evidence of soil formation may have been removed, it is more probable that the processes were too weak and of too short a duration to result in marked soil profile development. Thus, soil development began on all sections of the surficial deposits on this ridge at approximately the same time due to the relatively short time period between the retreat of the ice sheet and the drainage of the Lake Ontario basin below its present water level (5). Consequently, this cross-section of soils represents only one ground-surface. Although both buried and surficial sedimentary and erosional breaks occur due to the differing relative ages of the deposits, they do not involve breaks in the soil layer.

Conclusions

Sombric Brunisols have formed on a granite till overlying Precambrian quartzite. With increasingly impeded drainage, and higher amounts of clay, these soils grade through Brunisolic Luvisols to Gleyed Sombric Brunisols and Gleysols. Orthic Gleysols have formed on lacustrine clays on the north of the ridge, where gleization and the removal of the carbonates from the upper horizons have been marked. Where the granitic till covering the ridge is thin, and at the lakeshore on lacustrine clays, Regosols are found. Marked alteration of the primary minerals has not occurred, except for the alteration of mica to vermiculite. The heterogeneous nature of the

parent materials thus dominates over the effects of soil formation.

The absence of buried soils indicates that the present ground surface has been forming since shortly after the final retreat of the ice sheet. A calcareous till was probably deposited during the last advance of the Wisconsin ice sheet. The influx of Lake Iroquois deposited coarse sand and gravel on the south side of the ridge, and later lacustrine clays both north and south of the ridge as the ice front retreated to the north. Erosion shortly after the emergence of the ridge resulted in the mixing of the granitic sands with lacustrine clays at the sides of the ridge.

ACKNOWLEDGEMENTS

The authors wish to acknowledge that help and interest shown by Dr. J. A. McKeague of the Canada Department of Agriculture who also criticized and read the manuscript; Professors Arnold and Milford of Cornell University and Dr. W. A. Gorman of Queen's University.

We would like to express our appreciation to the work of the late Mr. J. K. Steele, a technician in the Department of Geography at Queen's University, and also to the National Research Council of Canada who provided a grant to carry out the project.

REFERENCES

1. Brewer, R. 1964. Fabric and mineral analysis of soils. John Wiley and Sons Inc., New York.
2. Butler, B. E., 1959. Periodic phenomena in landscapes as a basis for soil studies. C.S.I.R.O., Aust. Soil Publ. No. 14, Melbourne, Australia.
3. Churchward, H. M. 1961. Soil studies at Swan Hill, Victoria, Australia. 1. Soil layering. J. Soil Sci. 12: 73-86.
4. Foss, J. E. and Rust, R. H. 1968. Soil genesis study of a lithologic discontinuity in glacial drift in western Wisconsin. Soil Sci. Soc. Amer. Proc. 32: 393-398.
5. Henderson, E. P. 1967. Surficial geology north of the St. Lawrence River, Kingston to Prescott. In Guidebook, Geology of Eastern Ontario and Western Quebec. Annu. Meet. Geol. Ass. Can. and Mineral. Ass. Can., Kingston, Ontario, pp. 199-207.
6. Inman, Douglas L. 1952. Measures for describing the size distribution of sediments. J. Sediment. Petrol. 22: 125-145.

7. Jackson, M. L. 1958. Soil chemical analysis. Prentice-Hall Inc., Englewood Cliffs, N.J.
8. Jackson, M. L. 1965. Chemical composition of soils. In Chemistry of the Soil, 2nd ed., A.C.S. Monogr. No. 160, Firman E. Bear, ed., Reinhold Publ. Corp., New York, pp. 71-141.
9. Kilmer, Victor J. and Alexander, Lyle T. 1949. Methods of making mechanical analyses of soils. Soil Sci. 68: 15-24.
10. McKeague, J. A. and Day, J. H, 1966. Dithionite- and oxalate-extractable Fe and Al as aids in differentiating various classes of soils. Can. J. Soil. Sci. 46: 13-22.
11. Mulcahy, M. J. 1961. Soil distribution in relation to landscape development. Ziet. Geomorph. 5: 211-225.
12. National Soil Survey Committee of Canada. 1968. Report of the seventh meeting of the National Soil Survey Committee of Canada. Can Dep. Agr., Ottawa.
13. Pettersson, M. 1968. Indications of provenance

in some Anglesey drift soils. J. Soil Sci. 19: 168-173.

14. Rosenblum, Sam. 1958. Magnetic susceptibilities of minerals in the Frantz iso-dynamic magnetic separator. Amer. Mineral. 43: 170-173.

15. Ruhe, Robert V. 1956. Geomorphic surfaces and the nature of soils. Soil Sci. 82: 441-455.

16. United States Department of Agriculture, Soil Survey Staff. 1951. Soil survey manual. U.S. Dep. Agr. Handbook No. 18, U.S. Government Printing Office. Washington, D.C.

17. United States Department of Agriculture, Soil Survey Staff. Soil classification, a comprehensive system. 7th Approximation, 1960 and revisions of 1964 and 1967.

18. Van Dijk, D. C. 1959. Soil features in relation to erosional history in the vicinity of Canberra. C.S.I.R.O., Australia, Soil Publ. No. 13. Melbourne, Australia.

19. Walker, P. H. 1962. Soil layers on hillslopes: A study at Nowra, N.S.W., Australia. J. Soil Sci. 13: 167-177.

20. Wynne-Edwards, H. R. 1961. Geology — Gananoque, Ontario. Geol. Surv. Can. Map 27-1962.

Finally, there are many cases where soil scientists have opened avenues of investigation into Canadian soils at the local scale which physical geographers could well use as a basis for investigations with similar goals in the future. For instance, Gillespie and Protz (1969) found two soils (one developing on granite, the other on limestone) near Peterborough, Ontario, whose morphological and mineralogical characteristics pointed to a residual origin. This led them to dispute the generally held belief that residual soils (those developing on *in situ* weathered bedrock) have not had sufficient time to develop in extensively glaciated landscape such as Ontario.

Chapter 13

Studies in the Vegetation of Canada

Physical geographers have long been preoccupied with studies of landforms and climate. However, with the current concern for the maintenance of environment quality there is a need for a greater emphasis to be placed upon vegetational and biogeographical problems. The physical geographer, with his ability to integrate knowledge of the landscape, has a unique opportunity to contribute to this cause. Unfortunately, the opening remarks of the previous chapter, which lamented the lack of interest shown by physical geographers on research into soils, are also applicable in the case of studies involving vegetation. Once again, a brief background on the different approaches to vegetation studies is provided before launching into a discussion of the literature.

Historical Background

Just as in the case of soils, geographers have long been interested in the areal distribution of vegetation patterns and the way that these patterns are related to other factors of the natural landscape. On account of the biological implications in the term vegetation, the analysis and categorization of plant distributions can be tackled from either the floristic or ecological viewpoints. In the case of the former, lists of species for various locations are prepared in order to delimit those areas where plant taxonomic patterns tend to be repeated and to provide information on the origin, spread, and range of different plant species. Owing to the automatic emphasis placed upon floristics together with the implied historical bias, the adaptability of this approach to geographical studies is limited.

According to Raup (1942), the first serious effort to discuss world plant forms in other than purely floristic terms was made in 1805 by Alex Von Humboldt, one of the founders of modern geography. In the course of extensive travels, he carefully documented coincidences between botanical features and environmental conditions, and the pattern he discerned enabled him to establish a classification scheme founded upon plant morphology or outward form rather than floristics. Humboldt's major objective was to describe the areal variation of the earth's plant cover; he made no pretence of being able to distinguish relationships between plants and their environment although, by grouping taxonomically diverse but structurally related species, he laid the foundation that subsequently led to the formulation and growth of ecological viewpoints in plant geography. Following the added stimulus provided by Darwin's views on natural selection and habitat adaptation caused by environmental stress, plant geographers quickly realized that the form or physiognomy of plants might reflect prevailing environments. As a result, attention was shifted to the analysis of the functional significance of plant structures and the physiological relationships of plants to their environment. These approaches were later modified and incorporated into what we now consider to be ecologically based plant studies. Pioneering studies by Cowles (1899, 1901) demonstrated that characteristic plant communities tend to invade a given area in a successive manner. Later, Clements (1916) advocated that the progressive changes taking place in the various communities caused them to converge towards a dominant vegetation unit over the entire region irrespective of habitat differences. This was designated the climax community and was distinguished by internal stability compared with the preceding successional stages. According to Clements, the climax is essentially a reflection of the climatic regime in the area, and thus, climax communities should coincide geographically with climate. He also equated the climatic climax with the term "formation" which he considered to be a climax community wherein the structure of the vegetation is sufficiently uniform to mask habitat variations. Climax formations could be further subdivided into associations, each of which was considered to be homogeneous with respect to its physiognomy, ecological structure, and floristic composition (Weaver and Clements 1938).

Over the last few decades the concept of the climatic climax has been the object of considerable debate and much criticism mainly because of its restrictive monoclimax connotations. Many ecologists and biogeographers have argued that vegetation patterns reflect a wide range of environmental factors of which climate is only one. Other important factors include the soil type and profile characteristics, the depth to bedrock, the parent material, the dominant landforms, and the water supply. Any one of the above can invoke a sufficient response in the vegetation that will give rise to a specific plant community. Usually, however, these factors elicit a combined effect making it difficult to isolate one factor as being dominant above all others. At this level climatic effects are significantly modified. While there is less adherence to strict monoclimax principles today, most biogeographers and ecologists agree upon the existence of relatively uniform belts of vegetation, characterized by certain combinations of species with ecologically compatible structural characteristics. Moreover each belt generally exhibits some degree of stability. These belts, at least at the sub-continental scale, tend to correspond with basic climatic regimes. However closer inspection invariably reveals that local and sometimes regional variations in other environmental factors impart a heterogeneous character to the vegetation.

The move to reduce the importance of the climatic factor is also reflected in the growing tendency in recent years to regard vegetation and climate as merely components in an ecosystem. This term was initially proposed by Tansley (1935) to depict the interaction between the physical environment and biological communities. Evans (1956) argued that the concept should be expanded to encompass the circulation and transformation of energy between organisms and the environment. The concept now enjoys considerable popularity amongst a wide audience of biologists, ecologists, and geographers, and its flexibility is illustrated by its application to all scales of biogeographical and ecological enquiry.

On the basis of the foregoing discussion, the term vegetation can be regarded as encompassing all the plants in a given area. Thus geographers, who are confronted with the task of identifying, describing, and interpreting regional vegetation patterns, have generally adopted the ecological as opposed to the floristic point of view. Although the former can at times be unsatisfactory, especially in the case of studies at the local scale, it has enabled the establishment of broad but distinctive vegetation units over large areas which in turn has facilitated landscape description.

Those who investigate the variable nature of the earth's vegetative cover are usually referred to as biogeographers. The definition of biogeography varies according to the interests and background of the definer. However, there is a growing tendency amongst geographers to consider the field as bridging the gap between ecology and physical geography. In a recent review paper, Crowley (1967) furnishes some highly pertinent comments on the current status of biogeography in Canada. He implies that geographers should not be intimidated by investigators in the biological sciences who regard biogeography as a branch of ecology. Nonetheless he points out that a situation has developed in Canada whereby the bulk of biogeographic research is conducted by non-geographers. When geographers do tackle biogeographical problems, they frequently resort to traditional methods of ecology. Thus Crowley anticipates only a gradual growth in the amount of biogeographical research undertaken by professional geographers in Canada. He concludes by advocating that biogeographers must always maintain a geographical viewpoint and as a consequence they may be better advised to direct their efforts towards the study of complete ecosystems instead of concentrating upon their individual components.

Vegetation at the Sub-Continental Scale

One of the earliest documented attempts at regionalizing vegetation patterns across Canada was made by Macoun in 1894. Although his paper has not been reprinted here, the following remarks serve to amplify statements made earlier on the status of plant geography at the end of the nineteenth century. With the development of plant ecology still in its infancy, Macoun assumed a standpoint that was essentially floristic but with a strong geographical bias. He simply divided up the country into eight regions, based mainly on provincial boundaries, listed the major species in each, and commented briefly on their distribution. However, he did provide a few speculative comments on what he felt were the controlling environmental parameters at the regional and local scale. He also appreciated the potential hazards in uncontrolled deforestation practices and used his paper as a platform to focus attention on the need to conserve the forest cover. His strongly worded statements, in fact, are almost as appropriate today as they were then; for instance, he argued that

> . . . It is truly appalling when the magnitude of the national interests at stake are considered to view the spoilation which has been carried on quite recklessly under the protection of permits and licenses. When one is soberly told that this destruction was necessary in the interests of trade and for the development of the country, one is forced to deny the truth of such statements and to enter a protest against the fallacy concealed in them.

The influence of the expanded ecological literature is evident in the approach employed by Halliday (1937) in his proposal for a forest classification of Canada. He relied heavily upon the ecological classification system outlined by Weaver and Clements (1938) who recognized the existence of four major climax formations or Great Plant Formations in North America. Three of these, the Tundra Climax Formation, the Forest Climax Formation, and the Grassland Climax Formation are to be found in Canada. According to Weaver and Clements, each of these plant formations can be subdivided into Formation Types, although Halliday substitutes the term Forest Region to identify this category. He lists a total of eight regions: the Boreal Forest Region, the Subalpine Forest Region, the Montane Forest Region, the Coast Forest Region, the Columbia Forest Region, the Great Lakes-St. Lawrence Forest Region, the Acadian Forest Region, and the Deciduous Forest

Region. All but two of Halliday's regions are counterparts of Weaver and Clement's original formation types; the Columbia Forest Region and the Acadian Forest Region have been added to comply more closely with Canadian conditions. He further subdivides each region into Forest Sections which reflect environmental controls other than those stemming from the regional climate. Thus Halliday's Forest Sections correspond approximately with Weaver and Clement's plant associations. The above example of the traditional ecological approach is perpetuated in the following reprinted introduction to Rowe's (1959) regionalization of forests in Canada. Basically, it is an up-dated version of Halliday's classification, although the terminology employed is redefined on account of the contemporary disenchantment with the concept of the climatic climax. He seeks to avoid the monoclimax implication by depicting forest regions as geographic entities that are more or less homogeneous in terms of their vegetational characteristics and floristic composition. While retaining all eight regions identified and elaborated on by Halliday, he does make a number of boundary changes on the accompanying map in the light of recent observations. He then proceeds to describe each region. Like Halliday, he also subdivides each region into forest sections each of which is discussed in detail in the full report.

FOREST REGIONS OF CANADA*

J. S. Rowe

Introduction

The first notable description of Canadian forests on a national scale was W. E. D. Halliday's "A Forest Classification for Canada" published in 1937, a work whose merit won it immediate acceptance and brought it into widespread use as a standard reference. The present bulletin is based on this earlier publication, incorporating with it additional knowledge gathered during the last two decades.

The framework of the original has been maintained in essence although some innovations have been introduced. For example, many changes in the placement of boundary lines have been made in an attempt to refine the forest divisions. In areas of forest diversity it has seemed advisable to recognize a number of new Sections, which have, however, been kept within the numbering system previously established by the addition of letter postscripts. On

the map a clearer distinction has been made between the boreal forest proper and its northern and south-central sub-regions, the subarctic forest and the aspen parkland, respectively. Also, the forests of Newfoundland are here described for the first time within the Boreal Forest Region. It might be noted that in the introduction to his 1937 bulletin, Halliday recognized that revision would be necessary as new information became available.

A treatise such as this, professing to describe the vegetation of the greater part of Canada, is likely to be used as a reference for many different purposes. Accordingly, a brief discussion of the underlying rationale may forestall some misconceptions and errors of interpretation.

First, it should be pointed out that the parent publication, like the present one, was devoted to a geographic description of the forests; it outlined their areal distribution and was not a classification in the usually accepted sense. The

* From "Forest Regions in Canada" by J. S. Rowe, pp. 7-12, Canadian Department of Northern Affairs and Natural Resources, Forestry Bureau Bulletin 123, Ottawa. Reproduced by permission of Information Canada.

phrase "Forest Classification" is apt to be mis-interpreted, suggesting as it does the provision of a system of purely vegetative categories based on consistently applied criteria, and the change of title to "Forest Regions" removes this implication while expressing more aptly the purpose.

Halliday's primary division of the forestland was into "Regions" which were equated with the "Climaxes" of F. E. Clements and were similarly described as stable, climatically con-trolled formations characterized by the pres-ence of certain tree species, the climax dominants. For the most part, the Regions are the obvious large units of forest description that all field workers recognize. They are geo-graphic entities, and their status is not weak-ened by the contemporary loss of faith in climatic climaxes. The sub-divisions of the Regions—the Forest Sections—were necessary for a refinement of the descriptions. As Halli-day expressed it: "More detailed study of the Forest Region shows, in many cases, the possi-bility and advisability of distinguishing within the whole very definite areas, often of consid-erable extent, which are marked by the con-sistent presence of certain associations, and which show, in the mass, a character differing from other parts of the Region." In the above quotation the term "association" is to be inter-preted simply as a recurring community of one or more tree species.

The general purposes of areal forest descrip-tion are satisfied by such an approach, although charges of arbitrariness cannot be denied. Observably different areas—the Sections—are blocked out, then described and defined relative to one another. Herein lies the reason for the notable lack of consistency in the use of cri-teria for the definition of the Sections; as more or less obvious geographical entities they have been identified first, then the criteria to define them have been selected secondarily. The approach is justified on practical grounds, for while division of the forest according to the strict application of selected criteria would provide a logical classification on paper, at-tempts to use such a system for a physical division of the forestlands over the length and breadth of Canada would certainly prove impractical.

An alternative to the provision of geographi-cal forest Sections by a process of division "from above" would be their synthesis "from below". If detailed information at a large scale were available concerning the forest vegeta-tion and the land it occupies, and if ecological knowledge were sufficient for an understanding of the relationships between the two, then it might be possible to combine similar or related forestland units, or patterns of forestland units, into larger wholes for purposes of generaliza-tion and description. Unfortunately, the infor-mation necessary to carry this method through to a successful conclusion does not exist, except perhaps locally, in Canada; the only feasible approach is "from above".

In this bulletin the units of forest description are much the same as those used by Halliday, though a redefinition of terms is required. The Forest Region is here conceived as a major geographic belt or zone, characterized vegeta-tionally by a broad uniformity both in physiog-nomy and in the composition of the dominant tree species. It corresponds to the concrete forest "formation" as generally understood, but without the classic ecological implications. Particularly is the idea of a direct cause-effect relationship between present climate and Forest Regions (or their segments) disavowed. The relationships that doubtless exist between cli-mate and vegetation remain to be demonstrated in any particular area, with full awareness of the inter-related influences of all other com-ponents of the environment, past as well as present.

The Forest Section is a subdivision of the Region, conceived as a geographic area pos-sessing an individuality which is expressed relative to other Sections in a distinctive pat-terning of vegetation and of physiography. It is a unit of convenience for forest description which, because of the scale of the map, is usu-ally of large areal extent and, inevitably, is more or less heterogeneous. Wherever the for-ests and land are extremely variable over short distances, one Section or another must include this variability. Hence many unions of quite different forestlands are forced which at a larger scale would not be maintained.

The criteria appropriate for description of the boundaries of the Sections are the macro-features of vegetation; the distribution and range of conspicuous tree species, their life-forms (broad-leaved or needle-leaved), the physiognomy and relative areal extent of the communities in which they are associated, and the patterning of the total vegetation. At the small scale, boundaries drawn on the basis of these criteria are often coincident with major physiographic features which therefore serve a useful purpose as supplemental criteria for delimiting the Forest Sections.

The core of each Section is described in terms of the major tree species and of the cover types that they form, relatively more emphasis being given to the stable associations than to those resulting from recent fires and from exploitation of the forest. As detailed descriptions of forest types and of their comparative areal extents are not compatible with the purpose (broad description) and scale (1 inch equals 100 miles), these should not be looked for in the text. An attempt has been made to indicate what forests were present historically as well as what prevails now. Finally, a few brief notes on topography, geology and soils are given.

Appendices I and II, contain, respectively, selected climatic data for stations taken as representative of the Sections, and a glossary of the most commonly used technical terms. The Bibliography consists of a number of publications which were found useful in preparing descriptions of certain forest areas, particularly those of the more remote parts of Canada.

In separating and describing the Sections, Halliday's "Forest Classification" has of course been basic. The revision of the map was largely the work of W. G. E. Brown, formerly with the Forestry Branch, and in this he received information and advice from many sources; also, numerous people have contributed to the material from which the present text was prepared. Of necessity the writer has had to place his own interpretation on much of the available data, and full responsibility for such

interpretation is assumed. The following organizations and persons gave assistance in numerous ways, and grateful acknowledgement is here made: to the Officers-in-Charge and the personnel of the Forest Biology Laboratories, Department of Agriculture, and to the member Companies of the Pulp and Paper Research Institute of Canada, for much general information; to the Canadian Wildlife Service, Department of Northern Affairs and National Resources, especially for data concerning the northernmost forests; to members of the Geological Survey of Canada and the Geographical Branch, Department of Mines and Technical Surveys, who checked the accuracy of the descriptions of landforms and geology in the Sections; and to the Pedology and Soil Survey Section, Department of Agriculture, who rendered a similar service for the descriptions of soils. Assistance in specific areas has been given by Messrs. R. H. Spilsbury, R. L. Schmidt and G. M. Wilson, Drs. E. W. Tisdale, V. J. Krajina and E. H. Moss, Messrs. D. I. Crossley, W. R. Parks, H. I. Kagis, Drs. R. D. Bird, K. A. Armson, W. G. Dore, B. Boivin, Messrs. E. Bonner and S. T. B. Losee, Drs. D. Löve, F. K. Hare and H. D. Long. Finally, appreciation is expressed for the major contributions made by the Forestry Officers of the Forest Research Division of the Forestry Branch.

Forest Regions

The eight Forest Regions (see Fig. 1) are the Boreal, Subalpine, Montane, Coast, Columbia, Deciduous, Great Lakes-St. Lawrence, and Acadian. These Forest Regions, with the Grassland, the Arctic, and the Alpine Tundra, are shown by shading on the Forest Classification Map. A brief description of the Regions follows.

Boreal Forest Region This region comprises the greater part of the forested area of Canada, forming a continuous belt from Newfoundland and the Labrador coast westward to the Rocky Mountains and northwestward to Alaska. The white and the black spruces (*Picea glauca* and

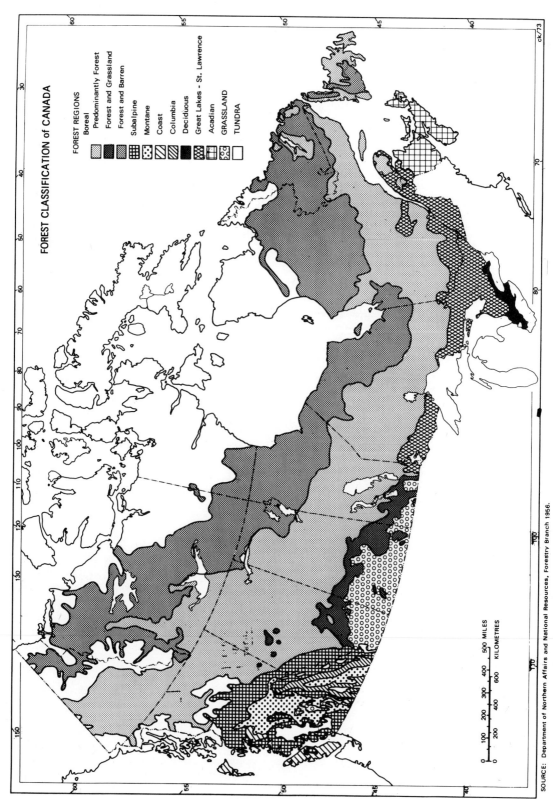

FOREST CLASSIFICATION of CANADA

FOREST REGIONS
- Boreal
- Predominantly Forest
- Forest and Grassland
- Forest and Barren
- Subalpine
- Montane
- Coast
- Columbia
- Deciduous
- Great Lakes - St. Lawrence
- Acadian
- GRASSLAND
- TUNDRA

500 MILES

0   100 200 300 400 500 MILES
0   100 200 300 400 500 600 KILOMETRES

Figure 1

SOURCE: Department of Northern Affairs and National Resources, Forestry Branch 1956.

ck/73

P. mariana)³ are characteristic species; other prominent conifers are tamarack (*Larix laricina*) which ranges throughout, balsam fir (*Abies balsamea*) and jack pine (*Pinus banksiana*) in the eastern and central portions, and alpine fir (*Abies lasiocarpa*) and lodgepole pine (*Pinus contorta* var. *latifolia*) in the western and northwestern parts. Although the forests are primarily coniferous, there is a general admixture of broadleaved trees such as the white birches (*Betula papyrifera* vars.) and poplars (*Populus tremuloides, P. balsamifera*); these play an important part in the central and south-central portions, particularly in the zone of transition to the prairie. In turn, the proportion of spruce and tamarack rises northward, and with increasingly rigorous climatic conditions the close forest gives way to the open lichen-woodland which finally merges into tundra. In the east there is along the southern border of the Region a considerable intermixture of species from the Great Lakes-St. Lawrence Forest such as the white and red pines (*Pinus strobus, P. resinosa*), yellow birch (*Betula lutea*), sugar maple (*Acer saccharum*), black ash (*Fraxinus nigra*), and eastern white cedar (*Thuja occidentalis*).

Subalpine Forest Region   This is a coniferous forest found on the mountain uplands in western Canada. It extends northward to the major divide separating the drainage of the Skeena, Nass, and Peace rivers on the south and to that of the Stikine and Liard rivers on the north. The characteristic species are Engelmann spruce (*Picea engelmanni*), alpine fir (*Abies lasiocarpa*), and lodgepole pine (*Pinus contorta* var. *latifolia*). There is a close relationship with the Boreal Region, from which the black and the white spruces (*Picea mariana, P. glauca*) and aspen (*Populus tremuloides*) intrude. There is also some entry of blue Douglas fir (*Pseudotsuga taxifolia* var. *glauca*) from the Montane Forest, and western hemlock (*Tsuga heterophylla*), western red cedar (*Thuja plicata*), and amabilis fir (*Abies amabilis*) from the Coast Forest. Other species found are western larch (*Larix occidentalis*), whitebark pine (*Pinus albicaulis*), and limber pine (*P. flexilis*), and on the coastal mountains yellow cedar (*Chamaecyparis nootkatensis*) and mountain hemlock (*Tsuga mertensiana*).

Montane Forest Region   The Region occupies a large part of the interior uplands of British Columbia, as well as a part of the Kootenay valley and a small area on the east side of the Rocky Mountains. It is a northern extension of the typical forest of much of the western mountain system in the United States, and comes in contact with the Coast, Columbia, and Subalpine Forests. Ponderosa pine (*Pinus ponderosa*) is a characteristic species of the southern portions. Blue Douglas fir (*Pseudotsuga taxifolia* var. *glauca*) is found throughout, but more particularly in the central and southern parts; lodgepole pine (*Pinus contorta* var. *latifolia*) and aspen (*Populus tremuloides*) are generally present, the latter being particularly well represented in the north-central portions. Engelmann spruce (*Picea engelmanni*) and alpine fir (*Abies lasiocarpa*) from the Subalpine Region become important constituents in the northern parts, together with white birch (*Betula papyrifera*). The white spruce (*Picea glauca*) though primarily boreal in affinity is also present here. Extensive prairie communities of bunch-grasses and forbs are found in many of the river valleys.

Coast Forest Region   This is part of the Pacific coast forest of North America. Essentially coniferous, it consists principally of western red cedar (*Thuja plicata*) and western hemlock (*Tsuga heterophylla*), with abundant sitka spruce (*Picea sitchensis*) in the north, and with the addition of Douglas fir (*Pseudotsuga taxifolia*) in the south. Amabilis fir (*Abies amabilis*) and yellow cedar (*Chamaecyparis nootkatensis*) occur widely, and together with mountain hemlock (*Tsuga mertensiana*) and alpine fir (*Abies lasiocarpa*), are common towards the timber-line. Western white pine (*Pinus monticola*) is found in the southern

---

³ Tree names are those used in "Native Trees of Canada". Bull. 61. Canada, Department of Northern Affairs and National Resources. 5th ed. 1956.

parts, and western yew (*Taxus brevifolia*) is scattered throughout. Broadleaved trees, such as black cottonwood (*Populus trichocarpa*), red alder (*Alnus rubra*), and broadleaf maple (*Acer macrophyllum*), have a limited distribution in this Region. Arbutus (*Arbutus menziesii*) and Garry oak (*Quercus garryana*) occur in Canada only on the southeast coast of Vancouver Island and the adjacent islands and mainland. These are species whose centers of population lie to the southward in the United States.

Columbia Forest Region   A large part of the Kootenay River valley, the upper valleys of the Thompson and Fraser rivers, and the Quesnel Lake area of British Columbia contain a coniferous forest closely resembling that of the Coast Region. Western red cedar (*Thuja plicata*) and western hemlock (*Tsuga heterophylla*) are the characteristic species in this interior "wet belt". Associated trees are the blue Douglas fir (*Pseudotsuga taxifolia* var. *glauca*), which is of general distribution, and, in the southern parts, western white pine (*Pinus monticola*), western larch (*Larix occidentalis*), grand fir (*Abies grandis*) and western yew (*Taxus brevifolia*). Engelmann spruce (*Picea engelmanni*) from the Subalpine Region is important in the upper Fraser valley, and is found to some extent at the upper levels of the forest in the remainder of the Region. At lower elevations in the west and in parts of the Kootenay valley the forest grades into the Montane Region and, in a few places, into prairie grasslands.

Deciduous Forest Region   A small portion of the deciduous forest, widespread in the eastern United States, occurs in southwestern Ontario between Lakes Huron, Erie, and Ontario. Here, with the broadleaved trees common to the Great Lakes-St. Lawrence Region, such as sugar maple (*Acer saccharum*), beech (*Fagus grandifolia*), white elm (*Ulmus americana*), basswood (*Tilia americana*), red ash (*Fraxinus pennsylvanica*), white oak (*Quercus alba*), and butternut (*Juglans cinerea*), are scattered a number of other broadleaved species which have their northern limits in this locality.

Among these are the tulip-tree (*Liriodendron tulipifera*), cucumber-tree (*Magnolia acuminata*), papaw (*Asimina triloba*), red mulberry (*Morus rubra*), Kentucky coffee-tree (*Gymnocladus dioica*), redbud (*Cercis canadensis*), black gum (*Nyssa sylvatica*), blue ash (*Fraxinus quadrangulata*), sassafras (*Sassafras albidum*), mockernut and pignut hickories (*Carya tomentosa, C. glabra*), and scarlet, black and pin oaks (*Quercus coccinea, Q. velutina, Q. palustris*). In addition, black walnut (*Juglans nigra*), sycamore (*Platanus occidentalis*), and swamp white oak (*Quercus bicolor*) are largely confined to this Region. Conifers are few, and there is only a scattered distribution of white pine (*Pinus strobus*), tamarack (*Larix laricina*), red juniper (*Juniperus virginiana*), and hemlock (*Tsuga canadensis*).

Great Lakes - St. Lawrence Forest Region   Along the Great Lakes and the St. Lawrence River valley lies a forest of a very mixed nature, characterized by the white and red pines (*Pinus strobus, P. resinosa*), eastern hemlock (*Tsuga canadensis*) and yellow birch (*Betula lutea*). With these are associated certain dominant broadleaved species common to the Deciduous Forest Region, such as sugar maple (*Acer saccharum*), red maple (*A. rubrum*), red oak (*Quercus rubra*), basswood (*Tilia americana*), and white elm (*Ulmus americana*). Other species with wide range are the eastern white cedar (*Thuja occidentalis*) and largetooth aspen (*Populus grandidentata*), and to a lesser extent, beech (*Fagus grandifolia*), white oak (*Quercus alba*), butternut (*Juglans cinerea*), and white ash (*Fraxinus americana*). Boreal species, such as the white and the black spruces (*Picea glauca, P. mariana*), balsam fir (*Abies balsamea*), jack pine (*Pinus banksiana*), poplars (*Populus tremuloides, P. balsamifera*), and white birch (*Betula papyrifera*), are intermixed, and in certain central portions as well as in the east, red spruce (*Picea rubens*) becomes abundant.

Acadian Forest Region   Over the greater part of the Maritime Provinces, exclusive of Newfoundland, there is a forest closely related to the Great Lakes-St. Lawrence Region and, to a lesser extent, to the Boreal Region. Red spruce (*Picea rubens*) is a characteristic though not exclusive species, and associated with it are balsam fir (*Abies balsamea*), yellow birch (*Betula lutea*) and sugar maple (*Acer saccharum*), with some red pine (*Pinus resinosa*), white pine (*P. strobus*) and hemlock (*Tsuga canadensis*). Beech (*Fagus grandifolia*) was formerly a more important forest constituent than at present, for the beech bark disease has drastically reduced its abundance in Nova Scotia, Prince Edward Island and southern New Brunswick. Other species of

wide distribution are the black and the white spruces (*Picea mariana, P. glauca*), red oak (*Quercus rubra*), white elm (*Ulmus americana*), black ash, (*Flaxinus nigra*), red maple (*Acer rubrum*), white birch (*Betula payrifera*), wire birch (*B. populifolia*), and the poplars (*Populus* spp.). Eastern white cedar (*Thuja occidentalis*) though present in New Brunswick is extremely rare elsewhere, and jack pine (*Pinus banksiana*) is apparently absent from the upper Saint John valley and the western half of Nova Scotia.

Vegetation at the Regional and Local Scale

While Halliday's original classification was designed to be continental in scope, by breaking down each forest region into sections, the scheme was able to represent some of the many variations in the forest pattern recognizable at the regional scale. Nonetheless, not all investigators have been in total agreement with Halliday's divisions. One region that has received considerable attention is that of Labrador-Ungava. Following along the lines of Halliday's scheme, the Finnish ecologist Hustich (1949a) developed a classification he felt was more appropriate on account of the attention it gave to the peculiar geologic and physiographic character of this area. On philosophical grounds, his classification appears to combine the physiognomic and floristic viewpoints from which he distinguishes eighteen phytogeographic sections. A year later, Hare (1950) also alluded to the variable nature of the northern boreal forests of Labrador-Ungava. He considered the forests to constitute a single formation but, on account of the comparative paucity of species represented and the glaciated nature of the terrain, he did not feel justified in dividing it into plant associations in an ecological sense. At the same time, he wished to dispel any misconceptions that the boreal forest comprised a broad unbroken cover of forest and muskeg. As a result, Hare constructed a zonal division of the forest on the basis of aerial photographs, over-flights, and the occasional ground traverse. He recognized three zonal divisions within the formation: the Forest-Tundra Ecotone, the Open Boreal Woodland, and the Main Boreal Forest. In an attempt to relate zonal boundaries to climatic parameters, he converted available climatic data into thermal efficiency and moisture indices (Thornthwaite 1948). He discerned a reasonably strong correspondence between thermal efficiency and the three basic zonal divisions but found no obvious correlation with respect to moisture provinces, a result supporting earlier contentions that temperature plays a more important role than moisture in governing the distribution of northern forests.

One of the most pressing problems facing scientists wishing to rationalize the distribution of natural phenomena such as vegetation and soils is the delimitation of meaningful boundaries separating identifiable regions. Utilizing the aerial photographic coverage of the Labrador peninsula and the physiognomic classification of vegetation established by Hustich (1949b), Hare and Taylor (1956) set out to determine the position of certain forest boundaries in a semi-statistical manner. By visually appraising the vegetation patterns from the aerial photographs, they recognized four dominant physiognomic classes or cover types: closed crown forests, lichen, woodlands, bog and muskeg, and bare rock. In this way they hoped to avoid the pit-falls associated with methods involving the concepts of plant succession and vegetation climax and to come to grips with the problem of objectively defining vegetation boundaries. Other investigators have since taken advantage of the increased aerial photographic coverage of northern regions for the purpose of mapping vegetation. The major physiognomic zonation of vegetation in northern Manitoba was determined from aerial photographs in a semi-quantitative manner by Ritchie (1960) and, while he conceded that the exact position of each boundary might require future revision, he was confident that he had successfully established the general configuration of each zone. He regarded this as an important prerequisite for later more intensive studies of vegetation structure and dynamics.

The papers discussed so far have dealt primarily with the forested areas of Canada. By contrast, Watts (1960) concentrates on the distribution of grasslands across the Prairie provinces. As with Halliday's scheme, Watts recognizes grasslands and forests as constituting the two primary formations in the southern Great Plains of Canada. His basic category is the vegetation "type" which supposedly can be equated with Weaver and Clements' association. He recognizes and describes six vegetation types within the grassland formation: the true prairie, the mixed-grass prairie, the short-grass prairie, the submontane mixed prairie, the aspen grove, and the aspen-oak grove. He also identifies three forest types, the mixed-wood forest, the Manitoba lowlands forest, and the subalpine forest, all of which correspond with Halliday's sections. Watts excludes the forest region altogether on the grounds that it is really too broad a category for such a limited area.

An example of the more traditional ecological approach to the analysis of vegetation patterns can be found in the work of McLean and Holland (1958). The authors set out to describe the vegetation zones along the Upper Columbia River Valley, British Columbia, between Canal Flats in the south and Bluewater Creek to the north. In the drier areas around Lake Windermere, they distinguished Dark Brown soils of the chernozemic variety. Occasionally these alternated with Brown Wooded soils which displayed a light-coloured A2 horizon developed as a result of weak podzolization. The trend away from soils associated with grasslands continued northward down the axis of the valley in response to increased precipitation. The effect of podzolization also became more pronounced giving rise to Brown Podzolic, Gray Wooded, and finally Podzol soils in the most northerly part of the study area. The authors also found the vegetation patterns to reflect the measured increase in precipitation both towards the north and with increasing elevation on the valley flanks. In the south they recognized a Douglas fir zone. One of the associations within this zone (the fescue/snowberry association) was identified as a climatic climax associated with Dark Brown soils. On wetter sites to the north and south, the above association gave way to a cover with a greater preponderance of Douglas fir. The Douglas fir/pinegrass association, found on Gray Wooded soils, was also assigned the status of climatic climax. Further north in the realm of the Podzols, and in areas where the precipitation exceeded 25 in., the Douglas fir zone merged with the cedar/hemlock zone. It was also characterized by a number of associations reflecting local variations in topography, soil type, and precipitation. A spruce/fir or subalpine zone was identified in the northern section of the study area and at elevations above 3600 ft. Van Ryswyk et al (1966) undertook a similar but considerably more detailed survey within a small grasslands area near Kamloops, British Columbia. From data collected at five weather stations and twenty-eight field sites, they established a series of vegetation zones and soil groups. They held climatic factors responsible for the rather abrupt change from one vegetation zone to the next.

The ecological literature is rich in studies dealing with plant succession. One method of studying the stages in the evolution of a stable plant cover is to find a freshly exposed natural land surface and carefully observe the chronological order in which different species or groups of species occupy the new surface. Opportunities for such an experience are comparatively rare on account of the scarcity of suitable surfaces and the lengthy period usually taken by plants in the colonization process. Nevertheless, Tisdale et al. (1966) were able to examine the early stages of plant succession across a series of recently constructed recessional moraines near Mount Robson, British Columbia, by comparing their observations with those made fifty years earlier by Cooper (1916).

The alternative procedure, which is the one normally followed in plant succession studies, involves the documentation of differences in the vegetation between adjacent stands of varying age. This was the approach employed by Kellman (1969) during his investigation of plant succession in a recently logged area near Haney, British Columbia. Kellman's paper is noteworthy for focussing attention on the issue of secondary succession* rather than adopting the more common practice of studying primary succession. His primary aim was to identify patterns of what he termed "interspecific associations" during the recolonization process back to the original forest cover. He chose four stands ranging in age from three to one hundred years. The oldest stand was designated the control site on the assumption that it represented the forest cover prior to logging and thus should contain all the "primary" elements. From the data collected at sites within each stand, he was able to recognize an abundant secondary flora which declined as the succession progressed. Further analysis of these data revealed that patterns of interspecific associations were not repeated in any way; nor was there any noticeable correspondence between the plant cover in each stand and the measured environmental parameters. Since the organization of plants appeared unique to each stand, Kellman felt it was not possible to predict the species likely to appear in association at any stage during the succession. Many investigations of primary succession have also led to the same conclusion.

The final paper reprinted in this section (Dale and Hoffman 1969) is basically an attempt to implement Crowley's recommendation that biogeographers should examine vegetation and soils within the context of the ecosystem. The authors contend that not only has there been a tendency for investigators to ignore disturbed sites as potential study areas but, in Ontario in particular, research into vegetation geography has been neglected. In view of the foregoing, Dale and Hoffman set out to identify and delimit ecosystems in a small area of southern Ontario by examining in detail the two major components of soil type and vegetative cover. From the vegetation pattern, the authors were able to distinguish six ecosystems within an area of approximately one square mile. They also discovered that, in all but two cases, each ecosystem was characterized by a distinctive

---

* Secondary succession is defined as ". . . the vegetation developed on sites subsequent to some event (usually catastrophic) which destroyed or severely disturbed the plant cover" (Kellman 1970). Primary succession refers to the various stages in the colonization of a fresh land surface where no plants have grown before.

organic soil type. Dale and Hoffman have shown that the method is sufficiently flexible to deal with disturbed or "semi-exploited" ecosystems. Thus the authors are concerned with the same environmental factors as McLean and Holland (1958), but their approach is quite different. This is partly a reflection of differences in scale. At the same time, however, by employing the ecosystem approach, Dale and Hoffman have been able to pursue their objectives without resorting to the controversial "climax" vegetation concept.

BOGLAND ECOSYSTEMS: SOME BIOGEOGRAPHICAL UNITS[a] *

H. M. Dale and D. W. Hoffman

In the review article on the status of biogeography in Canada, Crowley has made the recommendation that "biogeographers should direct more attention towards the study of the geography of the whole ecosystem rather than only to its components, such as vegetation, soil and so on."[1] This admirable suggestion should be implemented. Valuable studies made in British Columbia and in Quebec[2] will assist locally towards this goal, but unfortunately, for large areas in Canada, there are few geographers, soil scientists, biologists, or micrometeorologists attempting to delimit, identify, or describe these natural units.

In Ontario two tendencies in botanical studies have contributed to this situation. Firstly, there has been emphasis on floristic plant geography rather than vegetational geography. This primary interest in distribution of *species* is incidental to the study of vegetation and therefore ecosystems. The abundance and spatial distribution of the flora within sample stands are the essence of vegetational study. The presence of species does give valuable information about a stand of vegetation, but these other data are necessary, especially for an understanding of the roles of vegetation within the ecosystem.

The second tendency of emphasizing the individuality of each stand is more serious, since the underlying philosophy suggests that each stand is unique and therefore that only ordination, not classification, is possible. The denial of recurrent patterns in vegetation precludes mapping and meaningful geographical studies of vegetation and ecosystems.

In the study of vegetation and soil in Canada there is also a bias towards the study of "climax conditions," "undisturbed sites," and "virgin stands." Recently some attention is being turned to semi-exploited ecosystems and areas of secondary succession, areas formerly considered slightly improper for serious consideration by soil scientists, phytogeographers, and zoogeographers.

This study demonstrates that vegetation may be used to delimit ecosystems which may be immature. Components of the six ecosystems are described in relatively simple terms. It also emphasizes that local studies of ecosystems are necessary. Hills[3] has clearly stated the necessity for this in the forest ecosystems of Ontario. As a result, he has divided the province into thirteen site regions, each of which has nine site types based on several factors, but using air temperature and soil moisture as indices. Field studies over several years have demonstrated that the plant associations of Ontario do not correspond to those described for Quebec or Wisconsin.[4] Older classifications of organic soils have also been found inadequate for correlation with vegetation.

a Technical assistance for the senior author was financed by a NRC grant in aid of research on vegetation and hydrology.

* Reproduced from *The Canadian Geographer* XIII, 2, 1969, pp. 141-159 with the permission of the authors and publisher.

Fig. 1. Aerial photograph of the study area.

Soils in which the filter is purely organic and in which the ground water circulates in the profile, show little horizon development compared to mineral soils. The horizons are the result of changes in vegetation rather than those which normally develop in mineral soil due to the movement of materials in percolating water. Classification of these organic soils has been neglected. Until recently all organic soils were grouped in two classes—muck or peat—depending on their state of decomposition. Now many more classes have been defined in various categories.[5] However the identification and mapping of organic soils are difficult because of high water-tables and the inaccessibility of ecosystems which have developed on many of these areas. This study was undertaken to determine whether some ecosystems in Southern Ontario might be classified using vegetation and soils. It has been ably argued that the same set of factors is operative in the development of soils and vegetation.[6] One of these factors, "time," suggests successional changes in soil profile and in vegetation.

Radforth[7] has reported on the concurrence of certain characteristics of organic terrain with plant material in the Boreal Forest and Tundra regions. In Maine, a similar study[8] has reported the use of aerial photographs in determining the depth of peat deposits in bogs and swamps.

Study Area

Study sites were located on a one-half square mile area containing a high water-table, southwest of the city of Guelph on lots 6, 7, 8, and 9, Concessions III and IV, Puslinch Township. The land surface lying between the 1,025- and 1,050-foot contours consisted of several distinct but contiguous ecosystems which could be distinguished by characteristic vegetation. Part was a raised bog in which the living surface had grown above the influence of drainage water, whereas in other areas considerable erosion of the organic material had occurred. Figure 2 is an outline map of the area showing location of sites.

Vegetational Analysis

Aerial photographs from flights in the summer of 1955 and the early spring of 1960 confirmed ground surveys as to the extent of each kind of

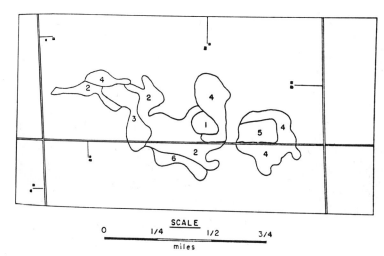

Fig. 2. Outline map of study area showing the location and extent of the six kinds of vegetation and soil types.

vegetation and the landscape was tentatively mapped using the vegetation units on the aerial photographs. Figure 1 is an aerial photograph of the area and Figure 2 is this map with some modifications. The six kinds of vegetation on this nearly level surface could be easily distinguished from stereo pairs. They are designated as follows: (1) dwarf bog forest of black spruce (*Picea mariana*), tamarack (*Larix laricina*), leatherleaf (*Chamaedaphne calyculata*), sphagnum (*Sphagnum* spp.); (2) raised bog of leatherleaf, labrador tea (*Ledum groenlandicum*), sphagnum, hair moss (*Polytrichum juniperinum*), white pine (*Pinus strobus*), tamarack; (3) eroding bog of cotton grass (*Eriophorum* sp.), huckleberry (*Gaylussacia braccata*), giant blueberry (*Vaccinium corymbosum*), mountain holly (*Nemopanthus mucronatus*), white pine, tamarack; (4) spring swamp of American elm (*Ulmus americana*), balsam fir (*Abies balsamea*), yellow birch (*Betula lutea*), foam flower (*Tiarella cordifolia*), moss (*Mnium* spp.); (5) summer swamp of dwarfed black ash (*Fraxinum nigra*), tamarack, white cedar (*Thuja occidentalis*), holly (*Ilex verticillata*); (6) paludified bog of willow (*Salix* spp.), meadow sweet (*Spiraea alba*), mountain holly, chokeberry (*Aronia melanocarpa*), sedge (*Carex* spp.). (All species were identified following Gleason and Cronquist, 1963, Manual of Vascular Plants.)

The vegetation was sampled quantitatively. All species were listed and subjectively rated for abundance (1, 2, or 3). Four transects were walked through the central part of each stand. Quantitative ratings were also made by identifying and counting the number of each species of tree in aerial photographs in five circles (¼ acre) in each of the six areas. The abundance of each low shrub and herb species was estimated by determining the proportion of the area covered by each species. A fifteen-metre (49-foot) line was run and the interceptions by the crown of each species was measured in turn. The sum of the squares of each length of interception was used as an estimate of relative cover of each species. Four lines were run in each type of vegetation.

Soil Investigation

A detailed soil survey of the area was conducted and the soils were mapped and classified on the basis of their morphology. The distribution of the soils is shown in Figure 2. A representative soil from each area was sampled and was analysed on a horizon basis. The bulk density was determined from dry weight at 105°C (221°F); pH was determined at field moisture conditions using 1N HCl at a soil-reagent ratio of 1:1 by volume with an equilibrium time of ten minutes. Degree of decomposition was determined by treating samples in sodium pyrophosphate solution. The

TABLE I
Vegetation Analyses*

	Area					
	1	2	3	4	5	6
Upper layer, no./ha.	1100	120	330	220	280	140
Upper layer height (m.)	3	10	10	12+	5	5
Upper layer cover (%)	60	10	30	90	80	15
Subordinate layer, height (m.)	.3	.6	1.6	.8	1.8	2.0
Subordinate layer, cover (%)	46	74	28	5	17	67
Ground layer,cover (%)						
ferns	0	0	+	3	1	3
grass and herbs	0	+	31	14	17	40
moss and lichen	33	61	27	7	6	10
TOTAL VEGETATION	33	61	58	24	24	53
Litter	61	38	21	67	12	21
Bare soil	0	0	13	7	50	18

* These include number of individuals in upper layer; the average maximum height in metres in the upper and subordinate layers; and cover expressed as a percentage of total area for each of the three layers.

colour of the extract was compared with Munsell colour charts to identify one of the three stages of decomposition.[9]

The depth of organic material and the type of basal layer were determined as well as the position of the water table in October. It was assumed that this would be the lowest level for the year.

Results

Vegetation In each area the vegetation consisted of plants arranged in an upper, a subordinate, and a ground layer. The subordinate or shrub layer dominated areas 1, 2, and 6 whose height was 0.3, 0.6 and 2.0 metres respectively. The dwarf trees of 12 and 5 metres in height dominated areas 4 and 5; whereas area 3 had a heterogeneous profile pattern without any clearly dominant layer.

Table II gives the composition of the tree component or upper layer by species. Those which are predominantly shrub height, ½ to 3 metres, are designated. It will be noticed that the bog areas 1, 2, and 3 have a depauperate tree flora whereas tree species are well represented in areas 4, 5, and 6. This table clearly shows differences in tree species composition

in the six areas. Table III demonstrates that the differences found in the tree composition in Table II are echoed when all layers of vegetation are considered.

For each area 1 to 6, species are found peculiar to that area (faithful species). In Europe faithful species are frequently used to characterize vegetation. Abundance ratings have not been included although these further emphasize differences in the six areas.

These analyses of the six types of vegetation do not exactly coincide with plant association descriptions given by Dansereau[10] for Quebec although his descriptions are not necessarily of climax vegetation. Area 1 corresponds to his *Piceetum ericaceum* association with less balsam fir present than described. Area 2 is a matrix of *Chamaedaphnetum calyculatae* association with islands of *Ledetum groenlandici* and *Laricetum laricinae*. However, in the latter one layer is missing—the very sparse or barren layer 3 of Alder (*Alnus rugosa* var. *americana*). Area 3 corresponds to the association *Nemopanthetum mucronatae* (again without alder in layer 3). Dansereau (*loc. cit.*) describes the situation in Area 3 in which the *Ledetum* and *Chamaedaphnetum* is an earlier

TABLE II
Floral Composition of 6 Areas as Percentage of Tree Species
(From Aerial Photographs)

	Area					
	1	2	3	4	5	6
Tamarack (*Larex laricina*)	30*	38	31		25*	30*
Spruce, white (*Picea glauca*)					+	+
Spruce, black (*Picea mariana*)	65*	4			+	
Pine, white (*Pinus strobus*)	1	58	54			1
Birch, paper (*Betula papyrifera*)	+	1*	10*	2	1*	15
Birch, yellow (*Betula lutea*)				5		1
Elm, American (*Ulmus americana*)				25	3	5
Ash, black (*Fraxinus nigra*)				2	60*	1
Maple, red (*Acer rubrum*)			+	5	7*	2
Fir, balsam (*Abies balsamea*)	3*			15*	2*	
Poplar, aspen (*Populus tremuloides*)			+	10		+
Cedar, white (*Thuja occidentalis*)				8	10*	5*
Hemlock (*Tsuga canadensis*)				3		
Willow (*Salix* spp.)			5*		2*	40*

* Shrub height ($\frac{1}{2}$ to 3m).

\+ less than one per cent.

successional stage. Birch and red maple appear as scattered individuals as the shade at lower levels thins out the lower shrub layer. However, no description of a plant association from Quebec (*loc. cit.*) corresponds with area 4, with area 5, or with area 6.

Soil

Five major kinds of soil profiles were found. Four of the soils are organic—three mesisols and one humisol—and one is the peaty phase of a humic gleysol. The soil in area 1 is classified as a Unic Mesisol; in area 2 as a Stratic Mesisol; in area 3 as a Terric Stratic Mesisol; and in area 6 as an Orthic Humic Gleysol. Areas 4 and 5 had similar soil and are classified as Terric Humisols.[12]

Area 1   0–28 centimetres   Pale brown (designated by Munsell chart as 10YR6/4) very pale brown (10YR7/4) when rubbed; relatively unaltered plant remains; fibre content 95 per cent chiefly from Sphagnum mosses; sodium pyrophosphate extraction white paper (10YR8/2); pH 4.6.

28–150 centimetres   Dark brown (10YR3/3); content of fibres 50 to 60 per cent; fibres disintegrate on rubbing; rubbing colour is brown (10YR4/3); woody layer at 75 centimetres; sodium pyrophosphate extract on white filter paper (10YR8/2); pH 5.6.
Classification. Order: Histosol; Great Group: Mesisol; Sub Group: Unic Mesisol.

Area 2   0–15 centimetres   Pale brown (10YR 6/3) no change in colour when rubbed; fibre content 95 per cent, chiefly from Sphagnum mosses; leaves and wood fragments from Labrador tea and Leather leaf; fibres resist disintegration when rubbed; pH 4.6.
15–40 centimetres   Dark brown (10YR4/3); partially decomposed organic materials with 50 per cent fibres; fibres easily disintegrate on rubbing; pH 4.9.
40–50 centimetres.   Dark brown (10YR3/3); decomposed organic materials with 15 to 20 per cent fibres; sodium pyrophosphate extract on white filter paper (10YR4/4); pH 5.4.
50–65 centimetres   Brown (10YR5/3); partially decomposed organic materials with 50

TABLE III
Composition Indicating Number* of Species Present in and Faithful to Each Area

	Area					
	1	2	3	4	5	6
Trees	4	4	5	9	8	12
Shrubs	6	9 (2)	11 (1)	9 (4)	6	5
Herbs	1	0	3	32 (10)	17 (5)	6 (2)
Grass and sedges	1	0	4	1	4 (1)	2 (1)
Ferns	0	0	1	10 (8)	3	2 (1)
Mosses	5 (1)	4	3	2 (1)	4 (2)	2
TOTAL	17 (1)	17 (2)	24 (1)	63 (23)	42 (8)	27 (4)

* In parentheses.

per cent fibres; woody; finely divided organic material between fibres; fibres disintegrate when rubbed; pH 5.6.

65 centimetres + Dark brown (10YR4/3); finely divided organic material between the fibres; fibre content about 40 to 50 per cent; most fibres disintegrate when rubbed; pH 6.2. Classification. Order: Histosol; Great Group: Mesisol; Sub Group: Stratic Mesisol.

Area 3 0–5 centimetres Black (10YR2/1); well-decomposed organic material with 10 to 15 per cent fibre; colour when rubbed wet, black (10YR2/1); pH 5.8.

5–55 centimetres Dark brown (10YR3/2); organic material with 50 per cent of fibre; rubbed colour wet is brown (10YR4/3); coarsely fragmental and compact; pH 6.1.

55–90 centimetres Brown (10YR4/3); well-decomposed organic material with 10 per cent fibre; glossy appearance in root channels; rubbed colour wet is brown (10YR4/3); pH 6.8.

90–115 centimetres Dark brown (7.5YR3/2); organic material with 50 per cent of fibre; fibre disintegrates when wet and becomes finely divided and structureless; darkens on rubbing (7.5YR2/2); pH 7.2.

115 centimetres + Gray (10YR5/1) sandy loam; no mottles; massive; friable; calcareous; pH 7.6. Classification. Order: Histosol; Great Group: Mesisol; Sub Group: Terric Stratic Mesisol.

Areas 4 and 5 0–15 centimetres Black (10YR

2/1) well-decomposed organic material; fibre content less than 10 per cent; granular; sodium pyrophosphate extract on white filter paper (10YR4/4); pH 6.7.

15–104 centimetres Dark brown (10YR3/2); no colour change when rubbed wet; fibre content 20 to 25 per cent; finely divided organic material between the fibres; some woody material; pH 7.0.

104 centimetres Gray (10YR6/1) sand; single grain; loose; calcareous; pH 7.8. Classification. Order: Organic; Great Group: Humisol; Sub Group: Terric Humisol.

Area 6 0–28 centimetres Very dark brown (10YR2/2) well-decomposed organic material; fibre content less than 10 per cent; no colour change when rubbed wet; pH 7.0.

28–40 centimetres Clay loam; dark gray (10YR4/1); massive plastic; pH 7.2.

40–65 centimetres Clay loam; gray (10YR 5/1); mottled with brownish yellow (10YR6/6) mottles; massive; plastic; pH 7.3.

65 centimetres Gravelly sand; gray (10YR 6/1); mottled with brownish yellow (10YR6/8) mottles; single-grain; loose; calcareous; pH 7.8. Classification. Order: Gleysolic; Great Group: Humic Gleysol; Sub Group: Orthic Humic Gleysol.

The data in Table IV indicate ash content of fibric layers in these soils to be less than 10 per cent, mesic to be from 10 to about 20 per cent, and humic to be greater than 25 per cent. In fibric layers bulk density is less than 0.2

TABLE IV
Analyses of the Three Soil Layer Types Showing Ranges in Four Properties

Layer type	Ash content (Wt. %)	Holding capacity (Wt. %)	Bulk density (g./cc.)	pH
Fibric	3.6–8.4	1520–1680	0.09–0.15	4.5–5.3
Mesic	12.2–16.6	610–650	0.21–0.24	5.5–7.6
Humic	30.4–53.3	193–384	0.33–0.40	6.5–7.5

grams per cubic centimetre, in mesic between 0.2 and about 0.3, and in humic greater than 0.3. The figures on water-holding capacity show a definite pattern with the greatest capacity in fibric layers, the least in humic, and intermediate in the mesic layers.

Probing indicated that the main part of the bog occupies a rather steep-sided pothole. The present surface is comparatively level with a slope of less than 0.5 per cent to the southeast. A close relationship exists between the kind of profile, the thickness of the deposit, and the height of the water table.

Discussion

In four cases there is a direct relationship between vegetative cover and soil profile, as far as the higher categories of organic soil classification are concerned.

In the areas 4 and 5, however, vegetative cover differs although the areas occur on the same soil. A comparison of the two areas (4 and 5) shows a soil surface height on area 5 to be at least 60 centimetres higher than area 4. It has been demonstrated that organic soils decrease in elevation when the water table is below the soil surface and oxidation is thought to be the chief factor involved.[13] In area 5 the soil surface is under 7 to 18 centimetres of water in July, whereas at this time the surface in area 4 has no standing water. The road between area 4 and 5 is a corduroy or log-based road and has acted as an effective dam for 80-odd years since its construction. This has been sufficient time to produce differences in vegetation between areas 4 and 5 but insufficient to be reflected in the soil profile. This difference is thought to be of value in delimiting two separate ecosystems. The small number of species common to these two areas is good indication that neither vegetation is an earlier successional stage to the other.

Vegetation can be classified by amount of cover and flora composition of tree layers alone or by utilization of data from all layers of vegetation. It is possible in this region to delineate the sub-groups of organic soils by kind of vegetation. An increased variety of plants and a larger number of trees occur with shallower organic soil and increased decomposition. There is a relationship between tree species and soil conditions.

Delimiting six ecosystems in this sample study area confirms the observation that Southern Ontario consists of a mosaic of ecosystems (each composed of distinct vegetation, microclimate, soil, microorganism, and animal populations). These are recurrent and can be mapped. However, Ontario has proved so diverse that, for forested ecosystems, delimited for productivity purposes, it has been divided into thirteen site regions each with nine physiographic site types (14). Similar division will be necessary for other types of ecosystem. Descriptions of bogland vegetation from Quebec and Wisconsin do not apply. A complete study of an ecosystem should embrace Tshulok's eight points of view—composition, structure, physiology, genetics, ecology, history, chorology, and taxonomy (15). This study involves the first two, and the authors hope that other aspects will be dealt with, including the chorology and biogeography of these and many other ecosystems.

REFERENCES

1. Crowley, J. M., "Biogeography," *Can. Geogr.,* 11, no. 4, (1967), 312-26.
2. Krajina, V. J., *Bioclimatic Zones of British Columbia* (Univ. of B.C. Bot. Ser. no. 1, 1959). Grandtner, M. M., *La végétation forestière du Québec méridional* (Quebec, 1966).
3. Hills, G. A., *Soil-Forest Relationships in the Site Regions of Ontario,* First N. Amer. Forest Soils Conf., East Lansing, Bull. Agr. Exp. Sta., Mich. State Univ. (1958).
4. Dansereau, P., *Phytogeographia Laurentiana II: The Principal Plant Associations of the Saint Lawrence Valley,* Inst. Bot. Univ. Montreal, Contrib. 75 (1959). Curtis, J. T., *The Vegetation of Wisconsin* (Madison, 1959).
5. Dawson, J. E., "Organic Soils," *Advan. Agr.* 8 (1956), 337-401. Farnham, R. S., and H. R. Finney, "Classification and Properties of Organic Soils," *Advan. Agr.* 18 (1965), 115-60. Nat. Soil Surv. Committee, Rept. on Classification of Organic Soils, Can. Dept. Agr., Ottawa (mimeo). Rept. on 6th meeting of NSSC (1966), 68-80.

6. Major, J., "A Functional, Factorial Approach to Plant Ecology," *Ecol.* 32 (1951), 392-412.
7. Radforth, N. W., "The Use of Plant Material in Recognition of Northern Organic Terrain Characteristics," *Trans.* Roy. Soc. Can. 47 (1953).
8. Kennedy, R. A., *The Relationship of Maximum Peat Depth to Some Environmental Factors in Bogs and Swamps in Maine,* Maine Agr. Exper. Sta. Bull. 620 (1963), 53 pp.
9. Atkinson, H. J., G. R. Giles, A. J. MacLean, and J. R. Wright, *Chemical Methods of Soil Analysis.* Can. Dept. Agric., Contrib. 169 (Ottawa, 1958). Nat. Soil. Surv. Comm., Rept. on Classification of Organic Soils (1966). *Munsell Color Charts* (Baltimore, 1946).
10. Dansereau.
11. Curtis, J. T., *The Vegetation of Wisconsin* (Madison, 1959).
12. Nat. Soil Surv. Comm., Rept. on Classification of Organic Soils. Can. Dept. of Agr., (Ottawa, 1966, mimeo). Rept. on the 6th meeting of the Nat. Soil. Surv. Comm. Can. (1966), 68-80.
13. Dawson.

SOILS AND VEGETATION BIBLIOGRAPHY

Atkinson, H. J., G. R. Giles, A. J. MacLean and J. R. Wright: 1958, *Chemical methods of soil analysis,* Cont. Chemy. Div., Sci. Serv., Can. Dep. Agric. 169, 90 pp.

Baldwin, M., C. E. Kellogg, and J. Thorpe: 1938, "Soil classification," in: *Soils and Men,* U.S. Dep. Agric. Year Book, Washington, D.C., pp. 979-1001.

Boughner, C. C.: 1964, *The distribution of growing degree days in Canada,* Canadian Meteorological Memoir 17, 40 pp.

Bostock, H. S.: 1970, "Physiographic subdivisions of Canada," in: *Geology and Economic Minerals of Canada,* R. J. W. Douglas (ed.) Geol. Surv. Can. Economic Geology Report 1, pp. 11-30.

Bowser, W.E.: 1960, "The soils of the prairies," *Agric. Inst. Rev.* 15, pp. 24-26.

Bowser, W. E., A. S. Kjearsgaard, T. W. Peters and R. E. Wells: 1962, "Soil survey of Edmonton sheet, (83-H)," *Alberta Soil Survey Report* 21, 66 pp.

Brasher, B. R., D. P. Franzmeier, V. Valassis, and S. E. Davidson: 1966, "Use of Saran resin to coat natural soil clods for bulk density and water retention measurements," *Soil Science* 101, p. 108.

Brewer, R.: 1964, *Fabric and mineral analysis of soils,* Wiley, New York, 470 pp.

Brydon, J. E., H. Kodama and G. J. Ross: 1968, "Mineralogy and weathering of the clays in Ortha Podzols and other Podzolic soils in Canada," *Trans. of the 9th International Congress of Soil Science 1968* 3, pp. 41-51.

Bryson, R. A., W. N. Irving and J. A. Larsen: 1965, "Radiocarbon and soil evidence of former forest in the southern Canadian Arctic," *Science* 147, pp. 46-48.

Canada Department of Agriculture: 1970a, "Outline of the system of soil classification for Canada," in: *The System of Soil Classification for Canada,* Queen's Printer, Ottawa, pp. 19-23.

Canada Department of Agriculture: 1970b, *Proceedings of the eighth meeting of the Canada Soil Survey Committee,* Canada Soil Survey Committee Report, Ottawa, 195 pp.

Chapman, L. J.: "Physiography, climate and natural vegetation of Canada," *Agric. Inst. Rev.* 15, pp. 15-19.

Chapman, L. J. and D. E. Putnam: 1937, "The soils of south-central Ontario," *Scient. Agric.* 18, pp. 161-197.

Clark, J.S.: 1965, "The extraction of exchangeable cations from soils," *Can. Jl. Soil Sci.* 45, pp. 311-313.

Clark, J. S. and A. J. Green: 1964, "Some characteristics of gray soils of low base saturation from northeastern British Columbia," *Can. Jl. Soil Sci.* 44, pp. 319-328.

Clements, F. E.: 1916, "Plant succession, an analysis of the development of vegetation," *Publ. Carnegie Instn.* 242, 512 pp.

Cooper, W. S.: 1916, "Plant succession in the Mount Robson region, British Columbia," *Plant World* 19, pp. 211-238.

Corte, A. E.: 1966, "Particle sorting by repeated freezing and thawing," *Biul. Peryglac.* 15, pp. 175-240.

Cowles, H. C.: 1899, "The ecological relations of the vegetation on the sand dunes of Lake Michigan," *Botanical Gazette (Bot. Gaz.)* 27, pp. 95-117, 167-202, 281-308, 361-391.

Cowles, H. C.: 1901, "The physiographic ecology of Chicago and vicinity," *Bot. Gaz.* 31, pp. 73-108, 145-181.

Crowley, J. M.: 1967, "Biogeography," *Can. Geog.* 11, pp. 312-326.

Cruickshank, J. G.: 1971, "Soils and terrain units around Resolute, Cornwallis Island," *Arctic* 24, pp. 195-209.

Curtis, J. T.: 1959, *The Vegetation of Wisconsin,* University of Wisconsin Press, Madison, 657 pp.

Dale, H. M. and D. W. Hoffman: 1969, "Bogland ecosystems: some biogeographical units," *Can. Geog.* 13, pp. 141-149.

Dansereau, P.: 1959, *Phytogeographia Laurentiana. II. The principal plant associations of the Saint Lawrence Valley,* Contribution of the Institute of Botany, University of Montreal 75, 147 pp.

Dawson, J. E.: 1956, "Organic soils," *Advances in Agronomy (Adv. Agr.)* 8, pp. 377-402.

Dudas, M. J. and S. Pawluk: 1969, "Chernozem soils of the Alberta parklands," *Geoderma* 3, pp. 19-36.

Dumanski, J. and R. J. St. Arnaud: 1966, "A micropedological study of eluvial soil horizons," *Can. Jl. Soil Sci.* 46, pp. 287-292.

Ehrlich, W. A., H. M. Rice and J. H. Ellis: 1955, "Influence of the composition of parent materials on soil formation in Manitoba," *Can. J. Agric. Sci.* 35, pp. 407-421.

Evans, F. C.: 1956, "Ecosystem as the basic unit in ecology," *Science* 123, pp. 1127-1128.

Farnham, R. S. and H. R. Finney: 1965, "Classification and properties of organic soils," *Adv. Agr.* 7, pp. 115-160.

Feustel, I. C., A. Dutilly and M. S. Anderson: 1939, "Properties of soils from North American Arctic regions," *Soil Science* 48, pp. 183-198.

Geological Survey of Canada, Department of Energy, Mines and Resources: 1967, *Permafrost in Canada,* Map. 1246A.

Gillespie, J. E. and D. E. Elrick: 1968, "Micromorphological characteristics of an Oneida soil profile," *Can. Jl. Soil Sci.* 48, pp. 133-142.

Gillespie, J. E. and R. Protz: 1969, "Evidence for the residual character of two soils, one in granite and the other in limestone in Peterborough County, Ontario," *Can. Jl. Earth Sci.* 6, pp. 1217-1225.

Grandtner, M. M.: 1966, *La végétation forestière du Québéc méridional,* Presses de l'Université Laval, Quebec, 216 pp.

Guertin, R. K. and G. A. Bourbeau: 1971, "Dry grinding of soil thin sections," *Can. Jl. Soil Sci.* 51, pp. 243-248.

Halliday, W. E. D.: 1937, *A forest classification for Canada,* Bulletin, Forestry Branch, Canada 89, 50 pp.

Hamelin, L. E. and F. A. Cook: 1967, *Le Periglaciaire par l'Image — Ilustrated Glossary of Periglacial Phenomena,* Presses de l'Universite Laval, Quebec, 237 p.

Hare, F. K.: 1950, "Climate and zonal division of the boreal forest formation in eastern Canada," *Geog. Rev.* 40, pp. 615-635.

Hare, F. K.: 1963, "Climate," in: *A Report on the Physical Environment of the Quoich River Area Northwest Territories, Canada.* Mem. Rand Corp. 1997, pp. 79-140.

Hare, F. K. and R. G. Taylor: 1956, "The position of certain forest boundaries in southern Labrador-Ungava," *Geog. Bull.* 8, pp. 51-73.

Hills, G. A.: 1958, "Soil-forest relationships in site regions of Ontario," *Bulletin of the Michigan Agricultural Experimental Station,* pp. 190-212.

Hoffman, D. W., B. C. Matthews and R. E. Wicklund: 1963, *Soil survey of Wellington County, Ontario,* Soil Survey Report, Ontario 35, 64 pp.

Hubbard, W. A.: 1950, "The climate, soils, and soil-plant relationships of an area in southwestern Saskatchewan," *Scient. Agric.* 30, pp. 327-342.

Hustich, I.: 1949a, "Phytogeographical regions of Labrador," *Arctic* 2, pp. 36-42.

Hustich, I.: 1949b, "On the forest geography of the Labrador Peninsula: a preliminary synthesis," *Acta Geogr.* 10, pp. 36-42.

James, P. A.: 1970, "The soils of the Rankin Inlet area, Keewatin, N.W.T., Canada," *Arctic and Alpine Res.* 2, pp. 293-302.

James, P. A.: 1972, "The periglacial geomorphology of the Rankin Inlet area, Keewatin, N.W.T., Canada," *Biul. Peryglac.* 21, pp. 127-151.

Joel, A. H.: 1926, "Changing viewpoints and methods in soil classification," *Scient. Agric.* 6, pp. 225-232.

Joel, A. H.: 1928, "Diversity of soil type in the Prairie Provinces and cause of the same," *Scient. Agric.* 8, pp. 651-664.

Kellman, M.C.: 1969, "Plant species inter-relationships in a secondary succession in coastal British Columbia," *Syesis* 2, pp. 201-212.

Kellman, M. C.: 1970, "On the nature of secondary plant succession," *Proc. Can. Assoc. Geogr.,* Dept. of Geography, Univ. of Manitoba, Winnipeg, pp. 193-198.

Kennedy, R. A.: 1963, "The relationship of maximum peat depth to some environmental factors in bogs and swamps in Maine," *Bull. Me. Agric. Exp. Stn.* 620, 53 p.

Kodama, H. and J. E. Brydon: 1965, "Interstratified montmorillonite-mica clays from subsoils of the Prairie Provinces, Western Canada," in: *Clays and Clay Minerals,* Ingerson, E. (Ed.), Proc. 13th Natl. Conf. Clays and Clay Minerals 1964, Pergamon, Oxford, England, pp. 151-173.

Krajina, V. J.: 1959, "Bioclimatic zones of British Columbia," *Univ. Br. Columb. Bot. Ser.* 1, 47 pp.

Larsen, J. A.: 1964, "An outline of material for a postglacial bioclimatic history of Keewatin, N.W.T., Canada," *Tech. Rep. Dep. Met. Univ. Wis.* 15, 79 pp.

Leahey, A.: 1946, "The agricultural soil resources of Canada," *Agric. Inst. Rev.* 1, pp. 285-289.

Leahey, A.: 1947, "Characteristics of soils adjacent to the Mackenzie River in the N.W.T. of Canada," *Proc. Soil Sci. Soc. Am.* 12, pp. 458-461.

Leahey, A.: 1954, "Soil and agricultural problems in subarctic and arctic Canada," *Arctic* 7, pp. 249-254.

Leahey, A.: 1961, "The soils in Canada from a pedological viewpoint," in: *Soils in Canada: Geological, Pedological and Engineering Studies,* Leggett, R. K. (Ed.), Spec. Publs R. Soc. Can. 3 pp. 147-157.

Leahey, A.: 1963, "The Canadian system of soil classification and the Seventh Approximation," *Proc. Soil Sci. Soc. Am.* 27, pp. 224-225.

Leahey, A.: 1969, "Soils of Canada," in: *Vegetation, Soils and Wildlife,* J. G. Nelson and M. J. Chambers (eds.), Methuen, Toronto, pp. 113-122.

Macoun, J.: 1894, "The forests of Canada and their distribution," *Trans. R. Soc. Can.* 12, Sect. IV, pp. 3-20.

Major, J.: 1951, "A functional factorial approach to plant ecology," *Ecology* 32, pp. 392-412.

McKeague, J. A., G. A. Bourbeau and D. B. Cann: 1967, "Properties and genesis of a bisequa soil from Cape Breton Island," *Can. Jl. Soil Sci.* 47, pp. 101-110.

McKeague, J. A. and J. E. Brydon: 1970, "Mineralogical properties of ten reddish brown soils from the Atlantic provinces in relation to parent materials and pedogenesis," *Can. Jl. Soil Sci.* 50, pp. 47-55.

McKeague, J. A. and D. B. Cann: 1969, "Chemical and physical properties of some soils derived from reddish brown materials in the Atlantic Provinces," *Can. Jl. Soil Sci.* 49, pp. 65-78.

McKeague, J. A. and J. H. Day: 1966, "Dithionite and oxalate extractable Fe and Al as aids in differentiating various classes of soils," *Can. Jl. Soil Sci.* 46, pp. 13-22.

McKeague, J. A., J. I. MacDougall, K. K. Langmaid and G. A. Bourbeau: 1969, "Macro and micro-morphology of ten reddish brown soils from the Atlantic provinces," *Can. Jl. Soil Sci.* 49, pp. 53-63.

McKeague, J. A., N. M. Miles, T. W. Peters and D. W. Hoffman: 1972, "A Comparison of Luvisolic soils from three regions in Canada," *Geoderma* 7, pp. 49-69.

McLean, A. and W. D. Holland: 1958, "Vegetation zonation and their relationship to the soils and climate of the upper Columbia Valley," *Can. Jl. Pl. Sci.* 38, pp. 328-345.

McMillan, N. J.: 1960, "Soils of the Queen Elizabeth Islands (Canadian Arctic)," *Jl. Soil Sci.* 11, pp. 131-139.

Mehra, O. P. and M. L. Jackson, 1958, "Iron oxide removal from soils and clays by a dithionite-citrate system buffered with sodium bicarbonate," in: *Clays and clay minerals,* Ingerson, E. (Ed.), Proc. 7th Nat. Conf. Clays and Clay Minerals, 1958, pp. 317-327.

Morwick, F. E.: 1933, "Soils of southern Ontario," *Scient. Agric. 13,* pp. 449-454.

Moss, H. C.: 1954, "Soil classification in Saskatchewan," *Jl. Soil Sci.* 5, pp. 192-204.

Moss, H. C. and R. J. St. Arnaud: 1955, "Grey Wooded (Podzolic) soils of Saskatchewan, Canada," *Jl. Soil Sci.* 6, pp. 292-311.

Nat. Soil Surv. Committee of Canada: 1945, *Proceedings of the first conference of the National Soil Survey Committee of Canada,* Can. Dep. Agric., Ottawa, pp. 22-24.

Nat. Soil Surv. Committee of Canada: 1966, "Rept. on Classification of Organic Soils," *Rept. on 6th meeting of N.S.S.C.,* Can. Dept. Agric., Ottawa, (mimeo).

Nicholson, N. L. and Z. W. Sametz: 1970, "Regions of Canada and the regional concept," in: *Regional and Resource Planning in Canada,* R. R. Kreuger et al. (eds.), Holt, Rinehart and Winston, Toronto, pp. 6-23.

Pawluk, S.: 1961, "Mineralogical composition of some Grey Wooded soils developed from glacial till," *Can. Jl. Soil Sci.* 41, pp. 228-240.

Pawluk, S. and J. D. Lindsay: 1964, "Characteristics and genesis of Brunisolic soils of northern Alberta," *Can. Jl. Soil Sci.* 44, pp. 292-303.

Putnam, D. F.: 1951, "Pedogeography of Canada," *Geog. Bull.* 1, pp. 57-85.

Putnam, D. F., B. Brouillette, D. P. Kerr and J. L. Robinson: 1963, *Canadian Regions: A Geography of Canada,* J. M. Dent and Sons, Don Mills, Ontario, 601 pp.

Radforth, N. W.: 1953, "The use of plant material in recognition of northern organic terrain characteristics," *Trans. R. Soc. Can.* 47, sect. V. pp. 53-71.

Raup, H. M.: 1942, "Trends in the development of geographic botany," *Ann. Assoc. Am. Geogr.* 32, pp. 319-354.

Rice, H. M., S. A. Forman and L. M. Patry: 1959, "A study of some profiles from major soil zones in Saskatchewan and Alberta," *Can. Jl. Soil Sci.* 29, pp. 165-177.

Richards, N. R.: 1961, "The soils of Southern Ontario" in: *Soils in Canada: Geological, Pedological and Engineering Studies,* Leggett, R. F. (Ed.), Spec. Publs R. Soc. Can. 3, pp. 174-182.

Ritchie, J. C.: 1960, "The vegetation of northern Manitoba, V. Establishing the major zonation," *Arctic* 13, pp. 211-229.

Rowe, J. S.: 1959, "Forest regions of Canada," *Bull. For. Brch. Can.* 123, 71 pp.

Ruhnke, G. N.: 1926, "The soil survey in southern Ontario," *Scient. Agric.* 7, pp. 117-124.

Rutherford, G. K. and D. K. Sullivan: 1970, "Properties and geomorphic relationships of soils developed on a quartzite ridge near Kingston, Ontario," *Can. Jl. Soil Sci.* 50, pp. 419-429.

Ryswyk, A. L. van, A. L. McLean, and L. S. Marchand: 1966, "The climate, native vegetation and soils of some grasslands of different elevations in British Columbia," *Can. Jl. Pl. Sci.* 46, pp. 35-50.

Schnitzer, M.: 1969, "Reactions between fulvic acid, a soil humic compound, and inorganic soil constituents," *Proc. Soil Sci. Soc. Am.* 33, pp. 75-80.

Shutt, F. T.: 1910, "Some characteristics of the western prairie soils of Canada," *Jl. Agri. Sci.* 3, pp. 335-357.

Skinner, S. I. M., R. L. Halstead, and J. E. Brydon: 1959, "Quantitative manometric determination of calcite and dolomite in soils and limestones," *Can. Jl. Soil Sci.* 39, pp. 197-204.

Soil Survey Staff: 1967, *Supplement to Soil Classification System* (7th Approximation), U.S. Dept. Agric., Washington, D.C., 191 pp.

St. Arnaud, R. J. and M. M. Mortland: 1963, "Characteristics of the clay fractions in a Chernozemic to Podzolic sequence of soil profiles in Saskatchewan," *Can. Jl. Soil Sci.* 43, pp. 336-349.

St. Arnaud, R. J. and E. P. Whiteside: 1964, "Morphology and genesis of a Chernozemic to Podzolic sequence of soil profiles in Saskatchewan," *Can. Jl. Soil Sci.* 44, pp. 88-99.

Stobbe, P. C.: 1952, "The morphology and genesis of the Gray-Brown Podzolic and related soils of Eastern Canada," *Proc. Soil Sci. Soc. Am.* 16, pp. 81-84.

Stobbe, P. C.: 1960, "The Great Soil Groups of Canada: their characteristics and distribution," *Agric. Inst. Rev.* 15, pp. 20-22.

Stobbe, P. C.: 1962, "Classification of Canadian Soils," in: *Transactions of joint meeting of Commission.* IV and V, Lower Hutt, New Zealand, Int. Soc. Soil Sci., pp. 318-324.

Tansley, A. G.: 1935, "The use and abuse of vegetational concepts and terms," *Ecology* 16, pp. 284-307.

Tedrow, J. C. F.: 1962, "Morphological evidence of frost action in arctic soils," *Biul. Peryglac.* 11, pp. 343-352.

Tedrow, J. C. F.: 1966, "Polar desert soils," *Proc. Soil Sci. Soc. Am.* 30, pp. 381-387.

Tedrow, J. C. F.: 1968, "Pedogenic gradients of the polar regions," *Jl. Soil Sci.* 19, pp. 197-206.

Tedrow, J. C. F. and J. E. Cantlon: 1958, "Concepts of soil formation and classification in arctic regions," *Arctic* 11, pp. 166-179.

Tedrow, J. C. F. and L. A. Douglas: 1964, "Soil investigations on Banks Island," *Soil Sci.* 98, pp. 53-65.

Tedrow, J. C. F., J. V. Drew, D. E. Hill and L. A. Douglas: 1958, "Major genetic soils of the Arctic Slope of Alaska," *Jl. Soil Sci.* 9, pp. 33-45.

Thornthwaite, C. W.: 1948, "An approach towards a rational classification of climate," *Geog. Rev.* 38, pp. 55-94.

Thorp, J.: 1949, "Effects of certain animals that live in soils," *Scient. Mon.* 68, pp. 180-191.

Thorp, J. and G. D. Smith: 1949, "Higher categories of soil classification: Order, Suborder, and Great Soil Groups," *Soil Sci.* 67, pp. 117-126.

Tisdale, E. W., M. A. Fosberg, and C. E. Poulton: 1966, "Vegetation and soil development on a recently glaciated area near Mt. Robson, British Columbia," *Ecology* 47, pp. 517-523.

Warkentin, J. (Ed.): 1966, *Canada: a geographical interpretation,* Methuen, Toronto, 608 pp.

Watts, F. B.: 1960, "The natural vegetation of the southern Great Plains of Canada," *Geog. Bull.* 14, pp. 25-43.

Weaver, J. E. and F. E. Clements: 1938, *Plant Ecology,* McGraw-Hill, New York, (2nd ed.), 601 pp.

Williams, B. H. and W. E. Bowser: 1952, "Gray Wooded soils in parts of Alberta and Montana," *Proc. Soil Sci. Soc. Am.* 16, pp. 130-133.

Willis, A. L.: 1949, "Clay minerals present in eight representative Ontario soils," unpubl. Ph.D. thesis, University of Wisconsin, 110 pp.

World Soil Resource Report 33, by Dudal, R.: 1968, "Definitions of soil units for the soil map of the world," FAO/UNESCO Project. FAO, Rome, Italy, 72 pp.

Section 5
Overview

Chapter 14

Physical Geography:
A Canadian Context*

J. Brian Bird
McGill University

Description, interpretation, and explanation of the many phenomena that compose
the surface environment of Canada have concerned "the geographer scientist" since
the earliest Europeans arrived in the country. The description of the physical
geography of Canada, using the term in its broadest sense, together with the discovery and evaluation of economic wealth, provided the main motives for exploration for over three hundred years. The early, unsophisticated accounts of the
physical environments slowly changed in methodology and purpose, until by the
second half of the twentieth century Canadian physical geography is involved in
theoretical research of international significance and applied studies for national
objectives of environmental enrichment, planning, and an appreciation of the
aesthetic and material values of the land.

Physical geography resembled the other branches of geography in that un-
complicated description provided the basis of the early accounts. From eastern
Canada they are brief, incomplete and not infrequently faulty, both in fact and
inference from the observations. The purpose of an account is usually self-evident.
Dièreville in describing the two dominant terrains in the southwestern Annapolis
Valley near Port Royal at the end of the seventeenth century, writes:

> those called Uplands, which must be cleared in the Forest, are not good, and
> the seed does not come up well in them; . . . to grow Wheat, the Marshes
> which are inundated by the Sea at high Tide, must be drained; these are called
> Lowlands. . . .[1]

He had identified the two major physiographic units but failed to develop significant
conclusions from their presence.

By the middle of the eighteenth century descriptions by government surveyors including C. Morris in Nova Scotia and Captain Holland on the Island of
St. John (Prince Edward Island) are longer and more precise. Often the accounts
concentrate on an immediate resource of the area, commonly the quality and extent
of the forest. Shortly afterwards, and especially in Upper Canada from 1820
onwards, reports of travellers from the Old World and from settlers became more
numerous; it is such material which is included in Section 1 of this volume.

These accounts contain a broad range of geographical information; several
studies of the historical geography of parts of eastern Canada have been published,
but no major reconstruction of the physical geography and particularly its perception in the eyes of the early settlers, using as evidence the early descriptions, has
been written. The situation in the Prairies was very different. European settlement
was at least a century later than in the East, and a serious attempt was made to

determine the quality and potential of the land before transportation lines were fixed and extensive agricultural settlement commenced. In consequence, there is a considerable articulate body of literature describing the environments of the Prairies in the second half of the nineteenth century and interpreting them in terms of potential settlement.

The accounts of Palliser (1857-60), surveyors' reports of the end of the nineteenth and early twentieth centuries, and contemporary evidence in the form of fragments of plant associations that survive to the present, have enabled Watts to reconstruct the natural vegetation of the Prairies prior to European settlement.[2] The processes by which Prairie landscape was changed and the surface environments modified have been examined in the last few years[3,4] using an historical physical geography approach.

The surveys of Palliser and shortly afterwards Hind, Dawson, and Macoun provided the first comprehensive physical geography of the Prairies. Their works were essentially descriptive physiography based on subjective, non-quantified views of the resources and potential use of the land. More than a century later, the direct successors of these studies are the land utilization maps which are being prepared in conjunction with maps of land capability and land value. The programs involve mapping on a large scale by ground traverses, so that individual fields may be identified; remote sensing from aircraft is used increasingly for this work and within a decade the data may be collected from orbiting satellites. The current land capability maps include much detailed information on the physical environment; some physiographies, notably on southern Ontario,[5] incorporate considerable land utilization data. However, in general, physiographic and land utilization studies have evolved in different directions.

The slow progress of geomorphology in Canada for nearly a century after Palliser's expedition has been described succinctly by Parry.[6] Similarly, developments in other branches of physical geography were extremely limited. The few major studies that were produced, notably the pioneer work of Goldthwait in Nova Scotia,[7] were confined to the southern, settled margin of the country; to the north stretched the scientifically unknown boreal forest, subarctic woodlands, and arctic tundras. Exploration and description of the southerly sectors of this vast unsettled region had been primarily to estimate forest resources. Farther north, beyond the exploitable forest, description of more than half of Canada in the first part of the twentieth century was restricted to exploratory studies initiated, or at least supported, by the federal government, in which the officers of the Geological Survey of Canada were conspicuous. More complete and sophisticated descriptions of the northern four-fifths of the country awaited the development of new techniques, especially aerial photography, and a national demand for these studies as part of a fuller appreciation of the resources of the country.

In the past quarter century, techniques for extensive surveys in physical geography have been refined and there has been an increasing demand for the results; earlier chapters of this book reveal some of the achievements. At the same time objectives and methods of theoretical and applied physical geography were evolving and the significance of these changes forms the basis for the remainder of this chapter.

By the middle of the twentieth century, physical geography in the United States and West European countries was placed in an increasingly ambiguous position as a scholarly discipline.[8] Many physical geographers found they were attracted by topics and methodology, wholly into the earth sciences, whilst at the same time they sought to make a viable contribution to geography as a whole. Interdisciplinary stresses and crosscurrents were certainly as great, if not greater, in Canada than in other countries partly because in Canada the world trends of more than half a century were telescoped into a single decade and partly because slow and inadequate development of other earth sciences presented attractive openings for "pure" physical geographers. Although there were methodological conflicts which produced slowed expansion of Canadian geography in the federal institutions and in many universities, compensation was forthcoming in the strong contribution physical geography was seen to offer to a full understanding of the natural environment and its exploitation and management by man. Physical geography research has tended to isolate increasingly detailed phenomena which are examined in great detail, whilst neglecting, or being unable to handle the complex conditions that prevail on the land surface; in general the more detailed the study the less direct relevance it has had to real world conditions. The need to take a broad view of the total surface environment has in the last two decades become evident to many physical geographers in Canada. This has been especially true in the North where, to give but two examples, research developments in permafrost and glacier studies have indicated the strength of multidisciplinary approaches.

The presence of permafrost, by definition the temperature condition in which the ground remains below 0°C continuously for a number of years, adds an important element to analyses of surface phenomena in northern Canada. Permafrost is by no means a minor component either in extent or effect. It underlies one-fifth of the world's land area, and continuous permafrost is found beneath a quarter of Canada's territory and in a further quarter, discontinuous and sporadic permafrost is preserved (Fig. 1). Its distribution, areally and in depth, and the response of the terrain to permanently frozen subsoil has become a major field of investigation. Before 1950 Canadian studies of permafrost were few and the knowledge that was available was derived from Russian experience. In the last two decades extensive theoretical and applied research, sponsored, and in part undertaken, by the National Research Council has formed a basis for future developments.[9]

A fundamental effect of permafrost is to modify the distribution of soil moisture; in silts where the effect is greatest, the ice content may exceed 50 percent of the total weight of the soil, and sheets and veins of ground ice are common. If permafrost melts either naturally or through man-induced disturbances, changes of state in geomorphology, hydrology, pedology and botany follow (Fig. 2). In the recognition and understanding of the complex linkages between the several phenomena, physical geography, and especially biogeography, has much to offer.

Considerable engineering, economic, and ecological penalties result from disturbances of northern terrains and associated thermal changes in the ground. The projected construction of natural gas and oil pipelines in the Mackenzie Valley has been preceded by terrain mapping and interpretation of the hazards along the routes, including permafrost and ground-ice conditions, the probable severity of environmental disturbance, and the engineering performance of newly thawed soils.

Developments that may be anticipated in permafrost studies include quantification of the relationships between several environmental parameters associated with permafrost. The boundaries of permafrost have been related to climate; initially these were expressed in terms of annual air temperature when it was recognized that the −2° to −4°C isotherms coincide roughly with the southern limit of discontinuous permafrost. Later derivations have used annual freezing

PERMAFROST DEGRADATION

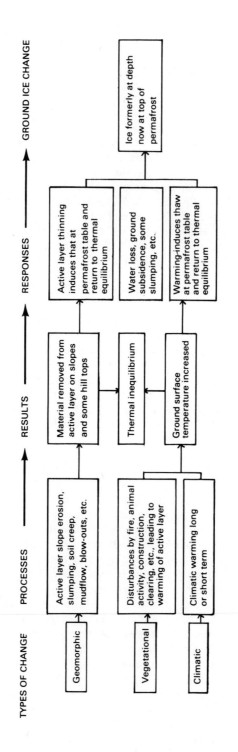

TYPES OF CHANGE   PROCESSES   →   RESULTS   →   RESPONSES   →   GROUND ICE CHANGE

Geomorphic → Active layer slope erosion, slumping, soil creep, mudflow, blow-outs, etc. → Material removed from active layer on slopes and some hill tops → Active layer thinning induces that at permafrost table and return to thermal equilibrium

Vegetational → Disturbances by fire, animal activity, construction, clearing, etc., leading to warming of active layer → Thermal inequilibrium → Water loss, ground subsidence, some slumping, etc.

Climatic → Climatic warming long or short term → Ground surface temperature increased → Warming-induces thaw at permafrost table and return to thermal equilibrium

Ice formerly at depth now at top of permafrost

after Mackay 10

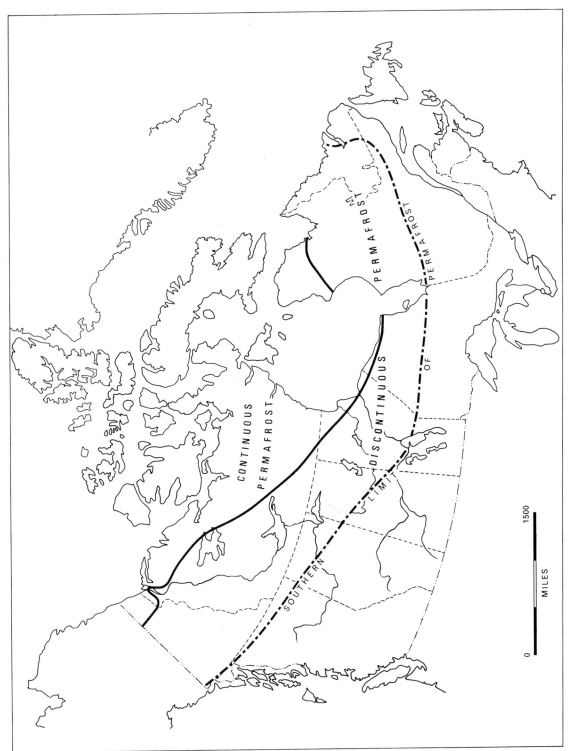

CONTINUOUS PERMAFROST

DISCONTINUOUS

PERMAFROST

PERMAFROST

SOUTHERN LIMIT OF PERMAFROST

0   MILES   1500

Figure 1

and thawing indices. Although they explain statistically the general regional distribution of permafrost, they do not attack the problem of process and this awaits analysis of heat and moisture exchange at the surface. The long-term objective is to predict the presence of permafrost and its depth. The latter is particularly variable near the southern margin where thicknesses may vary from less than 1 m to over 200 m. Reliable and rapid means of mapping the margins of the permafrost are needed using aerial photography. The recognition of permafrost indicators are important; however the validity of identifying permafrost conditions from the presence of wooded peat mounds (palsa) and peat plateaus, string bogs, and collapse scars has yet to be tested.

Snow has a special place in the physical geography of Canada. Causes of snowfall and its geographical distribution examined as part of meteorology and climatology form an important study. In a normal year no part of the country fails to receive some snow and all except small areas of coastal British Columbia have snow lying for some period of time. The influence of snow cover on surface phenomena is also significant: in geomorphology, snow is a source of moisture for mass movement, especially solifluction; it is the source of seasonally retarded run-off in hydrology; and it provides moisture and acts as a protective agency for the plant cover. Normally, when examining the responses of the surface environment and the biota to snow, it is impossible to deal effectively with less than the total system. As with permafrost programs, snow cover studies which ultimately have direct human relevance can most effectively be developed through the total, integrating approach of physical geography.

When winter snowfall (and associated frozen forms of precipitation, rime, hail, etc.) is of sufficient depth that it fails to melt in the succeeding summer, conditions are favourable for the development of glaciers; this situation has been widely studied as have the physical changes of a snow crystal through recrystallization into granular snow (firn) and then by compaction and expulsion of air into glacier ice. The duration of time for the changes to occur depends on snowfall and temperature and varies from a few years (under relatively warm conditions) to several centuries in the high Arctic.

Investigations in the last two decades of glaciers in northern Canada including the Penny and Barnes ice caps on Baffin Island, the Devon Island Ice Cap, glaciers on Axel Heiberg Island, and the ice caps of northern Ellesmere and Meighen islands have provided an insight into the processes acting with "cold" glaciers where the temperature beneath the surface of the glaciers is between -15 and $-20°C$. It is clear that the intense cold of these glaciers is responsible for considerable refreezing of meltwater, and that penetration of water into the interior of the glacier, a common phenomenon in temperate glaciers, is unusual. Detailed data not only indicate modified regimes of the glaciers but also differences in the processes by which the glaciers modify the rock surface over which they are passing.

A prominent landform associated with glaciers in high arctic environments is the push moraine. Although described from contemporary and Pleistocene glacial environments in many parts of the world, their extent in Canada is only now being appreciated with their recognition in the Queen Elizabeth Islands and analyses of their development* and their widespread presence in the southern

* An account of the distribution of push moraines and the processes acting on them is found in Kalin, M., "The Active Push Moraine of the Thompson Glacier, Axel Heiberg Island, Canadian Arctic Archipelago," *Axel Heiberg Research Reports, no. 4,* McGill University, Montreal 1971.

Prairies where they contribute to the height and morphology of many prominent landforms including the Manitoba Escarpment, the Missouri Coteau and several isolated uplands including the Neutral Hills.[11]

Glacier studies have not been restricted to analyzing geomorphic process during contemporary and Pleistocene glaciations; an important objective of glaciology has been the evaluation of glaciers as part of the total water resources of the country; this is especially true of western Canada where the sources of many of the British Columbia and Prairie rivers are ice fields. Glaciers, through evidence of changes in volume and areal extent, provide a link in physical geography between contemporary events and climates of the past. Cool periods in the eighteenth and nineteenth centuries in western Canada and other mid-latitude areas were associated with glacier maxima (Table 1).

Table 1
Fluctuations of Glaciers in the Rocky Mountains

Glacier	Beginning date of recession	Approximate rate of retreat mid-twentieth century
Angel (Mt. Edith Cavell)	1733	—
Athabasca	1731	35 m/yr
Dome	1875	10 m/yr
Peyto	—	34 m/yr
Saskatchewan	1854 (1st max) 1893 (2nd max)	35 m/yr
Yoho	1855	34 m/yr

The relationship between glacier advance and climatic deterioration is not simple and the delay between a change of climate and the response at the glacier snout may be several decades; some advances that are independent of climatic deterioration may occur through glacier "surging."

It is rarely possible to reconstruct past climate from glacial evidence without close checking between several of the biological and atmospheric sciences and extended analysis of contemporary conditions. The importance of a multidisciplinary approach has been widely appreciated in Canadian climatic-glaciological studies and the methods of physical geography have made notable contributions in this field. Successful research of this type requires the cooperation of many scientists, a semi-permanent base adjacent to a glacier to be occupied for several years, and strong scientific leadership to maintain the primary objective, for, as F. Müller and his colleagues have shown on Axel Heiberg Island, there is no way to evaluate glaciers as climatic indicators until the detailed processes are understood.

Glaciological research is also making significant contributions to the major unsolved problems facing Quaternary scientists in Canada, notably the growth and disappearance of the Pleistocene ice sheets. The solution of the problem lies in the reconstruction of the physical geography, or if you will, the natural systems of certain phases of the Pleistocene from the evidence and constraints supplied from the stratigraphy, palaeobotany, geomorphology, climatology, and geophysics. There is in fact an infinite series of historical physical geographers which has the objective of synthesizing and explaining past environments.

Although glaciers are a spectacular form of ice, other forms of terrestrial ice have also attracted scientific attention in northern environments. These include ground ice which occurs as massive sheets and wedges of subsurface ice. It is particularly prominent in northwestern Canada where it produces polygonal ground, pingos and conspicuous ice cliffs around the shores of the Beaufort Sea, and many other features;[12] ice acts as an important disruptive element in rock weathering in cold climates whilst geomorphic erosion of ice on rivers, lakes, and the sea shore, as well as its role in transporting sediments, has received wide attention. The modifying action of non-glacial ice on the surface of the land cannot be analysed by any single scientific discipline and has been most successfully investigated with the skills and techniques of several subjects under the general heading of periglacial studies.[13]

Periglacial studies together with permafrost and botanical investigations form the basis for examination of northern terrains, and the success of this work provides an outstanding example of the strength of a broadly based approach to environmental problems.

By 1945 three major environmental regions remained in Canada which had been only slightly influenced by western industrial technology and were still scientifically little-known. The three regions, the Arctic tundra, the boreal forest (north of commercial exploitation), and the northern part of the Western Cordillera occupied three-quarters of the country and were the last (with Alaska) surviving great wilderness areas of the North American continent. Within the next three decades, all three zones were threatened ecologically by the penetration of economic activity in the search for metallic ores, oil, and natural gas. By the early 1970s, the complexity and vulnerability of the ecosystems were recognized and processes within the systems were being investigated. Some of the resulting glaciological and periglacial studies within the Arctic have already been described; farther south the boreal forest was the first of the three zones to be analyzed geographically when Hare,[14] using aerial photography for mapping, recognized three main zonal vegetation subdivisions in eastern Canada, divisions that were later extended west of Hudson Bay.[15] This was the first major application of extensive, remote sensing techniques to Canadian geography, and it marked an important development in biogeography.

Table 2[16]
Zonal Division of the North American Boreal Forest

Zonal Division	Characteristic Structure
Arctic Treeline	
Forest-Tundra	Scattered or isolated trees, often deformed or prostrate, set in a landscape with extensive tundra patches.
Open Woodland	Lichen-floored, open forest or woodland often with discontinuous shrub layer.
Closed Forest	Closed-crown forest with moist, deeply-shaded floor, often with abundant moisture.

Just as the Arctic studies underscored the linkages between the physical components of the terrestrial surface system in the far North, so the boreal forest research has emphasized increasingly the interaction of the main parameters related to climate, one set dealing with the storage of heat and moisture in the several media, and the other set with the rates of energy exchange between them.

The dual nature of physical geography also became evident in the studies of the boreal region. They lead notably to a more sophisticated knowledge of process in climatology through measurement and calculation of the radiation inputs; they also provided an insight into the mechanisms of past climates, and tentative reconstructions were attempted for the Wisconsin glacial and postglacial climates.[17,18]

The skills of the physical geographer have been applied to the interpretation of many of the surface phenomena in the boreal forest biome; in some cases scientists of other disciplines, especially ecologists, have developed independently similar approaches. This is evident in studies of the peat lands which form a significant part of the total area of the northern forest terrains. The catena of conditions where drainage is restricted varies from shallow ponds with totally aquatic vegetation to wet, hummocky sedge meadows commonly with standing water during the summer and slightly higher, extensive areas of hummocky, spongy sphagnums with some ground birch and stunted spruce. The latter may be interspersed with peat ridges, mounds, and plateaus 1-2 m above the lowest parts which support rather better developed forest. Peat thickness is extremely variable; it often averages about 1 m over large areas, although 3 m and more is not unusual.

Here again there is clearly a close relationship between the postglacial drainage on the mineral soils and the growth of the peat which in turn modifies the drainage in the flattest areas and at the same time changes the ground thermal regime. Throughout much of the northern boreal forest zone, permafrost occurs in isolated patches; it is rarely found beneath the high, well-drained, forested areas, nor is it common beneath the ponds and the wettest ground. However, it is not unusual beneath the thickest peat plateau sectors. Similar linked observations have been made on delta development in the boreal zone where an intimate relationship between mineral soils, ground temperatures (particularly annual and perennial soil frost), ground water, vegetation and surface processes has been clearly demonstrated in recent studies.[19,20]

Progress in northern physical geography has been considerable although there continue to be large areas, topically and geographically, that remain only partially known. A case in point, of considerable economic importance, is the water content of precipitation in Arctic and Subarctic regions. In the USSR, regional values of annual precipitation have recently been adjusted by an increase of 30 to 50 percent[21] and similar changes have been proposed for northern North America.[22]

These examples of research in northern Canada illustrate that there is one set of studies within physical geography which is essentially time-oriented, and which, if the time-scale is long in terms of man's experience, becomes closely linked with the even more general discipline of Quaternary studies — the reconstruction of geography and physiography of the past — and the processes of change through time. The other face of physical geography is essentially contemporary and is concerned with process today both of isolated parameters and within the terrestrial system as a whole. Both sets, and particularly the latter, can lead to model building, prediction, and consequently the application of physical geography to the modification and exploitation by man of the natural environment.

The dichotomy of time-oriented and process studies is evident notably in general physical geography but also in its individual components. It is clear in geomorphology where there is a fundamental assumption that process and land-form are interrelated. The search for clarification of this relationship and the basic nature of geomorphic processes has led in the past two decades to intensive and continuing studies of slope evolution, studies that are undoubtedly complicated by the polygenetic character of many slopes in both time and process modes. This is especially true in Canada where the events of the Wisconsin glaciation either created new materials for slope development (i.e., surficial, glacial sediments) or left conditions of marked disequilibrium when the glaciers withdrew. The necessity for some scientists to concentrate further and further away from the central theme of physical geography in order to understand the details of process and the response of materials on which they are acting is shown by studies at the University of Guelph on the origin and particularly the clay content of glacial materials. [23, 24]

Most proponents of physical geography in the past two decades have recognized the need for a rigorous treatment of the fundamental processes at work in the landscape. This theme has been stated clearly in geomorphology at the symposia on geomorphology held at the University of Guelph in 1969 and again in 1971. The appeal has not gone unanswered, and examination of process, whether it is in the transfer of energy at the surface of the ground as examined in climatology or in the detailed mechanism and rates of change involved in mass wasting, are being energetically tackled by many research groups across the country.

Analyses of slope profiles and geomorphic processes have been extended in recent years with ever-increasing sophistication of the techniques. Many mass wasting processes are extremely slow and are close to the limit of physical field measurement; even in mountainous areas where processes are spectacular and rapid, changes in the regional landscape, as distinct from individual slopes, are to be measured on a geological time scale.

It is evident in the early part of this chapter that the emphasis of physical geographical studies in Canada has been on the North. This is most conspicuous in geomorphology, but it applies also to climatology and to some extent hydrology, if glaciology is seen as a branch of the subject. It may well be asked whether similar clusters of coordinated studies will develop in other major physio-geographical regions of Canada. Studies in which the terrains and the processes operating on them are related to the vegetation cover, both original and in a man-modified form, and with climate. This may already be occurring in the Western Cordillera, where for several years, geomorphologists in particular have been working on a number of linked problems. The recent publication of a collection of papers on the geomorphology of the mountains of western Canada, to which the majority of the field workers active in the field have contributed, shows the advances that have already been made towards the ultimate objective of an explanatory description of the evolution, energy balance, and future development of high mountain areas.[25]

Studies in the western mountains point to high rates of slope retreat and denudation which contrast with the low-energy situation that exists in the plains farther east. However, mountain geomorphology in western Canada, as elsewhere, frequently stresses the non-equilibrum slope conditions that exist at the present time, due to the presence of glacially modified slopes which are potentially unstable in the present environment. Although much of the high energy situation is a consequence of the considerable available relief in mountainous areas, it is evident that the rapid mass wasting measured in many parts of the Canadian West occurs in over-deepened valleys which have been recently occupied by ice.

A reasonably typical rate of bedrock slope retreat in mountains is 5 cm/1000 years. Wide variations of the order of several magnitudes may exist where rocks are particularly vulnerable to weathering processes or where a physical process such as avalanches is concentrated; where strong slope retreat and active downcutting occur, total denudation may be considerably greater.

The validity of extrapolating contemporary process rates to describe the changes of the last million years, let alone for the much longer period during which the mountains were forming, both endogenetically and exogenetically, is questionable. A conservative estimate of denudation in mountains has been obtained recently from radioisotope dating based on ^{230}Th/^{234}U ratios in calcite deposits from ancient caves. It showed that Crowsnest Pass, which is today incised 1200 m into the mountains of southern Alberta, has been deepened by not more than 100 m in the last 200,000 years.[26] Assuming this to be correct and reasonably typical, the broad landform elements of the Rocky Mountains in this area existed more than a million years ago.

A complementary approach to the rates of landscape change is to be obtained through hydrology by measuring annual sediment loads in river basins. In the Canadian stream basins that drain into Lake Ontario it has been found that the quantity of dissolved sediments is nearly twelve times greater than the suspended sediments, the calculated mean rate of denudation being about 5 cm/1000 years. This figure seems unusually high for an area of low relief when it is compared with the rate in the western mountains. However, it is similar to rates in areas of the United States with comparable precipitation, and it almost certainly reflects the dramatic change of vegetation cover in the last two centuries during which southern Ontario has been cleared and settled by Europeans.[27] The vegetation and land use changes induced major alterations in the hydrologic system with enhanced run-off, increase of sediment load, modification of slopes, and minor but nonetheless real changes in local climate — the latter especially since the development of large urban communities. Similar methods for estimating overall denudation rates and some of the implications of European settlement and changes of land use are also becoming available from southern British Columbia.[28]

A valid question in physical geography, and one of considerable significance, is how rapidly do the principal components of the natural environment including the plant cover, soils, slopes, and hydrology respond to change, whether man-induced or linked to variations of energy input and hence to climate. A related question is to determine how important inherited events are in the present characteristics of a given area. Answers to such questions are difficult to give, especially when so little is known quantitatively about climatic change. The importance of glaciers in this respect has already been mentioned. Another source of information for climatic change in southern Canada has been through analyses of pollen preserved in bogs. It is assumed that the pollen represents the species and relative density of tree species and some non-arboreal plants when the pollen rained down on the bog. From this information an estimate of the climate can be made. However problems of dating the particular pollen horizons and interpreting the results are many and usually only general observations are possible.

In the Canadian North, it is possible to reconstruct part of the postglacial climatic change although evidence is limited. The final disappearance of glacier ice from an area at the close of the Wisconsin glaciation was accompanied by a rapid rise in temperature; depending on the locality, this rise of temperature began 6,000 to 9,500 years ago. The environmental conditions were then the most favourable in the North in postglacial time and are correlated with the hypsithermal interval elsewhere in the world. Between 3,500 and 4,000 years BP the climate deteriorated becoming cooler and probably wetter. It was accompanied by glacier advances, the growth of ground ice in the Northwest and the establishment of the Ellesmere ice shelf off northern Ellesmere Island. Several minor warm phases followed this deterioration (one dated at perhaps 1,000 years ago was arctic-wide), but the amelioration of climate did not achieve the same magnitude as the climatic optimum. The climate deteriorated once more in the thirteenth and fourteenth centuries and with oscillations, lower temperatures, and snowier conditions continued through until the middle or end of the nineteenth century; since then the climate has once more improved, although there is some evidence that in the last two decades there has been a renewed downturn. Process studies which will provide linkages between observed climatic change and the response of the associated ecosystems are now urgently required.

Process studies in geomorphology and climatology have also been linked successfully with the perception and response to them by man. This is most evident when the process is rapid, disruptive, and potentially catastrophic. All too frequently examples have been provided by landslides in the St. Lawrence Lowlands. Typically they occur when Champlain Sea clays, which were deposited under marine or estuarine conditions, experience increased sensitivity and decreased shear strength as a result of the leaching of enclosed salts. If one face of the clay is unsupported as along a riverbank, a disturbance, either man-made or natural such as an earth tremor, may induce an earthflow. Flow commences as successive failure of a steep slope which is followed by liquefaction of the clays. Over seven hundred flows have been mapped; a prehistoric earthflow in Alfred Township (Ontario) covered fifteen square miles and another near Kenogami (Quebec), dated at 1400-1500 AD, engulfed eight square miles. Three disastrous flows occurred in 1971 with the destruction of buildings and land at Gatineau, Quebec, and Casselman, Ontario, and the death of thirty-one persons at Saint-Jean-Vianney in the Saguenay Valley. Notwithstanding the dangers of building in susceptible localities, people continue to locate in these areas. The possible adjustments to the hazards of the danger zones are summarized in Table 3.[29]

The magnitude-frequency of physical, damaging events and the human response to environmental hazards in general, bring the physical and human branches of geography into juxtaposition. Recent studies in southern Ontario have illustrated the difficulties of generalization in this field whilst at the same time indicating rational decision criteria for responding to hazards.[30]

Increasing emphasis on process is also evident in hydrology. The value, real and potential, of water as a renewable resource in a country which has an abundance and indeed an overabundance of water in most sectors is only now becoming obvious. Everywhere the demand for water increases: water for irrigation purposes in southwestern Ontario, the Prairies, and interior British Columbia; water for energy production, for urban and industrial areas in all parts of the country; water for recreational purposes; and potentially, in the future, water for export to the United States.

Physical geographers have already made significant contributions to the hydrological sciences especially in studies of fluvial processes, the problems of run-off and the quality of water. The close relationship between vegetation, snowfall, snow cover, spring run-off, and the annual water budget in hydroelectric storage basins, has been demonstrated by the McGill University Subarctic Laboratory group in northern Quebec/Labrador in a series of closely-linked experiments begun in 1970.

Table 3
Alternative Adjustments to Clay Landslide Hazards
in the Ottawa, St. Lawrence and Saguenay Valley Region

Affect the cause	Modify the hazard	Modify the loss potential	Adjust to losses
Stabilize the clay by installing a system which maintains an optimal amount of moisture in the clay within desirable limits.	Engineering controls e.g., construct reinforcing buttresses or retaining walls, slope stabilization	Design buildings to withstand slides or float on mud slides	Bear losses
			Insurance
		Regulate land use	Public relief and rehabilitation
		Prepare warnings and forecasts	
Stabilize by controlling supply of moisture to the clay, e.g., remove or cover any sand or other porous materials on the surface near the clay, control vegetation.		Temporary evacuation	
		Permanent evacuation	

Hydrology may be considered a subsystem of the general system of terrestrial environmental science. As a result, it shares with physical geography many of the same methods and problems; it also shares the difficulties of obtaining advanced training as contributions are required from civil engineering, geomorphology, agrometeorology, agroclimatology, and from other sciences. The identification of water as a resource involves inputs in training from economic systems studies, and the social value of water involves inputs from the whole area of social studies. As with physical geography as a whole there is a broad spectrum of knowledge which has to be integrated before a total picture of water and its value to man can be understood and managed. To these studies the geographer and particularly the physical geographer has brought special skills, and it must be expected that increasingly some physical geographers will specialize in a "total approach" to water development.

Inevitably, in view of the historical development of physical geography, many of the separate chapters of the book have stressed the contributions of specializations within physical geography to a particular problem, rather than the contribution of physical geography as a whole. This underlies once again the fundamental problem in physical geography. Ultimately its field is the terrestrial environment of the real world but it did not take physical geographers long to discover the complexity of the real world; they responded by isolating individual parts and analyzing extremely small units under the simplest possible conditions. While this has enabled individual processes, whether on the earth's surface or in the atmosphere to be examined and explained in physical terms, the detailed analysis that is required has moved the investigator away from the generalist concept of physical geography and particularly from the real world in which it is a matter of constant observation that the many separate components are closely linked. The dichotomy in physical geography that resulted has already been described. At an elementary level physical geography provides a simple, descriptive overview of the physical conditions at the earth's surface and has offered a limited amount of information on request to the human geographer, who, however, rejected geographical determinism long ago and has felt less and less the relevance of a physical basis for his subject.

In Canada the negative response of the human geographer to physical geography has never been as extreme as it has been in the United States — if only, one must assume, because, despite the high degree of urbanization, the physical environment visibly dominates the life of the nation for periods each winter and is the most obvious reason for the slow expansion of settlement into the boreal sector. Whilst these trends were developing in geography, other changes were occurring first in the physical and biological sciences and later in man's perception of his world. Many physical geographers have specialized in various subdisciplines of the subject and have subsequently concentrated their attention in adjacent sciences. The cost to physical geography has been high, and increasingly arguments have been presented for the return of physical geography to a unified examination of the whole landscape in research and teaching. Meanwhile other sciences of the earth's surface were discovering the interdependence of the separate environmental parameters and had recognized that the whole is both greater and more complex than the separate units. With this realization came the concept of the interaction of the biotic, physical, and chemical environments in a single ecosystem. The development of these conceptual insights coincided with a growing public awareness of pollution problems, anxiety about the state of the environment, and the increasing recognition that man has to become involved in environmental management, initially in cities, locally in rural areas, and ultimately for the whole world.

Whilst physical geography has no unique claim to the study of the natural environment it must be clear that it has a great deal to offer in measuring, interpreting, and advising on its use and management. The type of generalist physical geographer most suited to this purpose, in contrast to the extreme specialist, has recently been described.[31]

Many problems of decision-making in environmental management lie at the intersection of socio-economic systems and the natural environment process-response systems of the physical geographer. It will, however, be recognized that commonly it is not the environment nor even certain elements of the environment that directly control man's reaction but rather the manner in which man perceives the environment and chooses to optimize his response, often apparently irrationally in the conditions that prevail.

The geographer, familiar with physical process-response environment systems, is able to extract benefits from the system without injurious secondary effects and degradation of the environment, as long as he understands the critical operation of the system in its natural form and the thresholds that exist in it. Western man is at a stage in which he can intervene in limited physical subsystems, including the environment, immediately adjacent to him without disaster. To a minor, and at present less certain level, he can intervene in larger mesosystems leading to the production of local rainfall from suitable clouds and the possible modification of hurricanes. At the macrosystem level embracing large parts of the earth's surface, the dangers of disturbing the quasi-equilibrium state of the environment are evident although the benefits are fascinating and will ultimately attract man's technological skills and resources; such changes as the melting of the sea ice cover of the polar seas, which, when a threshold is passed, may lead to a permanently open ocean, and the possibility of modifying the circulation and sea ice of Hudson Bay are problems for the future. The complexity of decision-making in environmental intervention is great and it is easier, although in many cases still potentially disastrous, to work within existing natural systems rather than to create artificial systems.

Physical geography in Canada has advanced greatly since the earliest descriptions of the landscape. Even that preliminary stage, in a sophisticated sense, is far from complete: now, in addition, the possibilities of both detailed studies and more general approaches to major problems of resource and environmental management offer intriguing potential for future development in physical geography.

REFERENCES

1. Quoted by Bird, J. B. "Settlement Patterns in Maritime Canada, 1687-1786," *Geographical Review,* vol. 45, 1955, pp. 385-404.
2. Watts, F. B. "The Natural Vegetation of the Southern Great Plains of Canada," *Geographical Bulletin,* no. 14, 1960, pp. 25-43.
3. Nelson, J. G. "Some Reflections on Man's Impact on the Landscape of the Canadian Prairies and Nearby Areas." In Smith, P. J. (ed.) *The Prairie Provinces,* University of Toronto Press, Toronto, 1972, pp. 33-50.
4. Nelson, J. G. and R. F. England. "Some Comments on the Causes and Effects of Fire in the Northern Grasslands Area of Canada and the Nearby United States, ca. 1750-1900," *The Canadian Geographer,* vol. 15, 1971, pp. 295-306.
5. Chapman, L. J. and D. F. Putnam. *The Physiography of Southern Ontario.* University of Toronto Press, Toronto 1966, 386 pp.
6. Parry, J. T., ch. 5.
7. Goldthwaite, J. W., "Physiography of Nova Scotia," *Geol. Surv. Can. Mem.,* 140, 1924, 179 pp.
8. Chorley, R. J. "The Role and Relations of Physical Geography," in *Progress in Geography,* vol. 3, 1971, Arnold, London, pp. 87-109.
9. Brown, R. J. E. *Permafrost in Canada.* University of Toronto Press, Toronto, 1970, 234 pp.
10. Mackay, J. R., in Brown, R. J. E. (ed.), Proceedings of a Seminar on the Permafrost Active Layer, May 4 and 5, 1971, *Nat. Res. Council Assoc. Com. Geot. Res. Tech. Mem.,* no. 103, 1971, pp. 26-30.
11. Bird, J. B. *The Natural Landscapes of Canada.* Wiley of Canada, Toronto, 1972, 191 pp.
12. Mackay, J. R. "The World of Underground Ice," *Annals Assoc. Amer. Geog.,* vol. 52, 1972, pp. 1-22.
13. An indication of the broad range of periglacial phenomena is presented in Hamelin, L.-E. and F. A. Cook, *Illustrated Glossary of Periglacial Phenomena,* Les Presses de L'Université Laval, Québec, 1967, 237 pp. and Péwé, T. L. (ed.), *The Periglacial Environment,* McGill-Queen's University Press, Montreal, 1969, 487 pp.
14. Hare, F. K. "Climate and Zonal Divisions of the Boreal Forest Formation in Eastern Canada," *Geographical Review,* vol. 40, 1950, pp. 615-635.
15. Ritchie, J. C. "The Vegetation of Northern Manitoba, V: Establishing the Major Zonation," *Arctic,* vol. 13, 1960, pp. 210-229.
16. Hare, F. K. and J. C. Ritchie. "The Boreal Bioclimates." *Geographical Review,* vol. 62, 1972, pp. 353-365.

17. Barry, R. G. "Meteorological Aspects of the Glacial History of Labrador-Ungava with Special Reference to Atmospheric Vapour Transport," *Geographical Bulletin,* vol. 8, 1966, pp. 319-340.
18. Bryson, R. A. and W. M. Wenland. "Tentative Climatic Patterns for some Late Glacial and Postglacial Episodes in Central North America," in *Life, Land and Water,* University of Manitoba Press, Winnipeg, 1967, pp. 271-298.
19. Dirschl, H. J. "Geobotanical Processes in the Saskatchewan River Delta." *Can. J. Earth Sci.,* vol. 9, 1972, pp. 1529-1549.
20. Gill, D. Modification of Levee Morphology by Erosion in the Mackenzie River Delta, Northwest Territories, Canada. *Inst. Brit. Geog. Spec. Publ.,* no. 4, 1972, pp. 123-138.
21. Bochkov, A. P., A. I. Chebotarev, and K. P. Voskresensky. "Water Resources and Water Balance of the USSR." *Symposium on the World Water Balance,* IASH, Brussels, vol. 2, 1970, pp. 324-330.
22. Hare, F. K. and J. E. Hay. "Anomalies in the Large-Scale Annual Water Balance over Northern North America." *The Canadian Geographer,* vol. 15, 1971, pp. 79-94.
23. Yatsu, E. and A. Falconer (ed.), *Research Methods in Pleistocene Geomorphology,* 2nd Guelph Symposium on Geomorphology, 1971:1972, 285 pp.
24. Falconer, A. and E. Yatsu. "Objectives and Methods in the Study of Glacier Depositional Materials." *International Geography, 1972,* University of Toronto Press, Toronto, 1972, pp. 19-21.
25. Slaymaker, O. and H. J. McPherson (ed). *Mountain Geomorphology,* Tantalus Research Ltd., Vancouver, 1972, 274 pp.
26. Ford, D. C., P. L. Thompson and H. P. Schwarz. "Dating Cave Calcite by the Uranium Disequilibrium Method: Some Preliminary Results from Crowsnest Pass, Alberta," *International Geography, 1972,* University of Toronto Press, Toronto, 1972, pp. 21-24.
27. Ongley, E. D. Contemporary Denudation in Lake Ontario Watersheds. *International Geography, 1972,* Toronto University Press, 1972, pp. 119-121.
    ——— Sediment Discharge from Canadian Basins into Lake Ontario. *Can. J. Earth Sci.,* vol. 10, 1973, pp. 146-156.
28. Slaymaker, H. O. "Physiography and Hydrology of Six River Basins," in Robinson, J. L. (ed), *Studies in Canadian Geography: British Columbia,* University of Toronto Press, Toronto, 1972, pp. 33-68.
29. Burton, I. and L. May. "An Ecological Approach to Landslide Hazard in the Ottawa, St. Lawrence and Saguenay Valley Region." *Revue de Géographie de Montréal,* vol. 26, 1972, pp. 199-202.
30. Hewitt, K. and I. Burton. "The Hazardousness of a Place: A Regional Ecology of Damaging Events." *University of Toronto Geography Research Publication,* no. 6, 1971, 154 pp.
31. Hare, F. K. and C. I. Jackson. "Environment: a Geographic Perspective," *Dept. Environment Geographical Paper,* no. 52, 1972, 16 pp.

APPENDIX

Comments on Using This Book

In the preparation of this book it was our intention to show that physical geography was both alive and relevant. This book is not intended to be used in the place of an introductory text; it is written to be used *with* an introductory text. We do not in this volume explain what the ice age is or how meanders form in streams. Nor do we attempt to describe cloud formation, air mass movement or the conditions under which a podzol is formed. All these things are ably done in the available introductory text books.

We provide "Canadian content" by drawing attention to Canadian work which aims to extend knowledge of physical geography. This work is of international repute and so not all the papers we have selected are easily understood by the student at the introductory level. This is intentional. As the students of introductory physical geography complete the first course this volume can be used to stretch their abilities a little and serve as an indication of the content of senior courses in the subject. In the early stages of a physical geography course students should use the introductory section of this book as light reading to give some very general framework for the material which follows. The paper by Kelly (Chapter 2 above) should give a clear guide to the importance of physical geography as an ingredient in the development of human geography.

The other sections of this book should be used with the appropriate sections of the introductory text. Parry's paper on geomorphology in Canada (Chapter 5 above) is an excellent summary of what happened in the development of that subject in Canada. Many of the institutions and people named in the later part of Parry's paper should be known to the instructors and some of the students in the introductory courses. This personal link between the class and the subject should not be ignored. Physical geography is still developing and the fact that the majority of introductory texts have been written by scholars in the United States should not obscure our local and national activity in physical geography. Oke and Fuggle's paper (see Chapter 11 above) is based on data gathered in the Montreal area, James writes about soils in the Rankin Inlet Area, McKeague *et al* compare soils across the country. Most of the authors of the papers reprinted in this volume are now distributed throughout Canada and thus in any major population centre it is possible to invite one of the Canadians currently working in physical geography to give a guest lecture. We suggest that this provides local interest and Canadian content and adds a personal dimension to introductory physical geography in university and college courses. It is our experience that students in introductory physical geography courses benefit from this demonstration that the contents of introductory physical geography books are of immediate local relevance.

Using This Book as a Supplement to the Available Texts

The majority of physical geography texts are very well designed volumes containing the materials required in the introductory physical geography courses. They are comprehensive volumes which provide an excellent introduction to the subject matter. However, from the Canadian user's point of view the volumes subtly imply that a majority of physical geography originates in the United States and furthermore they suggest that vegetation and soils of the U.S. are the key to an understanding of the world's vegetation and soils. Climate does not suffer so much from this nationalistic flavour because the global atmosphere, by definition, envelopes the earth, and the classification of climates is undertaken on a world-wide basis. The geomorphology sections of these books do have an abundance of U.S. examples used to illustrate the text. These comments are not intended to be criticisms of the calibre of the illustrations or their suitability for the texts. In Canada, however, the students should realize that there are many Canadian examples of the major types of landforms. In most areas a short field trip will provide numerous examples of landforms created by glaciers, and many of the major rock types may also be examined either as hand specimens from glacial deposits or in place in the shield and mountain areas. Specific comments on the use of this book follow.

Section One

Any directions on the use of this book must re-state the link between section 1 and the introductory part of the available texts. We provide examples of the importance of physical geography to the explorers of the sixteenth and seventeenth centuries and to the nineteenth century settlers of Ontario. The exploration of the Canadian west also provides insight into man's perception of his environment as recently as 1907. It should also be noted that standard texts do Canada a great disservice by failing to note that world time zones, a product of an international conference held in Washington in 1883, were formulated by Sir Sandford Fleming, a Canadian railroad employee who saw the need to rationalize time zones for many international purposes, particularly rail travel. Fleming also proposed the use of the "twenty-four hour clock" for use in timetables. In addition to these links with the growth of Canada we include a chapter which comments on the present context of physical geography; we hope this chapter will stimulate discussion and provide food for thought.

Section Two

The second section of this book relates to the landforms section of the physical geography texts. The texts are concerned to convey information about rock types, landforms and landforming processes. We attempt to illustrate the way in which men have reached their present level of understanding of these processes. Parry's paper (see Chapter 5 above) is a comprehensive account of the development of geomorphology in Canada. The contrast between the work of Lyell and Dreimanis in the early part of Chapter 6 serves to emphasize the importance of changes in perception and technology in any understanding of the landscape. These two papers and the paper by Andrews and Sim (Chapter 6 above) relate to the glaciation sections of the text-books in that all three papers are concerned with the effects of glaciation on the land. The remaining three papers in Chapter 6, papers by Carson, Yatsu, and Goodchild and Ford all relate to the behaviour of landform materials. Carson provides a comprehensive review of work on slopes extending beyond the text-book introduction to slopes. Yatsu draws attention to the lack of understanding of the materials which are moulded into landforms and proposes that more attention should be paid to studies of the properties of landform materials. Goodchild and Ford investigate a specific form related to a specific rock-type and provide a detailed and interesting study of the process using laboratory and field data to reach their conclusion. The geomorphology section of this book does not directly integrate with the specific process and description content of the text-books; it provides a series of comments and a history which, we hope, provide some insight into the present state of geomorphology in Canada.

Section Three

The third section of this book extends the Weather and Climate sections of the introductory texts. The texts tend towards a uniform system of presenting a generalized discussion of atmospheric processes. This is related to the view that the understanding of physical processes of the atmosphere is essential in an interpretation of the physical environment. Weather is presented as a systematic study of global processes; heat balance, temperature, pressure, winds, moisture and air mass stability all being considered in detail. Climate is the regional component of these systematic studies. It is characteristically handled by the use of one or other of the several available modifications of the Koppen classification system.

   The content of the texts is sound. They present a factual account of the ingredients essential for the informed study of weather and climate but they neglect the need to relate these to the functioning of a nation. We present a view of Weather and Climate which is strongly oriented to Canadian climates. This view is based on research which strives to solve problems of national importance. The development of Canada's north requires a clear and detailed understanding of the Arctic climate. Similar needs are generated by the modification of climate which man himself creates either by clearing land for agriculture or by creating large cities.

   This book addresses the problems of Urban Climatology in a Canadian context (Chapter 11 above) and provides material which is an important extension of the usual material presented in the physical geography texts. Similar remarks apply to the section on Agroclimatology (Chapter 10 above), the specialized chapters on the Canadian Arctic (Chapter 8 above) and the climatology of Snow and Ice (Chapter 9 above). The introductory review of Canadian Climatology (Chapter 7 above) provides a useful background for students in Canadian universities and attempts to indicate the growth, achievements and present status of this subject area.

Section Four

Section 4 of this book is designed to offer a Canadian context for the material presented in the available introductory texts. An examination of contemporary texts in Physical Geography shows that those sections dealing with soils and vegetation normally have two functions; the first is a general overview designed to acquaint the student with fundamental concepts and viewpoints. This serves as the basis for a subsequent section or chapter discussing the distributions on a world scale. Thus, the scope in terms of geographical extent and depth of introductory comment is considerably greater than that offered here. We do not aspire to fulfill the needs of students requiring basic instruction in soils or biogeography; rather, we aim to emphasize the elements of the discipline within the context of Canadian studies. On occasion it is necessary to introduce certain concepts or terms in order that discussions focussing on Canadian contributions appear more meaningful, but as stated in our earlier remarks this volume is a companion to introductory texts.

   There is a tendency for introductory texts to generalize soils information and present material on classification and distribution of soils linked to the United States Department of Agriculture 7th approximation. This link, however, is not always very strong and the traditional use of the subdivision of soils into pedalfer and pedocal groups and then into the Great Soil Groups is often found. Because Canadian soil surveyors do not use the 7th Approximation this material is of a marginal significance for the Canadian student. The paper by Hoffman in Chapter 12 above should provide a more valuable source of information for many students.

The discussion in Chapter 12 above attempts to provide some material about the changes in classification systems with time. Soil survey is a comparatively recent undertaking and it is still establishing itself. Van Riper does discuss this type of material well in his Chapter 16 and students using this book should find that Van Riper's Chapter 16 and Chapter 12 above will blend together very well.

Vegetation is a difficult topic to deal with adequately within a physical geography course. The basic principles of botany cannot easily be incorporated into the course and, as a result, world vegetation zones represent the prime focus of the vegetation sections of introductory physical geography texts. The authors of these texts provide a broad regional overview which, when applied to Canada, does not adequately represent the primary vegetation units of the Canadian landscape. We provide a Canadian orientation to vegetation studies in Chapter 13 above, and we are aware that it is a very limited contribution to the subject. It is, however, material which extends the usual content of the introductory text and we recommend that students interested in this topic pursue their studies with reference to Dansereau's *Biogeography* and a basic training in botany.

Section Five

Section 5 offers a summary of Canadian physical geography by one of the best known physical geographers in Canada. Professor Bird's work on the Canadian Arctic and northern Canada together with his recently published work, *The natural landscapes of Canada* (Bird 1972), provides the best source of material on these topics for students of physical geography. This overview is a useful statement of the present status of physical geography and should be relevant to all students in physical geography courses.

Key to Use with Some Standard Texts

The following table correlates the chapters of this book with the sections and chapters of five texts commonly in use in physical geography courses for first and second year students. We hope it provides a useful guide for instructors and students. The five books used in the following table are

C.R.M. Books (1974) *Physical Geography Today: A Portrait of a Planet,* Del Mar, California, 518 pp.

Patton, C. P., C. S. Alexander and F. L. Kramer (1970), *Physical Geography.* Wadsworth, Belmont California, 408 pp.

Trewartha G. T., A. H. Robinson and E. H. Hammond (1967), *Physical Elements of Geography* (5th Edition) McGraw-Hill, New York, 544 pp.

Strahler, A. N. (1969), *Physical Geography* (Third Edition), John Wiley and Sons, New York, 757 pp.

Van Riper, J. E. (1971), *Man's Physical World* (Second Edition), McGraw-Hill, New York, 712 pp.

Chapters in *Physical Geography: The Canadian Context*

	INTRODUCTION				GEOMORPHOLOGY			WEATHER AND CLIMATE				SOILS & VEGETATION		OVERVIEW
	1	2	3	4	5	6	7	8	9	10	11	12	13	14
Physical Geography Today CRM Press	Appendix II Graphic Essay III	1, 2 All Geographic Essays Appendices II, IIII		*	11-16 inc.	14,15 11,12, 13,16	*	2-7 inc., 10 3,7	3,7	9,10	3	CHAPTERS 8, 9, 10 8	10	*
Physical Geography Patton et al	1,2	CHAPTERS 1 and 2	*	*	CHAPTERS 3-12 inc. *	3-10	*	20	6	21	*	CHAPTERS 21, 22, 23 22,23	21	*
Physical Geography A. N. Strahler	1,2,3,5	PART 1	*	*	PART IV *	22,23,24 29,32,33 34	*	PART II AND CHAPTERS 13-17 Inc. 13,17	*	14,18,20	*	CHAPTERS 18-21 inc. 18,19 20,21	18,19 **	*
Physical Elements of Geography Trewartha et al	1,2	INTRODUCTION	*	*	SECTION B *	13-17 inc.	*	SECTION A1 AND SECTION A2 11	11	*	*	SECTION C 22*	21*	*
Man's Physical World J. E. Van Riper	1,2,3	PREFACE, PART 1	*	1	PART 3 *	10,12	*	PART 2 8,14*	13,15	*	*	PART 5 PART 4 15,16*	13,14*	*

* indicates that we provide new material which is not contained in the physical geography texts.

Author Index

Ackerman, B., 192
Adams, W. P., 29, 36, 192
Adams, P. W., 56
Adkins, C. J., 192
Ager, B., 166, 192
Ahrnsbrak, W., 138-9, 141, 192
Allen, J. R. L., 107
Andreev, V. N., 192
Andrews, J. T., R. G. Barry and L. Drapier, 166, 192
Andrews, J. T., 48, 57, 70, 71, 107
Andrews, R. H., 161, 192
Angstrom, A., 193
Antevs, E., 107
Arnold, K. C., 193
Ashwell, A. W., 107
Atkinson, H. J., G. R. Giles, A. J. MacLean and J. R. Wright, 234, 279

Bach, W. and W. Patterson, 193
Badgley, F. I., 193
Baier, W., 172, 179, 193
Baird, P. D., 41, 55, 161, 193
Baker, F. C., 107
Baldwin, M., C. E. Kellogg and J. Thorpe, 208, 279
Barrett, H. M., 193
Barry, R. G., 118, 119, 193
Baynton, H. W., 176, 193
Beall, T., 18
Bishop, M., 8, 36
Blakadar, R. G., 107
Blanchard, R., 40, 41, 55
Boughner, C. C., 279
Boville, B. W., 119, 193
Bowen, 140
Bowser, W. E., 230, 279
Brasher, B. R., D. P. Franzmeier, V. Valassis and S. E. Davidson, 234, 279
Bretz, J. H., 107
Brewer, R., 234, 258, 279
Brown, D. M., 174
Brown, R. J. E., 47, 56, 140, 141, 142, 194
Brunger, A. J., 36
Brydon, J. E., H. Kodama

and G. J. Ross, 234, 247, 279
Bryson, R. A., 119, 194
Bryson, R. A., W. N. Irving and J. A. Larsen, 226, 279
Budyko, M. I., 194
Burbridge, F. E., 118, 194
Brazel, A. J., 138, 193
Boyd, D. W., 152, 153, 193
Bornstein, R. O., 193
Bird, J. B., 3, 34, 36, 45, 52, 57, 69, 107, 141, 193, 287, 301, 302
Benninghoff, W. S., 193

Carden, A. C. and H. M. Hennig, 171, 194
Careless, J. M. S. and R. D. Brown, 36
Carson, M. A., 45, 56, 77, 78, 91, 107, 108
Chalmers, 108
Champ, H., 176, 194
Chandler, T. J., 194
Chapman, L. J. and D. E. Putnam, 18, 36, 42, 55, 230, 280
Chorley, R. J., 91, 108
Chorley, R. J., A. J. Dunn, R. P. Beckinsale, 36
Chorley, R. J. and B. A. Kennedy, 36
Clark, A. H., 41, 55
Clark, J. F., 194
Clarke, J. S., 280
Clarke, J. S. and A. J. Greene, 232, 233, 247, 280
Claypole, E. W., 108
Clements, F. E., 261, 262, 280
Coachman, L. K., 194
Coleman, A. P., 108
Connor, A., 159, 194
Connor, A. J., 168, 194
Cook, F. A., 45, 47, 56
Cooper, W. S., 271, 280
Corte, A. E., 227, 280
Cowles, H. C., 260, 280
Crawford, C., and G. Johnston, 140, 141, 195
Crowe, B. W., C. L. Kuchuta

and W. A. Ross, 177, 195
Crowley, J. M., 262, 280
Cruickshank, J. G., 220, 280
Curl, R. L., 108
Curtis, J. T., 280

Dahl, E. R., 195
Dale, H. M. and D. W. Hoffman, 271, 272, 280
Dansereau, P., 280
Davies, R. N., 99, 108
Davis, W. M., 40, 79, 92
Dawson, G. M., 39, 40, 54
Dawson, J. E., 280
Dean, W. G., 43
DeBoer, H. J., 195
de Quervain, M., 195
Derikx, L. and H. Loijens, 165, 195
Dobson, M. R., 43, 55
Doronin, Yu. P., 195
Dorsey, H. G., 195
Doughty, J., et al, 195
Douglas, M. C. V. and R. N. Drummond, 43, 55
Douglas, R. J. W., 69, 108
Dreimanis, A., 61, 62, 63, 65, 66, 69, 70, 108
Drummond, R. N., 55
Dube, J. C. and L. E. Hamelin, 56
Dudas, M. J. and S. Pawluk, 230, 280
Dumanski, J. and R. J. St. Arnaud, 234, 280
Dunbar, M. and K. R. Greenway, 47, 56
Durnford, F., 19, 23
Dury, 77, 93, 108, 152, 179
Duthie, H. C. and Mannada Rani, R. G., 108
Dutton, C. E., 39
Dzerdzeyevakii, B. L., 195

East, C., 195
Ehrlich, W. A., H. M. Rice and J. H. Ellis, 233, 246, 280
Einarsson and Lowe, 177, 195, 322
Ells, R. W., 39

Emslie, J. H., 195
Evans, F. C., 261, 280
Eyre, J., 109

Farb, P., 36
Farnham, R. S. and H. R.
  Finey, 280
Fairchild, H. L., 109
Ferguson, H. et al, 195
Ferland, 176, 195
Ferrians, O. J., 195
Feustel, I. C., A. Dutilly, and
  M. S. Anderson, 219, 220,
  280
Findley, B. and G. McKay,
  160, 195
Fletcher, J., 120, 195
Flint, R. F., 41, 55, 109
Ford, D. C., 46, 56, 109
Ford, T. D., 109
Forsythe, J. L., 109
Foster, H. D., 98, 109
Frazer, E. M., 195
Frith, R., 195
Fuggle, R. E., 195
Fyles, J. G., 31, 36

Gadd, N. R., 109
Garnier, B., 195
Gavrilova, M. K., 195
Gilbert, G. K., 39
Gillespie, J. E. and D. E.
  Elrick, 233, 280
Gillespie, T., and K. King,
  173, 174, 196, 310, 313
Gillespie, J. E. and R. Protz,
  280
Glennie, E. A., 109
Godson, W. L., 196
Gold, L. W., 153, 166, 196,
  293
Goldthwait, J. W., 40, 55
Goldthwaite, R. P., A.
  Dreimanis, J.L.
  Forsythe, P. E. Karrow and
  G. W. White, 109
Goodchild, M. F. and D. C.
  Ford, 98
Goodison, B., 165, 196
Graham, W. and K. King,
  196
Grandtner, M. M., 280
Grigor'ev, A. A., 196
Guertin, R. K. and G. A.
  Bourbeau, 234, 280

Hack, J. T., 109
Hale, M. E., 161, 196
Halliday, W. E. D., 262, 280
Hamelin, L. E., 45, 56, 57
Hamelin, L. E., and F. A.
  Cook, 227, 280
Hannell, F. G., 138, 196
Hare, F. K., 116, 117, 119,
  120, 196, 197, 222, 269,
  281
Hare, F. K., and R. G.
  Taylor, 269, 281
Hattersley–Smith, G., 166
Hattersley–Smith, G. and H.
  Serson, 197
Havens, J. M., 197
Hayden, F. V., 39
Head, C. G., 6, 7, 36
Hector, J., 39, 54
Hills, G. A., 281
Hind, 39, 54
Hobbs, 117
Hobson, C. D. and J.
  Tarasmae, 109
Hoffman, D. W., B. C.
  Matthews and R. E. Wick-
  lund, 281
Holmes, R. and G. Robertson,
  197
Holmgren, B., 138, 161, 162,
  197
Hopkins, 174, 197
Hough, J. L., 109
Hubbard, W. A., 281
Hustich, I., 197, 269, 281
Hutton, W., 18

Ives, J. D., 47, 162, 197
Ives, J. D. and J. T. Andrews,
  109
Idso, S. B. and R. D. Jackson,
  197

Jackson, I. C., 160, 197
Jaiin, A. and M. Klapa, 109
James, P. A., 220, 281
Jeness, J. L., 41, 55, 140, 197
Joel, A. H., 281
Johnston, G. H., R. J. E.
  Brown and D. N. Pickers-
  gill, 197
Jopling, 97, 109

Karrow, P. F., 62, 63, 64, 65,
  67, 109

Keeler, E. W., 161, 197
Kellman, M. C., 271, 281
Kelly, K., 9, 10, 17, 18, 29
Kennedy, R. A., 281
King, 172, 197
Klein, C. J., 197
Kodama, H. and J. E.
  Brydon, 246, 281
Koerner, R., 161, 167, 197
Krajina, V. J., 281
Kratzer, P., 198
Kunkle, G. R., 109

Lachenbruch, A., 140, 198
Lajtoi, E. Z., 109
Larsen, J. A., 226, 281
Laycock, A., 169, 174, 198
Leahey, A., 210, 219, 281
Leahey, D. M., 176, 198
Lee, R., 119, 198
Lettau, H., 118, 198
Leverett, F., 110
Lewis, C. F. M., et al, 110
Lewis, E. L., 167, 198
Lindsay, D. G., 167, 198
Lister, H., 162, 198
Logan, L., 160, 198
Logan, W., 39
Loken, O. H. and J. T.
  Andrews, 57
Loken, O. H. and R. Sagar,
  165, 198
Lord, C. S., 55
Low, A. P., 39, 40, 54
Lyell, C., 58, 69, 70, 110
Lyon, T. L., et al, 110

Mack, A. R. and W. S.
  Ferguson, 172, 198
Mackay, J. R. and J. K.
  Stager, 46, 56
Macoun, J., 262, 281
Major, J., 281
Malone, T. F., 198
Marshunova, M. S. and N. T.
  Chernigovskiy, 198
Mateer, C. L., 176, 198
Mather, J. R. and C. W.
  Thornthwaite, 198
Maxson, J. H., 110
McConnell, R. G., 39, 54
McInnes, W., 25
McKay, G., 160, 198
McKay, G. and H. Thomp-
  son, 160, 198
McKeague, J. A., G. A.

Bourbeau and D. B. Cann, 233, 281

McKeague, J. A. and J. E. Brydon, 233, 234, 281

McKeague, J. A. and D. B. Cann, 233, 238, 247, 282

McKeague, J. A. and J. H. Day, 259, 282

McKeague, J. A. and J. I. MacDougall, K. K. Langmaid and G. A. Bourbeau, 282

McKeague, J. A., N. M. Miles, T. W. Peters and D. W. Hoffman, 230, 231, 282

McLean, A. and W. D. Holland, 270, 272, 282

McMillan, N. J., 219, 282

Mehra, O. P. and M. L. Jackson, 234, 282

Monteith, J. L., 140, 198

Moore, G. W., 110

Morison, S. E., 36

Morwick, F. E., 209, 230, 282

Moss, H. C., 230, 282

Moss, H. C. and R. J. St. Arnaud, 233, 282

Mukammal, E., K. King and H. Cork, 173, 199

Muller, E. H., 110

Muller, F., 45, 56, 61, 199

Munn, R. E., 176, 199

Munn, R. E. and M. Katz, 176, 199

Munn, R. E., M. Hirt, and B. Findlay, 199

Namais, J., 119, 199

Nicholson, N. L., 47, 56

Odell, N. E., 42, 55

Ogden, J. G. and R. J. Hay, 110

Oke, T. R., 176, 199

Oke, T. R. and C. East, 177, 199

Oke, T. R. and R. Fuggle, 180, 199

Oke, T. R. and F. G. Hannell, 176, 199

Oke, T. R., D. Yap and R. Fuggle, 179, 199

Ommanney, C., 165, 166, 199

O'Neil, A. and D. Gray, 160, 199

Orvig, S., 119, 161, 199

Ostrem, G., 162, 200

Palliser, J., 168

Parry, J. T., 39, 58, 110

Pawluk, S., 233, 246, 282

Pawluk, S., and J. D. Lindsay, 230, 282

Pluhar, A. and D. C. Ford, 56

Potter, J., 160, 177, 200

Pounder, E. R., 167, 200

Prest, V. K., 69, 110

Price, C. A., 110

Putnam, D. F., 209, 282

Putnam, D. F., B. Brouillette, D. P. Kerr and J. L. Robinson, 282

Putnam, D. F. and L. J. Chapman, 116, 200

Psyklywec, D., K. Davar and D. Bray, 200

Radcliffe, T., 18

Radforth, N. W., 47, 56, 282

Raup, H. M., 260, 282

Reed, R. J. and B. A. Kunkel, 119, 200

Reiche, P., 110

Rice, H. M., S. A. Forman and L. M. Patry, 247, 282

Richards, N. R., 230, 282

Richards, T. L. and V. S. Derco, 160, 200

Ritchie, J. C., 270, 282

Robertson, G. W., 174, 200

Robertson, G. W. and R. Holmes, 170, 200

Robitaille, B., 47, 56

Rouse, W. and J. McCutcheon, 176, 177, 200

Rouse, W. and R. B. Stewart, 138, 140, 200

Rowe, J. S., 263, 282

Rudnicki, J., 110

Ruhnke, G. N., 209, 282

Russel, R. J., 110

Rutherford, G. K. and D. K. Sullivan, 248, 282

Ryswyk, A. L. van, A. L.

McLean and L. S. Marchand, 282

Sagar, R. B., 161, 162, 200

Sanderson, M., 168, 200

Schaerer, P. A., 159, 200

Schnitzer, M., 282

Schwerdtfeger, P., and E. R. Pounder, 167, 201

Selirio, I. S. and D. M. Brown, 172, 201

Sharp, R. P., 42, 55

Shaw, R. H. and G. Thurtell, 173, 201

Shirreff, P., 13, 18

Shutt, F. T., 230, 282

Skinner, S. I. M., R. L. Halstead and J. E. Brydon, 283

Sim, V. W., 70, 71, 110

Sissons, J. B., 110

Spencer, J. W. W., 39, 64, 65, 110

Stanley, A. D., 165, 201

St. Arnaud, R. J. and M. M. Mortland, 233, 247, 282

St. Arnaud, R. J. and E. P. Whiteside, 233, 283

Stewart, E., 23, 25

Stobbe, P. C., 210, 233, 283

St. Onge, D. A., 47, 56

Strahler, A. N., 3, 36, 84, 93, 110

Stupart, R. F., 168, 201

Summers, P. W., 176, 177, 201

Sverdrup, H. U., 162, 201

Tanner, V., 42, 55

Tansley, A. G., 261, 283

Taylor, G., 40, 41

Tedrow, J. C. F., 219, 220, 222, 282

Tedrow, J. C. F. and J. E. Cantlon, 219, 283

Tedrow, J. C. F. and L. A. Douglas, 220, 283

Tedrow, J. C. F., J. V. Drew, D. E. Hill and L. A. Douglas, 226, 282

Terasmae, J., 64, 110

Thomas, M. K., 116, 160, 177, 201

Thompson, F. D., 160, 201

Thompson, S., 18

Thornthwaite, C. W., 168, 169, 269, 283
Thorp, J., 247, 283
Thorp, J. and G. D. Smith, 208, 283
Tisdale, E. W., M. A. Fosberg and C. E. Poulton, 283
Tissot, J., 119, 201
Traill, C. P., 12, 18

Unstead, J., 168, 202
Upham, W., 40, 55

Vavilov, N. I., 30, 36
Villimov, J., 118, 168, 202
Vowinkel, E., 119, 139, 202

Vowinkel, E. and B. Taylor, 119, 202
Vowinkel, E. and S. Orvig, 118, 119, 202

Wagner, P. L. and M. W. Mikesell, 30, 36
Wall, R. E., 110
Ward, W. and S. Orvig, 161, 202
Warkentin, J., 3, 36, 54, 207, 271, 283
Washburn, A. L., 42, 55
Watts, F. B., 270, 283
Weaver, J. E. and F. E. Clements, 261, 283
White, G. W., 110

Williams, B. H. and W. E. Bowser, 233, 283
Williams, G. D., 166, 168, 174, 203
Willis, A. L., 233, 246, 283
Witson, J. T., 42, 43, 55

Yatsu, E., 46, 56, 93, 94, 110, 111
Yatsu, E. and A. Falconer, 36
Yatsu, E., F. A. Dahms, A. Falconer, A. J. Ward, J. S. Wolfe, 36

Zacks, M. B., 168, 203

Subject Index

Agriculture, Northland, 21, 22, 23, 27, 28
    Ontario settlement, 10-17
AGROCLIMATOLOGY, 168-174
    Climate and crop types, 174
    Energy balance, 168-174
    Moisture flux, 173
    Soil moisture measurement, 171
    Soil moisture modelling, 172
    Water balance, 168-174
Air Pollution, 175-177
    Radiation balance, 177, 180-192
Arctic, Atmospheric Circulation, 116-118, 119, 121-129
    THE CLIMATE OF THE CANADIAN ARCTIC, 119-151
    Climate, 120-138
    Energy balance, 119-120, 131-138, 138-140
    Floating ice, 133-138, 166-167
    Geomorphology, 43-45, 46, 47, 71-77, 221-222, 296-299
    Land ice, 130, 131, 161-166, 293
    Permafrost, 131, 140-151, 289
    Soils, 219-229
    Vegetation, 129-130
Atmospheric Circulation, Arctic, 116-118, 119, 121-129

Biogeography, 261-262, 289
Bogland Ecosystem, 272-279
Classification, Soils, 207-210, 214, 216
    Vegetation, 262-265, 269-270
Climate, THE CLIMATE OF THE CANADIAN ARCTIC, 119-151
    Change, 15, 297-298
    History, 116-118
    Northland, 22, 28
    Permafrost, 131, 140-141, 142-151
    Settlers' interpretation, 14-16
    Soils, 214
    Vegetation, 129-130, 260-261
Climatology, AGROCLIMATOLOGY, 168-174
    THE STATUS OF CANADIAN CLIMATOLOGY, 115-118
    THE CLIMATE OF THE CANADIAN ARCTIC, 119-151
    THE CLIMATOLOGY OF SNOW AND ICE, 152-167
    URBAN CLIMATOLOGY, 175-192
Climatonomy, 118
Ecosystems, Bogland, 272-279
    General, 261-262
Energy Balance, Arctic, 119-120, 131-138, 138-140
    Crops, 172-174

Floating ice, 167
Land ice, 161-166
Permafrost, 140-151
Snow, 152-158
Urban areas, 179, 180-192
Exploration, 1-9, 19-29, 39-41
Forest, General, 294
Northlands, 21, 24, 25
Regions, 263-269
Geology, Soils, 212, 214
Geomorphology, Canadian perspective, 48
General, 298-299
GEOMORPHOLOGY IN CANADA, 39-57
Government research, 46-48
Great Lakes region, 58-61, 62-69
History, 39-57
Limestone scallops, 98-106
Northlands, 23-24, 25-27, 288-294, 296-301
Rock control, 94-97
Soils, 248-259
University research, 44-66
Glacial Deposits, Great Lakes region, 58-61, 62-69
Interpretation, 58-78
Northland, 20, 21, 23, 24, 26, 27
Perception by settlers in Ontario, 11, 12, 16
Rankin Inlet, 221-222
Hillslope Development, 78-93
History, Climate, 116-118
Geomorphology, 39-57
Soils, 207-210
Vegetation, 260-261
Ice, Characteristics, 292
Energy balance, 161-166, 167
Floating, 133-138, 166-167
Fluctuations of glaciers, 292-294
Land, 130-131, 161-166
Mass balance, 161-166
Land, Classification Southern Ontario, 16-17
Landform Material Science, 94-97
Microclimatology, Crops, 172-174
Ice, 161-165, 167
Permafrost, 143-150
Snow, 152-154
Tundra, 138-140
Urban areas, 178, 180-192
Micrometeorology, See Microclimatology
Models, Experimental, 98-106, 172
Process–response, 80
Statistical, 82
Northlands, Agriculture, 21, 22, 23, 27, 28
Climate, 22, 28

General, 295-296
Geomorphology, 23-24, 25-27, 288-294, 296-301
Glacial deposits, 20, 21, 23, 24, 26, 27
Ontario, Bogland ecosystems, 271-279
Climate, 14-16
Knowledge of soil depletion, 10-14
Land classification, 16-17
PRACTICAL KNOWLEDGE OF PHYSICAL GEOGRAPHY IN SOUTHERN ONTARIO DURING THE NINETEENTH CENTURY, 10-18
Soil erosion, 13-14
Soils, 10, 214
Organic Soils, 226-227, 273, 274, 276-278
Permafrost, Distribution, 131, 140-141, 289-290
Energy balance, 142-151
Physiography, Soils, 212, 213
Plant Succession, 12-13, 260-261, 271
Pleistocene, Deposits, 58-61, 62-77
Great Lakes region, 58-61, 62-69
Rock Control, 94-97
Snow, Avalanche control, 159
Energy balance, 152-158
Significance, 152, 292
Soil, Arctic, 219-229
Classification, 207-210, 214-216
Climate, 214
Distribution, 210-220
Erosion, 13, 14, 17
Geology, 212-214
History, 207-210
Limitations to use, 217-219
Luvisolic, 231-238
Nomenclature, 211
Organic, 226-227
Patterns, 10-12
Physiography, 212
Relations to geomorphology, 248-259
SOILS IN THE CANADIAN LANDSCAPE, 207-259
Terminology, 211
Vegetation, 214
Soil Moisture, measurement, 171
modelling, 172
Time Zones, Origin and use, 4
URBAN CLIMATOLOGY, 175-192
Heat island, 177-178, 188
Pollution, 175-177, 188, 189
Radiation, 180-192
Water vapour, 188-189
Vegetation, Arctic, 129-130

Bogland ecosystem, 272-279
Classification, 262-265, 269, 270
Climaxes, 261, 262, 264
Cover types, 10-12
Deforestation, 13
Distribution, 262-278
Ecosystems, 261, 262, 271-279
Forest regions, 263-269
Grasslands, 270

History, 260-261
Patterns, 10
Plant succession, 12-13, 260-261, 271
Reforestation, 14
Soils, 214, 270-271, 272-278
STUDIES IN THE VEGETATION OF CANADA, 260-283
Water Balance, 168-174
Turbulent transfer of moisture, 173

Index of Contributing Authors

J. Brian Bird, Professor and Chairman, Department of Geography, McGill University, Montreal, Quebec: "Physical Geography: A Canadian Context," pages 287-302.

D. W. Hoffman, Professor, Department of Land Resource Science, Director, Centre for Resources Development, University of Guelph, Guelph, Ontario: "The Canadian Soils System," pages 211-219.

K. Kelly, Associate Professor and Chairman, Department of Geography, University of Guelph, Guelph, Ontario: "Practical Knowledge of Physical Geography in Southern Ontario During the Nineteenth Century," pages 10-18.

J. T. Parry, Associate Professor, Department of Geography, McGill University, Montreal, Quebec: "Geomorphology in Canada," pages 39-57.

Index of Reprinted Papers Listed by Author

Andrews, J. T. and V. W. Sim, "Examination of the Carbonate Content of Drift in the Area of Foxe Basin, N.W.T.," pages 71-77.

Brown, R. J. E., "Influence of Vegetation on Permafrost," pages 142-151.

Carson, M. A., "Models of Hillslope Development Under Mass Failure," pages 78-93.

Dale, H. M. and D. W. Hoffman, "Bogland Ecosystems: Some Biogeographical Units," pages 272-279.

Dreimanis, A., "Late-Pleistocene Lakes in the Ontario and Erie Basins," pages 62-69.

Durnford, F. G., "Evidence Delivered Before the Select Committee," extracts only, pages 19-23.

Gold, L. W., "Micrometeorological Observations of the Snow and Ice Section, Division of Building Research, National Research Council," pages 153-159.

Goodchild, M. F. and D. C. Ford, "Analysis of Scallop Patterns by Simulation Under Controlled Conditions," pages 98-106.

Hare, F. K., "The Arctic," pages 120-138.

James, P. A., "The Soils of the Rankin Inlet Area, Keewatin, N.W.T., Canada," pages 220-229.

Lyell, C., "On the Ridges, Elevated Beaches, Inland Cliffs, and Boulder Formations of the Canadian Lakes and Valley of St. Lawrence," pages 58-61.

McInnes, W., "Evidence Given Before the Select Committee," extract only, pages 25-28.

McKeague, J. A., N. M. Miles, T. W. Peters and D. W. Hoffman, "A Comparison of Luvisolic Soils from Three Regions in Canada," pages 231-248.

Oke, T. R. and R. F. Fuggle, "Comparison of Urban/Rural Counter and Net Radiation at Night," pages 180-192.

Rowe, J. S., "Forest Regions of Canada," introduction only, pages 263-269.

Rutherford, G. K. and D. K. Sullivan, "Properties and Geomorphic Relationships of Soils Developed on a Quartzite Ridge, near Kingston, Ontario," pages 248-259.

Stewart, E., "Evidence Before the Select Committee," extracts only, pages 23-25.

Yatsu, E., "Landform Material Science—Rock Control in Geomorphology," pages 94-97.